George Greaves

Sieves
in Number
Theory

 Springer

George Greaves
School of Mathematics
University of Wales
Senghennydd Road
Cardiff CF24 4YH
Wales, U.K.
e-mail: greaves@cf.ac.uk

Library of Congress Cataloging-in-Publication Data applied for

Die Deutsche Bibliothek - CIP-Einheitsaufnahme

Greaves, George:
Sieves in number theory / George Greaves. - Berlin ; Heidelberg ;
New York ; Barcelona ; Hong Kong ; London ; Milan ; Paris ; Singapore ;
Tokyo : Springer, 2001
  (Ergebnisse der Mathematik und ihrer Grenzgebiete ; Folge 3, Vol. 43)
  ISBN 3-540-41647-1

Mathematics Subject Classification (2000): 11N35

ISSN 0071-1136
ISBN 3-540-41647-1 Springer-Verlag Berlin Heidelberg New York

This work is subject to copyright. All rights are reserved, whether the whole or part of the material is concerned, specifically the rights of translation, reprinting, reuse of illustrations, recitation, broadcasting, reproduction on microfilms or in any other ways, and storage in data banks. Duplication of this publication or parts thereof is permitted only under the provisions of the German Copyright Law of September 9, 1965, in its current version, and permission for use must always be obtained from Springer-Verlag. Violations are liable for prosecution under the German Copyright Law.

Springer-Verlag Berlin Heidelberg New York
a member of BertelsmannSpringer Science+Business Media GmbH

http://www.springer.de

© Springer-Verlag Berlin Heidelberg 2001
Printed in Germany

Typeset by the author using a Springer T$_E$X macro package.
Printed on acid-free paper   SPIN 10746941   44/3142LK - 5 4 3 2 1 0

Preface

Slightly more than 25 years ago, the first text devoted entirely to sieve methods made its appearance, rapidly to become a standard source and reference in the subject. The book of H. Halberstam and H.-E. Richert had actually been conceived in the mid-1960's. The initial stimulus had been provided by the paper of W. B. Jurkat and Richert, which determined the sifting limit for the linear sieve, using a combination of the λ^2 method of A. Selberg with combinatorial ideas which were in themselves of great importance and interest. One of the declared objectives in writing their book was to place on record the sharpest form of what they called Selberg sieve theory available at the time.

At the same time combinatorial methods were not neglected, and Halberstam and Richert included an account of the purely combinatorial method of Brun, which they believed to be worthy of further examination. Necessarily they included only a briefer mention of the development of these ideas due (independently) to J. B. Rosser and H. Iwaniec that became generally available around 1971.

These combinatorial notions have played a central part in subsequent developments, most notably in the papers of Iwaniec, and a further account may be thought to be timely. There have also been some developments in the theory of the so-called sieve with weights, and an account of these will also be included in this book.

The author's aim is to present a self-contained account of the "small" sieve method associated with the names of Brun, Selberg and their successors in a form that will be suitable for university graduates making their first acquaintance with the subject, leading them towards the frontiers of modern researches and unsolved problems in the subject area.

This book could not have reached its present form without assistance and guidance that has been received from a number of sources. In particular the author is indebted to Henryk Iwaniec for making available valuable unpublished material in the form of lecture notes prepared in connection with courses given to graduate students at Rutgers University.

Notation and Layout

Theorems, lemmas and displays will be numbered consecutively and independently within sections, and referred to from outside the section in which they appear by prefixing their reference number with their section number. For example Theorem 1 in Section 3.2 is referred to as Theorem 1 from within Section 3.2 and as Theorem 3.2.1 from elsewhere. Since section numbers appear in the running heads the reader will be able to locate any numbered entry quickly and without ambiguity.

There are several notations and conventions that are used so regularly throughout the book that explicit reference to them every time they are used would become tedious. The pages on which some of these are introduced are indicated in the following table.

List of Sieve Notations

The Function $S(a, P)$	13
The Set $\mathcal{A}_d$	14
The Sum $S(\mathcal{A}, P)$	14
The Remainders $r_\mathcal{A}(d)$	19
The Function ρ	19
The Level D	20
The Product $V(P)$	22
The Sifting Density κ	28
The Constants K and L	28
The Sifting Limit $\beta(\kappa)$	37

Other notations that are standard in Number Theory are freely employed: μ for the Möbius function, ϕ for Euler's function, and so on. Frequently these are also explained on their first use. The total number of prime factors of n has, as usual, been denoted by $\Omega(n)$, but we write $\nu(n)$ for the number of distinct prime factors.

Bibliographic information, historical remarks, supplementary information and the like are for the most part given in the notes at the end of each chapter where they will not interrupt the flow of the main text. The author wishes to make the usual disclaimer is respect of some of his historical remarks: he

VIII   Notation and Layout

is not a historian, and has not consulted the primary source in connection with all assertions made. If an intermediate secondary source has made an erroneous statement there is a possibility that it is repeated somewhere in this book.

The "author, year" system of references to published material is used. A year given in brackets will refer to a date of publication.

Table of Contents

Introduction 1

1. The Structure of Sifting Arguments 7
    1.1 The Sieves of Eratosthenes and Legendre 7
        1.1.1 The Contribution of Eratosthenes 7
        1.1.2 Legendre's Sieve 8
        1.1.3 An Estimate for $\pi(X)$ 10
        1.1.4 The Distribution of Primes 12
    1.2 Examples of Sifting Situations 13
        1.2.1 Notations 13
        1.2.2 The Integers in an Interval $(Y-X, Y]$ 14
        1.2.3 Numbers Given by Polynomial Expressions 15
        1.2.4 Arithmetic Progressions 16
        1.2.5 Sums of Two Squares 17
        1.2.6 Polynomials with Prime Arguments 17
    1.3 A General Formulation of a Sifting Situation 19
        1.3.1 The Basic Formulation 19
        1.3.2 Legendre's Sieve in a General Setting 23
        1.3.3 A Generalised Formulation 25
        1.3.4 A Further Generalisation 26
        1.3.5 Sifting Density 27
        1.3.6 The Sifting Limit $\beta(\kappa)$ 37
        1.3.7 Composition of Sieves 37
    1.4 Notes on Chapter 1 39

2. Selberg's Upper Bound Method 41
    2.1 The Sifting Apparatus 42
        2.1.1 Selberg's Theorem 42
        2.1.2 The Numbers $\lambda(d)$ 46
        2.1.3 A Simple Application 49
    2.2 General Estimates of $G(x)$ and $E(D, P)$ 50
        2.2.1 An Estimate by Rankin's Device 51
        2.2.2 Asymptotic Formulas 54
        2.2.3 The Error Term 60

X   Table of Contents

    2.3 Applications 63
        2.3.1 Arithmetic Progressions 63
        2.3.2 Prime Twins and Goldbach's Problem 64
        2.3.3 Polynomial Sequences 66
    2.4 Notes on Chapter 2 68

3. Combinatorial Methods 71
    3.1 The Construction of Combinatorial Sieves 72
        3.1.1 Preliminary Discussion of Brun's Ideas 72
        3.1.2 Fundamental Inequalities and Identities 73
        3.1.3 Buchstab's Identity 77
        3.1.4 The Combinatorial Sieve Lemma 78
    3.2 Brun's Pure Sieve 79
        3.2.1 Inequalities and Identities 80
        3.2.2 The "Pure Sieve" Theorem 81
        3.2.3 A Corollary 83
        3.2.4 Prime Twins 84
    3.3 A Modern Edition of Brun's Sieve 85
        3.3.1 Rosser's Choice of χ 86
        3.3.2 A Technical Estimate 87
        3.3.3 A Simplifying Approximation 88
        3.3.4 A Combinatorial Sieve Theorem 89
        3.3.5 Applications 94
    3.4 Brun's Version of his Method 96
        3.4.1 Brun's Choice of χ 96
        3.4.2 The Estimations 97
        3.4.3 The Result 98
    3.5 Notes on Chapter 3 100

4. Rosser's Sieve 103
    4.1 Approximations by Continuous Functions 105
        4.1.1 The Recurrence Relations 106
        4.1.2 Partial Summation 108
        4.1.3 The Leading Terms 110
    4.2 The Functions F and f 113
        4.2.1 The Difference-Differential Equations 113
        4.2.2 The Adjoint Equation and the Inner Product 115
        4.2.3 Solutions of the Adjoint Equation 117
        4.2.4 Particular Values of $F(s)$ and $f(s)$ 124
        4.2.5 Asymptotic Analysis as $\kappa \to \infty$ 129
    4.3 The Convergence Problem 134
        4.3.1 The Auxiliary Functions 135
        4.3.2 Adjoints and Inner Products 137
        4.3.3 The Case $\kappa \leq \frac{1}{2}$ 141

4.4	A Sieve Theorem Following Rosser	144
	4.4.1	The Case $\kappa > \frac{1}{2}$: a First Result	145
	4.4.2	Theorem 1 when $\kappa \leq \frac{1}{2}$	148
	4.4.3	An Improved Version of Proposition 1	150
	4.4.4	A Two-Sided Estimate	156
4.5	Extremal Examples	158
	4.5.1	The Linear Case	158
	4.5.2	The Case $\kappa = \frac{1}{2}$	164
4.6	Notes on Chapter 4	167

5. The Sieve with Weights 173

5.1	Simpler Weighting Devices	176
	5.1.1	Logarithmic Weights	176
	5.1.2	Modified Logarithmic Weights	179
	5.1.3	Some Applications	184
5.2	More Elaborate Weighted Sieves	185
	5.2.1	An Improved Weighting Device	186
	5.2.2	Buchstab's Weights	188
5.3	A Weighted Sieve Following Rosser	193
	5.3.1	Combining Sieving and Weighting	194
	5.3.2	The Reduction Identities	202
	5.3.3	An Identity for the Main Term	205
	5.3.4	The Estimate for the Main Term	210
5.4	Notes on Chapter 5	216

6. The Remainder Term in the Linear Sieve 223

6.1	The Bilinear Nature of Rosser's Construction	226
	6.1.1	The Factorisation of χ_D	227
	6.1.2	Discretisations of Rosser's Sieve	229
	6.1.3	Specification of Details	234
	6.1.4	The Leading Contributions to the Main Term	238
	6.1.5	Composition of Sieves	241
	6.1.6	The Remainder Term	243
6.2	Sifting Short Intervals	245
	6.2.1	The Smoothed Formulation	246
	6.2.2	The Remainder Sums	250
	6.2.3	Trigonometrical Sums	252
6.3	Notes on Chapter 6	256

7. Lower Bound Sieves when $\kappa > 1$ 259

7.1	An Extension of Selberg's Upper Bound	260
	7.1.1	The Integral Equation and the Function $\sigma(s)$	261
	7.1.2	The Estimation of $G_z(s)$	265

7.2	A Lower Bound Sieve via Buchstab's Identity	....	270
	7.2.1 Buchstab's Iterations	....	271
	7.2.2 The Buchstab Transform of the λ^2 Method	....	273
	7.2.3 The Sifting Limit as $\kappa \to \infty$	....	277
7.3	Selberg's $\Lambda^2 \Lambda^-$ Method	....	285
	7.3.1 An Identity for the Main Term	....	286
	7.3.2 The Improved Sifting Limit for Large κ	....	288
7.4	Notes on Chapter 7	....	291

References 297

Index 303

Introduction

Sieve Methods in Number Theory have roots which can be traced back to antiquity, but the modern era may be said to have begun with the papers of Viggo Brun, in particular with his article "Le Crible d'Eratosthène et le Théorème de Goldbach" in 1920. Some of the theorems in this paper provided a significant step towards some of the most difficult and tantalising questions in arithmetic. Among Brun's results we find the following (the numbering, as well as the translation, is ours).

Theorem 1. *If n is sufficiently large, then between n and $n + \sqrt{n}$ there is a number with at most eleven prime factors.*

Theorem 2. *Every sufficiently large even number can be expressed as the sum of two numbers each of which has no more than nine prime factors.*

Theorem 3. *There exist infinitely many pairs of numbers, having difference 2, of which the number of prime factors does not exceed nine.*

Theorem 4. *For sufficiently large x, the number of prime twins not exceeding x does not exceed $100x/\log^2 x$.*

Correspondingly, it had long been conjectured that

Conjecture 1. There is a prime between n and $n + \sqrt{n}$, if n is large enough.

Conjecture 2. Every even number greater than 2 is expressible as a sum of two primes.

Conjecture 3. There are infinitely many prime twins: numbers $p, p+2$ both of which are prime.

Conjecture 2 is of course the well-known hypothesis of Goldbach communicated in a letter to Euler. Conjecture 1 is essentially that contained in the snappier assertion: there is always a prime between consecutive squares.

The theorems listed were not the first recorded by Brun in this subject area. Already in 1915 he had obtained a weaker version of Theorem 4, a corollary of which was that the series $\sum 1/p$, where p is one of a pair of twin primes, is convergent or finite.

The mathematical community of the time did not immediately give Brun's results the recognition they later received. E. Landau left Brun's 1920 paper unread for a decade, apparently because he was not predisposed to believe that elementary methods as used by Brun could penetrate problems such as Goldbach's to the asserted extent. It may be added that Brun's paper is written in a way which even today is not easy to read, although (as we hope to demonstrate) with the benefit of hindsight the method appears straightforward enough. Landau was however quite happy with Brun's earlier result from 1915, and included an account of it in Volume I of his "Vorlesungen" in 1927.

Theorem 4, an example of what is now called an upper bound sieve result, indicated a way in which Brun's methods would have a particular influence in Number Theory. It has often transpired that some quantity, which must not be too large if a certain argument is to succeed, can be successfully bounded above by Brun's methods, in situations where the finer analytic tools of Number Theory will not work. An early example of this situation arose in the divisor problem of Titchmarsh, where one seeks to estimate $\sum d(p-1)$ summed over primes $p \leq x$, $d(n)$ being the number of divisors of n. Even after assuming the Riemann Hypothesis (although this is not necessary now) there remains a range of p around $\sqrt{x}$ where Brun's methods were required.

Examples of the utility of Brun's methods soon proliferated, especially as the result of the activities of P. Erdős. An important instance occurred in the work of Snirel'man in 1933 on representing numbers as a sum of (a rather large number of) primes. It was as a result of this work that Landau's scepticism about Brun's methods was overcome, and he included an account of them in his Cambridge tract in 1937. We will give a description of Brun's methods in Chapter 3, which will also incorporate some of the ideas later used by J. B. Rosser, H. Iwaniec, and others.

A major milestone was the appearance of the sieve method of A. Selberg in 1947. Selberg's method is simpler to understand and to implement than Brun's, and in many circumstances (though not all) leads to better results. In the first instance it delivers only an upper bound sieve. From the point of view of those needing applications elsewhere in Number Theory in the style of by Titchmarsh, Erdős and others this was perhaps not a disadvantage. Selberg himself subsequently indicated ways in which his ideas could be used to obtain lower bound sieves, as are needed for results such as Brun's Theorems 1, 2 and 3. However, as Selberg wrote, it seemed that "... we have not yet found the right approach to the problem of the lower bound." This is one area in which major open problems remain.

The simple upper bound method of Selberg is described in Chapter 2.

The problem of how one might combine the upper bound method of Selberg with ideas of a more combinatorial nature as used by Brun and his successors was addressed by W. B. Jurkat and H.-E. Richert in a paper appearing in 1965. They worked with the sieve of density (or "dimension") 1, in a technical sense which we describe in Chapter 1. Actually their paper worked with a restricted case of this "linear sieve" context; a general version using the same approach appeared in 1974 in the book of H. Halberstam and Richert. The paper of Jurkat and Richert gave a lower bound result which, in respect of the asymptotic expression for its main term, was (in view of an extremal example due to Selberg and discussed in Chapter 4) best possible.

Having seen the paper of Jurkat and Richert, H. Iwaniec was inspired to take up the topic and showed, in a Warsaw dissertation submitted in 1968, that combinatorial ideas alone, of the level of sophistication appearing in the paper of Jurkat and Richert, were sufficient to prove a slightly improved version of their theorem.

At a conference held in Stony Brook in 1969, Selberg described essentially the same construction, the essence of which had appeared in a manuscript of J. B. Rosser dating from the 1950's. Rosser's manuscript was never published, and even today very few people have seen it; the present author is not one of them. The proceedings of the Stony Brook meeting contain the account given by Selberg. We will describe a version of the Rosser-Iwaniec sieve method in Chapter 4 (though some ideas drawn from this method will also appear in the simpler discussions of Chapter 3).

The sieve method of Brun and his successors works with a "suitable" collection $\mathcal{A}$ of numbers not exceeding N, say, and estimates (from above or from below) the number of such numbers that are not divisible by sifting primes not exceeding a certain size N^c. It would be extremely satisfactory to be able to work satisfactorily with $c = \frac{1}{2}$ (thereby counting primes by a sieve method) but it transpires that there are excellent reasons why this is, in general, beyond reach. The power of a sieve method, if unaided by supplementary input from elsewhere, is limited by the examples, provided by Selberg, of which we give an account towards the end of Chapter 4. Thus Brun's Theorems 1, 2 and 3 refer not to primes, but to numbers having a bounded number of prime factors (all of which exceed N^c, for a certain $c < \frac{1}{2}$).

P. Kuhn observed in 1941 that in this situation one could obtain improved bounds for the number of prime factors by relaxing the restrictions on their size. This was done by "weighting" the sieve in a certain way which we discuss in Chapter 5. The weighting device of Kuhn was developed by N. C. Ankeny and H. Onishi in 1964, by A. A. Buchstab in 1965 and by Richert in 1969, and has also been the subject of more recent work which we describe in Chapter 5. It will become clear that the theory of the sieve with weights is in a less satisfactory state than is the theory of the standard unweighted sieve, and remains an area where important problems remain outstanding. In particular, there is a limitation on what one might hope to prove, again set by an example due to Selberg, but in the case of the sieve with weights the quality of the results attained to date does not approach the limitation set by Selberg's example. It may be conjectured that results attaining this limiting quality ought to be achievable, but establishing this conjecture remains an important unsolved problem.

In 1975 J.-R. Chen established that between n and $n + \sqrt{n}$ (sufficiently large) there lies a number having at most two prime factors. This depended not only on the sieve with weights but on a non-trivial treatment of associated error terms (see the discussion in Sect. 1.4.1). A more general treatment in this direction (the "bilinear form of the error term") appeared in a paper by Iwaniec in 1980, which has become a essential weapon in the armoury of those seeking to obtain good results in the applications of sieves to problems in Number Theory. There is an introduction to the bilinear error term and its implications in Chapter 6.

For the lower bound sifting problem, the methods of Rosser and of Iwaniec are not very effective when the sifting density (or dimension) exceeds 1. On the other hand the upper bound method of Selberg is very effective in these situations. The problem of how to harness Selberg's ideas to the lower bound sifting problem is one which still seems to be rather imperfectly solved.

One approach uses an iterative method used by Buchstab in the 1930's to improve the power of Brun's sieve, actually in a way that converges towards the results of Rosser and of Iwaniec. In this case Buchstab's iterative method is again used, but starting from Selberg's upper bound method. The first iteration of this process was undertaken by Ankeny and Onishi in 1964, and already gives results which are rather strong for sifting density of moderate size, say 2. We will describe this process in some detail. This will involve us in extending Selberg's upper bound method beyond the form found satisfactory in Chapter 2 for dealing with its direct applications.

For the purpose of applications in number theory one needs a numerical determination of the result achieved by the method of Ankeny and Onishi, and in particular of the sifting limit, a measure of the size of the set of sifting primes with which success is achieved. This question is not covered in this book, but some numerical values are quoted from the literature. As the sifting density κ tends towards infinity, however, a more mathematical analysis is possible. Ankeny and Onishi found that the sifting limit attained in this method is asymptotic to $2C\kappa$, where C is a constant close to 1·22. We will give a derivation of this result, though not by the method used by Ankeny and Onishi.

The result of the limiting case of Buchstab's iterative process, starting from Selberg's upper bound, has been determined in a series of papers by Halberstam, Richert and H. G. Diamond. This gives improved results for moderate κ, but appears to converge towards the Ankeny-Onishi result as $\kappa \to \infty$. For this work, we refer the reader to the literature.

A significant result in the theory of sieves was described by Selberg at a symposium held in his honour in Oslo in 1987. Selberg's result refers to the value of the sifting limit in the limiting case where the sifting density tends to infinity, and thus is rather removed from the usual applications in number theory where the sifting density is either equal to or is not very different from 1. Selberg showed that a rather different method of applying his ideas to the lower bound sifting problem, suggested by him in his early papers on the subject, gives a sifting limit asymptotic to 2κ as $\kappa \to \infty$. We will describe a somewhat simplified edition of this process, as already given by Selberg in Volume 2 of his Collected Works. This simple version is already sufficient to give the same asymptotic value for the sifting limit.

This result highlights some tantalising questions. Having identified the fact that the methods of Ankeny and Onishi, and of Diamond, Halberstam and Richert are not the best for large κ, the challenge is to incorporate more of Selberg's ideas in a way that leads to improvements for smaller κ. This challenge remains one of the major unsolved problems of the subject.

The other major outstanding problem in the theory of sieves relates to the sieve with weights, where important unresolved questions remain even in the case of the linear sieve of sifting density 1. Here, useful guidance can be drawn from the details of the extremal example first pointed out by Selberg in connection with the usual sieve without weights.

The author hopes that others will be inspired to attack and perhaps satisfactorily solve these problems.

The art of applying sieve methods remains a thriving field in which exciting new developments have recently appeared. The extremal example of Selberg discussed in Chapter 4 raises a certain "parity obstacle" which prevents the detection of prime numbers in suitable sequences by unaided methods of the type discussed in this volume. In conjunction with new ideas which break the parity obstacle, and which are rather beyond the scope of this book, further progress can be made. We cannot do better than mention recent results by Iwaniec and J. Friedlander, who established the infinitude of primes of the form $x^2 + y^4$, and by D. R. Heath-Brown, who showed that there are infinitely many primes represented by the binary cubic form $x^3 + 2y^3$.

1. The Structure of Sifting Arguments

This chapter provides an extended introduction to the rest of this book. It includes a discussion of the earliest sieve arguments, those of Eratosthenes and Legendre, but makes no explicit mention of the 20th century sieve methods to which this volume is devoted. It does however provide a number of examples of contexts in which sifting arguments can be employed. It also provides a description of the framework within which our account of the subject will be placed, and discusses several technical questions which make repeated appearances later on. These include the concept of *sifting density* (also known as the *dimension* of a sieve). An account of the well-known process of partial summation also appears here, more particularly in the somewhat specialised forms that will frequently be useful.

1.1 The Sieves of Eratosthenes and Legendre

We begin with the oldest argument recognisable as a sieve, the well-known argument due to Eratosthenes. Here one deletes multiples of primes from the positive integers not exceeding a specified number x. Rather closer to modern sieve ideas is that of Legendre, who analysed the process when x is a real variable rather than a fixed pre-assigned number. We will give an account of both of these processes, together with a recap of some essential items from 19th century prime number theory.

1.1.1 The Contribution of Eratosthenes

Eratosthenes's sieve was created in the third century B.C. According to the earliest surviving accounts, it appears that he began with a list of odd numbers; he deleted 3^2 and every third number thereafter, then deleted 5^2 and every fifth number thereafter, and so on. It is not clear from these accounts that the purpose of the sieve was to tabulate primes. It may have been to generate a factor table. For this purpose one would note not only that a number had been deleted, but of what primes it was a multiple that caused its possibly multiple deletion.

8   1. The Structure of Sifting Arguments

The awareness that a table of primes (up to N, say) could be generated in this way may date from much later, around the 13th century A.D. The process now generally referred to as the "Sieve of Eratosthenes" depends also on the observation that a number n, if not prime, has a prime factor not exceeding $\sqrt{n}$.

We may begin with a list of all the integers between 1 and N (avoiding Eratosthenes's "pre-sieving" with respect to the prime 2). Delete the multiples of all the primes p between 1 and $\sqrt{N}$. Then the survivors of this sifting process are the number 1 and the primes between $\sqrt{N}$ and N. Observe (it is significant) that some numbers have been cancelled more than once. Indeed, in this process a number n is cancelled k times, where k is the number of prime factors of n that do not exceed $\sqrt{N}$. The process is illustrated below.

The Sieve.

1	~~2~~	~~3~~	~~4~~	5	~~6~~
7	~~8~~	~~9~~	~~10~~	11	~~12~~
13	~~14~~	~~15~~	~~16~~	17	~~18~~
19	~~20~~	~~21~~	~~22~~	23	~~24~~
~~25~~	~~26~~	~~27~~	~~28~~	29	~~30~~
31	~~32~~	~~33~~	~~34~~	~~35~~	~~36~~

1.1.2 Legendre's Sieve

In 1808, A.M. Legendre provided a more analytic formulation of the principle of Sect. 1.1.1. This idea can be used to provide information, as Legendre did, about the number $\pi(X)$ of primes p for which $p \leq X$. Legendre's process was rather a weak one, essentially because of the difficulties involved in keeping track of the effect of multiple deletions arising when a number is a multiple of several small "sifting" primes.

Let $\mathcal{A}$ denote the set of integers in the interval $[1, X]$:

$$\mathcal{A} = \{n \in \mathbb{Z} : 1 \leq n \leq X\},$$

and let

$$P = \prod_{p<z} p \tag{2.1}$$

denote the product of the primes less than z. Then (using a notation to be redefined more generally in Sect. 1.2.1) let

$$S(\mathcal{A}, P) = \left|\{n \in \mathbb{Z} : 1 \leq n \leq X, \ (n, P) = 1\}\right|$$

1.1 The Sieves of Eratosthenes and Legendre

denote the number of integers between 1 and X that have no prime factors smaller than z. We will be forced to work with a value of z that is very considerably smaller than the value $\sqrt{X}$ that we would prefer.

For simplicity suppose X and z are not integers, so that the inequalities $n \leq X$ and $n < X$ are synonymous. For any z we obtain

$$S(\mathcal{A}, P) \geq \pi(X) - \pi(z) + 1 , \qquad (2.2)$$

since $S(\mathcal{A}, P)$ counts the number 1, does not count the primes p with $p < z$, but does count the larger primes p with $z \leq p \leq X$. It also counts the composite numbers that are products of primes exceeding z. Thus an estimation of $S(\mathcal{A}, P)$ will yield an upper bound for the number $\pi(X)$.

With a view to our requirements in Chapter 3 let us perform the thought-experiment of tracing the effect of successive steps in the sifting process. Start with the $[X]$ integers n with $1 \leq n \leq X$, where, as usual, $[X]$ denotes the greatest integer n such that $n \leq X$, so that $[X] + 1 > X$ and

$$X - 1 < [X] \leq X . \qquad (2.3)$$

For each prime p_1 with $p_1 < z$ we delete its multiples mp_1 for which $mp_1 \leq X$, $[X/p_1]$ in number. At this stage the count of numbers minus deletions would be $[X] - \sum[X/p_1]$. However, this process would have mishandled those n divisible by two such primes (p_1 and p_2, say) by counting two deletions, so we should increase our count by 1 in respect of these $[X/p_1p_2]$ numbers for each pair $\langle p_1, p_2 \rangle$ (ordered, with $p_1 > p_2$). Our count has now reached

$$[X] - \sum_{p_1 < z} \left[\frac{X}{p_1}\right] + \sum_{p_2 < p_1 < z} \left[\frac{X}{p_1 p_2}\right] , \qquad (2.4)$$

but of course this mishandles those n divisible by three primes, as they have so far been counted with a weight

$$1 - \binom{3}{1} + \binom{3}{2} = 1 ,$$

and so our count needs to be reduced appropriately.

By continuing this reasoning we arrive at the identity

$$S(\mathcal{A}, P) = [X] - \sum_{p_1 < z}\left[\frac{X}{p_1}\right] + \sum_{p_2 < p_1 < z}\left[\frac{X}{p_1 p_2}\right] - \sum_{p_3 < p_2 < p_1 < z}\left[\frac{X}{p_1 p_2 p_3}\right] + \cdots$$

$$= \sum_{d | P} \mu(d) \left[\frac{X}{d}\right] , \qquad (2.5)$$

where P is as in (2.1) and the Möbius function μ is (as usual) defined as is implied by (2.5):

$$\mu(1) = 1$$
$$\mu(p_1 p_2 \ldots p_r) = (-1)^r \text{ if } p_1, \ldots p_r \text{ are distinct primes} \qquad (2.6)$$
$$\mu(d) = 0 \qquad \text{if } p^2 | d \text{ for some prime } p .$$

That is to say, μ is the multiplicative function defined at prime-powers p^α by

$$\mu(p) = -1, \qquad \mu(p^\alpha) = 0 \text{ if } \alpha \geq 2 . \tag{2.7}$$

All this material is of course to be found in many elementary texts, as is in particular a proof of (2.5). Indeed, the right side of (2.5) is

$$\sum_{d|P} \mu(d) \sum_{1 \leq m < X/d} 1 = \sum_{n < X} \sum_{d|(n,P)} \mu(d) ,$$

where we have relabelled dm as n, and the condition that d divides the highest common factor (n, P) is a convenient expression of the simultaneous conditions $d|n$, $d|P$. The equation (2.5) now follows from the characteristic property of the Möbius function μ:

$$\sum_{d|A} \mu(d) = \begin{cases} 1 & \text{if } A = 1 \\ 0 & \text{if not.} \end{cases} \tag{2.8}$$

This in turn follows by observing, for example, that the expression on the left is a multiplicative function of A, since μ is a multiplicative function, and its value at prime-powers p^α for $\alpha \geq 1$ is $1 - 1 = 0$:

$$\sum_{d|A} \mu(d) = \prod_{p|A} (1 + \mu(p)) ,$$

and (2.8) follows from (2.7).

1.1.3 An Estimate for $\pi(X)$

We discuss the details of the application of the general ideas of Legendre's approach, as outlined in Sect. 1.1.2, to the specific question of bounding (from above) the number $\pi(X)$ of primes not exceeding X.

Start from the expression (2.5) for the quantity $S(\mathcal{A}, P)$ appearing in the inequality (2.2), where P is as in (2.1). Use the inequality (2.3) in the simplified form

$$[X] = X - \{X\} = X + O(1) ,$$

the bounded quantity denoted by $O(1)$ in fact lying in the interval $(-1, 0]$. Using this in (2.5) gives

$$\begin{aligned} S(\mathcal{A}, P) &= X \sum_{d|P} \frac{\mu(d)}{d} + O\left(\sum_{d|P} 1\right) \\ &= X \prod_{p < z} \left(1 - \frac{1}{p}\right) + O(2^{\pi(z)}) , \end{aligned} \tag{3.1}$$

1.1 The Sieves of Eratosthenes and Legendre

the product over p being a convenient expression of the sum

$$\sum_{d|P} \frac{\mu(d)}{d} = 1 - \sum_{p_1 < z} \frac{1}{p_1} + \sum_{p_2 < p_1 < z} \frac{1}{p_1 p_2} - \cdots,$$

while $2^{\pi(z)}$ counts the number of possible d, since each prime $p < z$ may or may not be included as a factor of d.

It is an elementary matter to estimate the product over p in (3.1) with sufficient accuracy for our purposes. For the application to (2.2) we need only to estimate it from above. Its reciprocal satisfies

$$\prod_{p < z}\left(1 - \frac{1}{p}\right)^{-1} = \prod_{p < z} \sum_{r=1}^{\infty} \frac{1}{p^r} > \sum_{n < z} \frac{1}{n} > \log z,$$

the last inequality following from a straightforward comparison of the sum with the corresponding integral. Thus (3.1) gives (if $z > 1$)

$$S(\mathcal{A}, P) < \frac{X}{\log z} + O\left(2^{\pi(z)}\right). \tag{3.2}$$

The great weakness of this argument, however, is the excessive size of the term $2^{\pi(z)}$ in terms of z. Indeed, in (3.2) we can do little better than to take $z = \log X$, so that

$$2^{\pi(z)} < 2^{\log X} = X^{\log 2},$$

and from (2.2) we now obtain (for $X \geq 3$, say)

$$\pi(X) = O\left(\frac{X}{\log \log X}\right). \tag{3.3}$$

Probabilistic Considerations. There are a few remarks that can be made about a heuristic line of reasoning that proceeds as follows. The probability of an integer not being divisible by a prime p is $1 - 1/p$, so the expected number of integers not exceeding X and divisible by no prime $p < z$ might be

$$X \prod_{p < z}\left(1 - \frac{1}{p}\right), \tag{3.4}$$

i.e. as in (3.1), but without the error term. There are two remarks to be made about this argument. In the first place, it gives a wrong answer. Suppose $z = X^c$, where $\frac{1}{2} \leq c < 1$. Then $\pi(X) + O(X^c)$ numbers would be counted by (3.4), as in (2.2). On the other hand Mertens' formula (see (4.2)) shows that (3.4) is asymptotic to $e^{-\gamma} X/c \log X$. This cannot be correct for every such c. In particular, reference to the Prime Number Theorem shows that it is not correct at $c = \frac{1}{2}$.

12   1. The Structure of Sifting Arguments

The failing in the argument is that it treats the probabilities of being divisible by different primes as being independent. But the probability of a positive number $n < X$ being divisible by d is

$$p_X(d) = \frac{1}{[X]}\left[\frac{X}{d}\right] = \left(\frac{X}{d} - \left\{\frac{X}{d}\right\}\right) \Big/ (X - \{X\}) , \qquad (3.5)$$

and thus $p_X(p_1 p_2) \neq p_X(p_1) p_X(p_2)$ when p_1, p_2 are distinct primes. This is because of the occurrence of the fractional parts $\{\cdot\}$ in (3.5).

In this connection note that the primary objective of sieve theory is to develop (in a wider context) constructions which do not allow these fractional parts to dominate the argument as did that leading to (3.1). On the other hand the dependence on $z = X^c$ exhibited by (3.4) shows that the effect of these fractional parts can not be expected to be totally negligible.

1.1.4 The Distribution of Primes

We give a brief survey of some classical results, not derived using sifting ideas, that we will need in the sequel.

As we shall see, Legendre's idea was potentially of rather general applicability, but for the specific question about the number $\pi(X)$ much better answers were later obtained by Chebyshev and his successors by more specialised methods involving, at the very least, the use of uniqueness of factorisation. The reader can find, in many standard texts, proofs (for $X \geq 3$, say) of Chebyshev's inequalities

$$c_1 \frac{X}{\log X} < \pi(X) < c_2 \frac{X}{\log X}$$

for explicit c_1, c_2 with $c_1 < 1 < c_2$. We shall refer to the related formula

$$\sum_{p \leq X} \frac{\log p}{p} = \log X + O(1) , \qquad (4.1)$$

from which the result

$$\sum_{p \leq X} \frac{1}{p} = \log \log X + C + O\left(\frac{1}{\log X}\right)$$

is easily deduced by partial summation. We shall also need Mertens' product

$$\prod_{p < z} \left(1 - \frac{1}{p}\right)^{-1} = e^\gamma \log z + O(1) , \qquad (4.2)$$

where γ is Euler's constant $0{\cdot}577\ldots$. In passing we shall also refer to the Prime Number Theorem

$$\pi(X) \sim \frac{X}{\log X} \quad \text{as} \quad x \to \infty ,$$

though this deeper theorem is not often needed in discussions confined strictly to the topic of Sieves in Number Theory.

It is, perhaps, to be regretted that the method of Legendre leads to a result as weak as (3.3). For, as we shall outline in Sect. 1.3, the process he developed is potentially applicable in a far wider range of circumstances than are the tools of analytic number theory used to prove the Prime Number Theorem. We shall, for example, see in Chap. 2 and elsewhere how developments growing out of Legendre's idea can show that the number of primes in any interval $(Y - X, Y]$ of length X is bounded from above by the inequality

$$\pi(Y) - \pi(Y - X) = O\left(\frac{X}{\log X}\right)$$

uniformly in Y. This result is quite out of reach of purely analytic methods based on the use of Riemann's Zeta-function as soon as Y is significantly larger than X^2, even if we assume the Riemann Hypothesis about the location of its complex zeros.

1.2 Examples of Sifting Situations

We will give a few instances of contexts in which the sieve methods to be described in subsequent chapters will be applicable. Legendre's method can be applied in these situations, but except in very special instances will lead, as before, only to very weak results.

We begin with a few of the notations that will be standard in this volume.

1.2.1 Notations

The identity (1.5) below and the ensuing remarks will indicate the significance of the quantity $\mathcal{A}_d$ defined in (1.3).

We generalise slightly on what was said in Sect. 1.1. Let P denote a product of distinct primes, not necessarily containing all the primes up to some specified bound. Frequently, however, we will take

$$P = \prod_{p<z} p. \tag{1.1}$$

Definition 1. Write $S(A) = \sum_{d|A} \mu(d)$. Further, abbreviate $S\big((a, P)\big)$ to $S(a, P)$, so that

$$S(a, P) = \sum_{d|(a,P)} \mu(d) = \begin{cases} 1 & \text{if } (a, P) = 1 \\ 0 & \text{otherwise.} \end{cases} \tag{1.2}$$

Here the second equality is the characteristic property (1.1.2.8) of the Möbius function μ. Observe that the condition $(a, P) = 1$ says that a is not divisible by any "sifting" prime factor of the product P.

Let $\mathcal{A}$ now denote a finite sequence of integers. We need a concise notation for the subsequence of those a in $\mathcal{A}$ that are divisible by d.

Definition 2. Denote

$$\mathcal{A}_d = \left\{ a \in \mathcal{A} : a \equiv 0, \bmod d \right\}. \tag{1.3}$$

Definition 3. With $S(a, P)$ is as in (1.2) define

$$S(\mathcal{A}, P) = \sum_{a \in \mathcal{A}} S(a, P). \tag{1.4}$$

Thus $S(\mathcal{A}, P)$ is the number of a in $\mathcal{A}$ not divisible by any prime from the product P. Then

$$S(\mathcal{A}, P) = \sum_{a \in \mathcal{A}} \sum_{d \mid a; d \mid P} \mu(d) = \sum_{d \mid P} \mu(d) |\mathcal{A}_d|, \tag{1.5}$$

which we may term "Legendre's Identity".

In the more efficient sieve methods to be discussed later the identity (1.2) will be replaced by an inequality in which the Möbius function μ is replaced by some other function λ. To make progress a suitable estimate is then required for the number $\mathcal{A}_d$ of a in $\mathcal{A}$ divisible by d. We will see below how such estimates are obtained in a few typical situations.

1.2.2 The Integers in an Interval $(Y - X, Y]$

Here we take $\mathcal{A} = \{a \in \mathbb{N} : Y - X < a \leq Y\}$, where $Y \geq X > 0$, and let P be as in (1.1). The discussion of Sect. 1.1.2 related to the case $Y = X$, and the general case proceeds similarly. For $\mathcal{A}_d$, as defined in (1.3), we obtain

$$|\mathcal{A}_d| = [Y/d] - [(Y - X)/d] = \frac{X}{d} + O(1), \tag{2.1}$$

in the usual integer part notation already used in (2.3). In fact the entry $O(1)$ arises as $-\{Y/d\} + \{(Y - X)/d\}$, and therefore lies in the interval $(-1, 1)$.

If we apply Legendre's identity (1.5), we obtain, as in Sect. 1.1.3,

$$S(\mathcal{A}, P) \leq \frac{X}{\log z} + O(2^z).$$

On choosing $z = \log X$ we find

$$\pi(Y) - \pi(Y - X) = O\left(\frac{X}{\log \log X}\right) \quad \text{if} \quad X \geq 3. \tag{2.2}$$

Here the rôle of the inequality $X \geq 3$ is merely to ensure $\log \log X > 0$.

The methods of Chapter 2 will lead to a sharper upper bound of the appropriate order $O(X/\log X)$ for the quantity on the left of (2.2).

1.2.3 Numbers Given by Polynomial Expressions

Here we take

$$\mathcal{A} = \{f(n) : Y - X < n \leq Y\},$$

where $f(n)$ is a polynomial over the integers such as $n^2 + 1$ or (as are relevant to Brun's investigations concerning the prime-twin and Goldbach problems) $n(n-2)$ or $n(n-C)$. Again, P will be as in (1.1).

To estimate $|\mathcal{A}_d|$ consider each residue class, mod d separately. We find

$$|\mathcal{A}_d| = \sum_{\substack{1 \leq l \leq d \\ f(l) \equiv 0, \bmod d}} \sum_{\substack{Y - X < n \leq X \\ n \equiv l, \bmod d}} 1 = \sum_{l} \left(\frac{X}{d} + O(1)\right). \tag{3.1}$$

Here, the inner sum over n was estimated by expressing n as $l + dm$, so that m is confined to an interval of length X/d almost exactly as in Sect. 1.2.2. Let $\rho(d)$ denote the number of roots of the congruence $f(l) \equiv 0$, mod d. Then

$$|\mathcal{A}_d| = X \frac{\rho(d)}{d} + O(\rho(d)). \tag{3.2}$$

From this, Legendre's identity (1.5) leads to

$$S(\mathcal{A}, P) = X \prod_{p < z} \left(1 - \frac{\rho(p)}{p}\right) + O\left(\sum_{d \mid P} \rho(d)\right). \tag{3.3}$$

There is a complication that can arise, for example in connection with Goldbach's problem of representing an even integer N as a sum of two primes. Brun's approach to this question is to sift the set

$$\mathcal{A} = \{n(N - n) : 1 \leq n < N\}.$$

Then (3.2) holds as before, but with $X = N$ and the function ρ depending on N:

$$\rho(p) = 2 \quad \text{if} \quad p \nmid N, \qquad \rho(p) = 1 \quad \text{if} \quad p \mid N. \tag{3.4}$$

This dependence on X of the function ρ is capable of causing a certain amount of difficulty if the situation is not handled with a certain degree of care.

1.2.4 Arithmetic Progressions

A special case of the situation in Sect. 1.2.3 arises when $f(n) = kn + l$, so that the numbers $m = f(n)$ lie in the arithmetic progression $a \equiv l$, mod k.

We may (at least initially) confine attention to the case when $(l, k) = 1$, since otherwise we can write

$$a = f(n) = (k, l)(k_1 n + l_1), \qquad (4.1)$$

where $k_1 = k/(k, l)$, $l_1 = l/(k, l)$. In (3.2), ρ is now the multiplicative function specified at primes p by

$$\rho(p) = \begin{cases} 1 & \text{if } p \nmid k \\ 0 & \text{if } p \mid k, \end{cases} \qquad (4.2)$$

so that

$$\rho(d) = \begin{cases} 1 & \text{if } (d, k) = 1 \\ 0 & \text{if } (d, k) > 1. \end{cases} \qquad (4.3)$$

In particular we obtain in (3.1) that $|\mathcal{A}_d| = 0$ if $(d, k) > 1$. Thus (3.2) becomes

$$|\mathcal{A}_d| = \begin{cases} \dfrac{X}{d} + O(1) & \text{if } (d, k) = 1 \\ 0 & \text{if } (d, k) > 1. \end{cases}$$

In (3.3) we now have

$$\prod_{p<z}\left(1 - \frac{\rho(p)}{p}\right) = \prod_{p<z;\, p\nmid k}\left(1 - \frac{1}{p}\right) = \prod_{p<z;\, p\mid k}\left(1 - \frac{1}{p}\right)^{-1} \prod_{p<z}\left(1 - \frac{1}{p}\right). \qquad (4.4)$$

If z exceeds the largest prime factor of k this becomes

$$\frac{k}{\phi(k)} \prod_{p<z}\left(1 - \frac{1}{p}\right),$$

where ϕ is Euler's function.

If we were to allow the case $(l, k) > 1$ then (4.2) should be replaced by

$$\rho(p) = \begin{cases} 1 & \text{if } p \nmid k \\ 0 & \text{if } p \mid k, \; p \nmid l \\ p & \text{if } p \mid k, \; p \mid l, \end{cases} \qquad (4.5)$$

In the case $(d, k) > 1$ the expression for $\rho(d)$ is now more elaborate than (4.3), actually (d, k) if $(d, k) \mid l$, and 0 otherwise. Observe however that in (4.4) the factor $1 - \rho(p)/p$ would reduce to 0 for primes p dividing both l and k. This illustrates the point that it is a rather unnecessary process to attempt to estimate the number of a in $\mathcal{A}$ not divisible by such primes p when the equation (4.1) already says that all a in $\mathcal{A}$ are divisible by p. Thus the value $\rho(p) = p$ arising in (4.5) when $p \mid (k, l)$ would not be relevant to a sensibly conceived sifting process.

1.2.5 Sums of Two Squares

Take $\mathcal{A}$ to be the integers between 1 and X:

$$\mathcal{A} = \left\{ n \in \mathbb{Z} : 1 \leq n \leq X \right\}.$$

Then $\mathcal{A}_d$ is as in (2.1), now with $Y = X$. Take

$$P = \prod_{p < z;\, p \equiv 3,\, \text{mod } 4} p$$

in place of (1.1).

If (for simplicity) we disregard the numbers divisible by squares of primes in P then the remaining "special" sums of two squares are the numbers not divisible by any prime in P. The number $N(X)$ of these "special" numbers that lie in $\mathcal{A}$ satisfies the inequality

$$N(X) = S(\mathcal{A}, P) = X \prod_{\substack{p < z \\ p \equiv 3,\, \text{mod } 4}} \left(1 - \frac{1}{p}\right) + O(2^z),$$

where we used Legendre's identity (1.5) and the expression in (2.1).

This example illustrates the point that a sifting process need not involve all (or even most) of the primes up to a given bound z.

1.2.6 Polynomials with Prime Arguments

Take

$$\mathcal{A} = \left\{ f(p) : 1 \leq p \leq x \right\}, \tag{6.1}$$

where as usual the symbol p denotes a prime. This example is of interest because, for instance, some better results can be obtained concerning the prime-twin problem by sifting the sequence (6.1) with $f(p) = p + 2$ than can be found by the earlier more elementary approach of sifting the sequence $n(n+2)$.

Proceeding as in Sect. 1.2.3 we obtain

$$|\mathcal{A}_d| = \sum_{\substack{1 \leq l \leq d \\ f(l) \equiv 0,\, \text{mod } d}} \sum_{\substack{1 \leq p \leq X \\ p \equiv l,\, \text{mod } d}} 1. \tag{6.2}$$

Approximations are available for the inner sum of the type

$$\sum_{\substack{1 \leq p \leq X \\ p \equiv l,\, \text{mod } d}} 1 = \frac{\operatorname{li} x}{\phi(d)} + E(x; l, d) \quad \text{if} \quad (l, d) = 1. \tag{6.3}$$

18   1. The Structure of Sifting Arguments

If $(l,d) > 1$ then this sum is at most 1, this arising only if (l,d) is prime. In (6.3), the error term $E(x;l,d)$ will be discussed briefly below, and $\phi(d)$ is Euler's function. The logarithmic integral $\operatorname{li} x$ is (for our purposes) given by

$$\operatorname{li} x = \int_2^x \frac{dt}{\log t} \sim \frac{x}{\log x} \quad \text{as } x \to \infty.$$

This asymptotic relation for $\operatorname{li} x$, and indeed a full asymptotic expansion, is easily established using integration by parts.

Let $\psi(d)$ denote the number of solutions of

$$f(l) \equiv 0, \bmod d, \quad (l,d) = 1, \quad 1 \leq l \leq d.$$

Then after substituting (6.3) in (6.2) we find

$$|\mathcal{A}_d| = \operatorname{li} x \frac{\rho(d)}{d} + r_A(d),$$

where

$$\rho(d) = \frac{d\psi(d)}{\phi(d)}, \quad |r_A(d)| \leq \left| \sum_{1 \leq l \leq d} E(x;l,d) \right| + \nu(d).$$

Here the entry $\nu(d) \leq \log d / \log 2$ bounds the number of times that (l,d) might be a prime factor of d, in which case this prime could contribute 1 to (6.2).

Thus ρ is a multiplicative function. Note in particular that in the prime-twin situation $f(p) = p - 2$ we have

$$\rho(p) = \frac{p}{p-1} \quad \text{if } p > 2.$$

For our purposes good information about $E(x;l,d)$ is supplied by the Bombieri-Vinogradov Theorem in the following form.

Lemma 1. *For each $\alpha > 0$ there exists $\beta > 0$ and x_0 such that*

$$\sum_{d \leq x^{1/2}/\log^\beta x} |E(x;l,d)| \leq \frac{x}{\log^\alpha x} \quad \text{if } x > x_0.$$

This lemma is sufficient to allow progress. We refrain from recording the effect of combining this deep theorem from the second half of the twentieth century with Legendre's identity from the first half of the nineteenth on the grounds that the combination would be bizarre.

1.3 A General Formulation of a Sifting Situation

We will set up a general framework within which we will discuss sifting problems of the type discussed in Sect. 1.2. Our immediate objective is to create an environment in which (1.5) below can be enunciated. This can be done very simply, as in Sect. 1.3.1, and this discussion will be adequate for almost all purposes required in this volume. There are, however, benefits to be obtained from considering more general situations, again leading to (1.5). We describe one such formulation in Sect. 1.3.3 and a more general one in Sect. 1.3.4.

Sect. 1.3.1 contains some further notations which we will adopt as standard, in addition to those appearing in Sect. 1.2.1.

Largely by way of an illustration, we note in Sect. 1.3.2 what Legendre's method gives in the context of Sect. 1.3.1. We take the opportunity to describe how, in a rather special "local" context, Legendre's idea can attain results that are distinctly stronger than those described in Sect. 1.2.

A concept of central importance in this volume is that of a *sifting density*, also known as the *dimension* of a sieve. This refers to the multiplicative function ρ introduced in (1.1), also appearing in the more general situations described in Sects. 1.1.3 and 1.1.4. The function ρ will be said to have a sifting density κ if a certain average of $\rho(p)$ at primes p does not exceed κ. There is a choice of ways in which this concept may be formulated, and some space is devoted to describing a particularly convenient two of these ways and the relationship between them.

Sects. 1.3.6 and 1.3.7 describe some other technical points which it is convenient to place in this section.

1.3.1 The Basic Formulation

The reader can verify that the examples given in Sect. 1.2 all conform to the description which follows.

Definition 4. When $\mathcal{A}_d$, X and $\rho(d)$ are specified define the "remainder" $r_A(d)$ by

$$|\mathcal{A}_d| = X \frac{\rho(d)}{d} + r_A(d) \quad \text{if} \quad d|P, \, d \leq D. \tag{1.1}$$

The equation (1.1) will be needed with ρ a multiplicative function, and D to be specified. For this to be useful in practice X and ρ will have to be chosen so that the remainder $r_A(d)$ is small, at least in some average sense as exemplified in Sect. 1.2.6. In the earlier examples in Sect. 1.2 these remainders were $O(1)$ (possibly with the implied constant depending on $\mathcal{A}$, as occurred in Sects. 1.2.3 and 1.2.4.)

1. The Structure of Sifting Arguments

In the sieves to which the bulk of this volume is devoted we construct functions λ_D^+, λ_D^- with the properties

$$\sum_{d|A} \lambda_D^-(d) \leq \sum_{d|A} \mu(d) \leq \sum_{d|A} \lambda_D^+(d) \quad \text{when} \quad A|P. \tag{1.2}$$

The objective is to select λ_D^+, λ_D^- in such a way that these inequalities become effective in a given sifting situation for relatively large products P, certainly substantially larger than is usable in the prototype method of Legendre. In this connection the label D will have the property

$$\lambda_D^\pm(d) \neq 0 \Longrightarrow d \leq D, \tag{1.3}$$

so that (1.1) is not needed when $d > D$.

Definition 5. *When (1.2) and (1.3) hold, say that λ^+, λ^- are (respectively) upper and lower sifting functions of level D for the product P.*

Here the term "level" is a contraction of "level of support".

In connection with the sieve with weights discussed in Chapter 5 there will be occasions when we will need to revise our objective, and replace (1.2) by a more general requirement

$$\sum_{d|A} \lambda_D^-(d) \leq \sum_{d|A} \mu(d)\psi(d) \leq \sum_{d|A} \lambda_D^+(d) \quad \text{when} \quad A|P, \tag{1.4}$$

in which ψ will be a suitable "weight" function. In this situation we will be dealing with expressions

$$S(a, P, \psi) = \sum_{d|(a,P)} \mu(d)\psi(d), \qquad S(\mathcal{A}, P, \psi) = \sum_{a \in \mathcal{A}} S(a, P, \psi)$$

in place of $S(\mathcal{A}, P)$ and $S(\mathcal{A}, P)$, to which these expressions reduce when $\psi(d) = 1$. In the interests of simplicity we will for the time being avoid further reference to this elaboration.

Now take $A = (a, P)$ in (1.2) and sum over $a \in \mathcal{A}$ to obtain Theorem 1, as follows.

Theorem 1. *When $\mathcal{A}$ satisfies (1.1) and $\lambda^\pm$ satisfies (1.2) we obtain*

$$XV^-(D, P) + R^-(D, P) \leq S(\mathcal{A}, P) \leq XV^+(D, P) + R^+(D, P), \tag{1.5}$$

where S is as in (1.2.1.2) and

$$V^\pm(D, P) = \sum_{d|P} \frac{\lambda_D^\pm(d)\rho(d)}{d}, \qquad R^\pm(D, P) = \sum_{d|P} \lambda_D^\pm(d) r_\mathcal{A}(d). \tag{1.6}$$

1.3 A General Formulation of a Sifting Situation

Theorem 1 is of course scarcely a theorem until a suitable function λ^+ or λ^- has been constructed, but it lays down a basic framework for many of the results in subsequent chapters. When use is made of (1.5) the level D will be selected so that the terms $R^\pm$ in (1.5) are suitably small compared to the terms $XV^\pm$.

The problem to be dealt with then becomes the estimations of the entries $V^\pm$ and $R^\pm$. Sects. 1.3.3 and 1.3.4 give alternative and more general versions of (1.1) and (1.2.1.2), which do however lead to the same Theorem 1 and the same problem.

We will usually normalise our constructions so that

$$\lambda_D^\pm(1) = 1.$$

It will then turn out to be the case in our constructions that

$$|\lambda_D^\pm(d)| \leq 1 \quad \text{for all } d. \tag{1.7}$$

Corollary 1.1 then follows be estimating the terms involving $r_\mathcal{A}(d)$ in (1.5) by the following "trivial treatment of the remainder term":

$$\left|\sum_{d|P} \lambda_D^\pm(d) r_\mathcal{A}(d)\right| \leq \sum_{\substack{d|P \\ d \leq D}} |r_\mathcal{A}(d)|.$$

Corollary 1.1. *Suppose* (1.2), (1.7) *and* (1.1) *hold. Then*

$$XV^-(D,P) - \sum_{\substack{d|P \\ d \leq D}} |r_\mathcal{A}(d)| \leq S(\mathcal{A},P) \leq XV^+(D,P) + \sum_{\substack{d|P \\ d \leq D}} |r_\mathcal{A}(d)|.$$

Frequently we shall refer to Corollary 1.1 rather than to Theorem 1 for a description of our results. In Chapter 6 we give a treatment of the remainder term which is non-trivial, in the sense that it does not rely on estimating each entry $r_\mathcal{A}(d)$ via its absolute value $|r_\mathcal{A}(d)|$.

Level of Distribution. As will be seen, this is a slightly vague concept, which has accordingly not been graced with the title "Definition". The idea is that it is a number D for which the r-terms in Corollary 1 are unimportantly small. Choose $\varepsilon > 0$, and say that $\mathcal{A}$ has level D if

$$\sum_{d < D} |r_\mathcal{A}(d)| < \varepsilon X$$

when (1.1) applies, with an appropriately chosen ρ.

If we choose $\varepsilon = 1$ then most of the examples in Sect. 1.2 have (with ρ as stated) level of distribution X, except for that of Sect. 1.2.6 where the level is rather smaller than $\sqrt{X}$.

The number of occasions when we will need to refer to this concept will be very small (there is one in the introduction to Chap. 5). The statements of our general theorems will leave the r-term unestimated, as in Theorem 1 or Corollary 1.1. Naturally, when an application is made to a specific situation appropriate estimates of the terms $R^{\pm}(D,P)$ arising from Theorem 1 will have to be made.

In the light of the discussion at the end of Sect. 1.2.4, and the fact that only those d with $d|P$ appear in (1.1), no real loss of generality is involved in making the following assumptions about the function ρ.

Hypothesis 1. Assume $0 \leq \rho(p) < p$ when $p|P$.

Hypothesis 2. Assume

$$\rho(p) = 0 \quad \text{whenever} \quad p \nmid P.$$

The convention that $\rho(d) \neq 0$ only when $d|P$ can simplify our notations, in that the condition $d|P$ may be omitted from sums where the summand has a factor $\rho(d)$. Where convenient, however, we will leave such conditions $d|P$ explicit.

We will adopt a standard notation for the products over primes of the type appearing in (1.2.3.3) and elsewhere. Here, the expression V is the prototype of the expressions $V^{\pm}$ considered in (1.6).

Definition 6. Define the expression $V(P)$ by

$$V(P) = \sum_{d|P} \frac{\mu(d)\rho(d)}{d} = \prod_{p|P} \left(1 - \frac{\rho(p)}{p}\right). \tag{1.8}$$

We will also denote

$$P(z) = \prod_{p|P,\, p<z} p. \tag{1.9}$$

A convenient convention will be that in contexts where no product P of primes is implied or specified then $P(z)$ denotes simply

$$P(z) = \prod_{p<z} p. \tag{1.10}$$

Information about individual values of $r_A(d)$ is not needed in Theorem 1 but it is frequently available. It often arises in situations of interest that

$$|r_A(d)| \leq \rho(d) \tag{1.11}$$

when an appropriate choice of $\rho(d)$ is made in (1.1). Most of the examples given in Sect. 1.2, in particular that in Sect. 1.2.3, possess this property. In this situation the remainder term in Corollary 1 satisfies

$$\sum_{\substack{d|P \\ d \leq D}} |r_{\mathcal{A}}(d)| \leq D \sum_{\substack{d|P \\ d \leq D}} \frac{\rho(d)}{d} \leq D \prod_{p|P} \left(1 - \frac{\rho(p)}{p}\right)^{-1} = \frac{D}{V(P)}. \quad (1.12)$$

We will put this estimate to use in Chapters 2 and 3.

1.3.2 Legendre's Sieve in a General Setting

For the purpose of comparison with the efficient sieve methods to be considered in subsequent chapters we record what Legendre's choice $\lambda_D^{\pm}(d) = \mu(d)$ gives in the context described in Sect. 1.3.1.

Theorem 2. *The sum S defined in Sect. 1.2 satisfies*

$$S(\mathcal{A}, P) = X V(P) + \theta \sum_{d|P} |r_{\mathcal{A}}(d)|,$$

for some θ with $|\theta| \leq 1$.

In this instance of Theorem 1 the expressions in (1.6) satisfy

$$V^{\pm}(D, P) = V(P), \quad |R^{\pm}(D.P)| \leq \sum_{d|P} |r_{\mathcal{A}}(d)|,$$

$V(P)$ being as in (1.8). In this theorem the level D of the function $\lambda_D = \mu$ is

$$D = P. \quad (2.1)$$

In the efficient sieve methods to be described later this (serious) constraint on the level D will not appear. Typically the main terms in Theorem 1 will be the subject of estimates of the shape

$$V^{\pm}(D, P) = c^{\pm}(D, z) V(P), \quad (2.2)$$

where $V^{\pm}$ is as in (1.6). The coefficient $c^{\pm}(D, z)$ will turn out to be essentially a function of $s = \log D / \log z$; in Legendre's method they were simply 1. It will then be necessary to select the level D in terms of X in such a way that the available estimates for the sums in (1.6) do not lead to an unduly weak statement in Theorem 1. Finally our selection of z in terms of D must have the property that $c^-(D, z)$ is not too small (certainly > 0), or alternatively that $c^+(D, z)$ is not too large.

Legendre's Sieve in a "Local" Context. The result in Theorem 2 can be significantly improved when we add an extra hypothesis

$$1 \leq a < X \quad \text{when} \quad a \in \mathcal{A}, \tag{2.3}$$

where X is the number appearing in (1.1). This would be of no significance if it were not coupled with a reasonable bound for $r_\mathcal{A}(d)$. In Theorem 3 we suppose

$$\rho(d) = 1, \quad |r_\mathcal{A}(d)| \leq 1, \tag{2.4}$$

the situation arising in the "short interval" context of Sect. 1.2.2. The method of proof in Theorem 3 would operate in a slightly wider context, for example if we assumed only $1 \leq a < Y$ with Y slightly larger than X, but would not allow Y to be as large as $X^{1+\delta}$ for any fixed $\delta > 0$ when X is large.

Theorem 3. *Suppose $\mathcal{A}$ satisfies (2.3), where the other entries in (1.1) satisfy (2.2). If $|P| < X^{1/(6 \log \log X)}$, then*

$$S(\mathcal{A}, P) = X \, V(P) \left(1 + O\left(\frac{1}{\log X}\right)\right).$$

This result is much stronger than that of Theorem 2, which would allow only $|P| < \log X$, as in Sect. 1.3.3. On the other hand the extra hypothesis (2.3) is quite restrictive.

An Application. Theorem 3 has been formulated so that we may take $\mathcal{A}$ to be the integers in $[1, X]$. Let P be the set of prime factors of an integer m, so that $|P| = \nu(m)$, the number of distinct prime factors of m. Theorem 3 then reduces to the following corollary.

Corollary 3.1. *Let $\phi(X, m)$ denote the number of integers between 1 and X that are coprime to m. Suppose $\nu(m) \leq X^{1/(6 \log \log X)}$. Then*

$$\phi(X, m) = X \frac{\phi(m)}{m} \left(1 + O\left(\frac{1}{\log X}\right)\right).$$

Proof of Theorem 3. We take advantage of the fact that $d < X$ when (2.3) holds and $d \mid a$. In Theorem 1, take $\lambda(d) = \mu(d)$ if $d < X$, and $\lambda(d) = 0$ otherwise. Then we find

$$V(D, P) = \sum_{d \mid P; d < X} \frac{\mu(d)}{d} = V(P) - \sum_{d \mid P; d \geq X} \frac{\mu(d)}{d}, \quad R(D, P) \leq \sum_{d \mid P; d < X} 1.$$

If $0 < \varepsilon < 1$, then

$$X \left| \sum_{d|P; d \geq X} \frac{\mu(d)}{d} \right| + \sum_{d|P; d < X} 1$$

$$\leq X^{1-\varepsilon} \sum_{d|P} \frac{1}{d^{1-\varepsilon}} \leq X^{1-\varepsilon} \prod_{p|P} \left(1 + \frac{1}{p^{1-\varepsilon}}\right). \quad (2.5)$$

Let z be any number with $\pi(z) > |P|$, and let $\varepsilon = 1/\log z$, so that $z^\varepsilon = e$ and $p^\varepsilon < e$ when $p < z$. Then, for given $|P|$, the product over p is maximised when P consists of the first $|P|$ primes, so that

$$\prod_{p|P}\left(1 + \frac{1}{p^{1-\varepsilon}}\right) \leq \prod_{p<z}\left(1 + \frac{e}{p}\right) \leq \exp \sum_{p<z} \frac{e}{p} = O((\log z)^e).$$

We may take $z = X^{1/5 \log \log X}$. Then

$$X^\varepsilon = X^{1/\log z} = e^{\log X/\log z} > e^{5 \log \log X} = \log^5 X.$$

This shows

$$\sum_{d|P; d<X} 1 = O\left(\frac{X}{(\log X)^{5-e}}\right) = O\left(\frac{X}{\log^2 X}\right),$$

with the same bound for the other entry on the left of (2.5). Theorem 3 now follows.

1.3.3 A Generalised Formulation

For some purposes it would be better to allow the members of the sequence $\mathcal{A}$ to be "weighted" (this will have nothing to do with the "Weighted Sieve" of Chapter 5). A context in which this idea is very useful occurs in Sect. 6.2. The formulation of Sect. 1.3.1 may be generalised slightly as follows.

Let the members n of the sequence $\mathcal{A}$ be assigned non-negative weights α_n. Then define

$$|\mathcal{A}_d| = \sum_{n \equiv 0, \bmod d} \alpha_n,$$

and assume (1.1). In this context we would replace (1.2.1.4) by

$$S(\mathcal{A}, P) = \sum_{n \in \mathcal{A}} \alpha_n S(n, P). \quad (3.1)$$

Then we obtain the inequalities (1.5) required in Theorem 1.

This formulation of the sifting situation avoids a minor problem of terminology that might otherwise occur. We were careful to refer to our sequences $\mathcal{A}$ as sequences as opposed to sets. This was not because their ordering has any significance, but merely because some member n of $\mathcal{A}$ might have occurred more than once. In the current formulation such n could be simply be assigned an appropriate weight $\alpha(n)$.

1.3.4 A Further Generalisation

There is a somewhat more far-reaching way in which the situation of Sect. 1.3.1 might be generalised.

In (1.2.1.2) we let $S(a, P)$ count those a which do not lie in the residue class 0, mod p, for each "sifting" prime dividing P. More generally, one might ask that a should not lie in any of a specified list of excluded residue classes

$$\mathcal{E}_p = \left\{e_1, e_2, \ldots, e_{\rho(p)}\right\}, \bmod p . \tag{4.1}$$

Particularly in the context of "large" sifting applications, where the number $\rho(p)$ is allowed to increase unboundedly with p, this would be the appropriate formulation. An instance important in some applications is where the excluded residue classes e_i are those which are not kth power residues, mod p.

In the problems considered in the body of this volume a bounded number of residue classes would be excluded for each sifting prime p. For example in the question of Sect. 1.2.3, where the sifting condition was $f(n) \not\equiv 0$, mod p, an equivalent formulation is that $n \not\equiv e_i$, mod p, where $e_1, e_2, \ldots$ are the residues mod p for which $f(e_i) \equiv 0$, mod p. For this question the formulation in terms of the residues e_i is not necessary, although it may nevertheless be illuminating.

We now wish to consider, in place of (1.2.1.2),

$$S'(n, P) = \begin{cases} 1 & \text{if } n \notin \mathcal{E}_p \text{ whenever } p | P \\ 0 & \text{otherwise} . \end{cases}$$

The description of $S'(n, P)$ in terms of the Möbius function μ is as follows. For squarefree d dividing P let $\mathcal{E}_d$ denote the set of "excluded" residue classes e, mod d given by the condition

$$e \equiv e_i, \bmod p \text{ for some } i \text{ and some } p \text{ dividing } d .$$

Then

$$S'(n, P) = \sum_{d | P; \, n \in \mathcal{E}_d} \mu(d) , \tag{4.2}$$

since the sum over d is

$$\prod_{p | P; \, n \in \mathcal{E}_p} (1 + \mu(p)) , \tag{4.3}$$

the product being taken over those p dividing P such that n lies in some excluded class e_i, mod p. Since $\mu(p) = -1$ this gives that the product in (4.3) is as required by (4.2) and (4.1).

In this formulation we can again let members n of a sequence (or set) $\mathcal{A}$ carry non-negative weights α_n. In this context replace (3.1) by

$$S(\mathcal{A}, P) = \sum_{a \in \mathcal{A}} \alpha_n S'(n, P) .$$

1.3 A General Formulation of a Sifting Situation

On summing over n and invoking (1.2) we obtain

$$\sum_n \alpha_n \sum_{d|P; n \in \mathcal{E}_d} \lambda_D^-(d) \leq S(\mathcal{A}, P) \leq \sum_n \alpha_n \sum_{d|P; n \in \mathcal{E}_d} \lambda_D^+(d),$$

whence

$$\sum_{d|P} \lambda_D^-(d) \sum_{n \in \mathcal{E}_d} \alpha_n \leq S(\mathcal{A}, P) \leq \sum_{d|P} \lambda_D^+(d) \sum_{n \in \mathcal{E}_d} \alpha_n . \qquad (4.4)$$

In place of (1.1) we now write

$$\sum_{n \in \mathcal{E}_d} \alpha_n = X \frac{\rho(d)}{d} + r_\mathcal{A}(d) . \qquad (4.5)$$

From (4.4) and (4.5) we again obtain the inequality (1.5) required in Theorem 1.

In a particularly important case of this situation there is a natural interpretation of the significance of the function ρ. Suppose

$$\sum_{n \equiv e, \bmod d} \alpha_n = \frac{X}{d} + R(e, d) . \qquad (4.6)$$

Then we obtain (4.5) with $r_\mathcal{A}(d) = \sum_{e \in \mathcal{E}_d} R(e, d)$ and $\rho(d) = |\mathcal{E}_d|$, the number of residue classes, mod d, excluded in the sifting process.

A situation where (4.6) arises is as follows. Suppose the weights α_n were the characteristic function of an interval of length X. Then (4.6) holds with $R(e, d) = O(1)$, as was observed in (1.2.2.1). In this case we would obtain $r_\mathcal{A}(d) = O(\rho(d))$, a conclusion already noted at (1.2.3.2) and used to derive (1.12).

1.3.5 Sifting Density

The function ρ introduced in (1.1) plays a central rôle in sifting arguments, in the first instance via its occurrence in the product $V(P(z))$ appearing in (1.8).

In the sifting situations we will encounter, of which some examples were provided in Sect. 1.2, the quantity $\rho(p)$ has an average value κ, in some appropriate sense. In many instances the following formula holds:

$$\sum_{p < z} \frac{\rho(p) \log p}{p} = \kappa \log z + O(1) . \qquad (5.1)$$

For example, in the prime twin problem, where we take $f(n) = n(n+2)$ in Sect. 1.2.3, we have $\rho(p) = 2$ for each prime $p > 2$, and (5.1) follows, with $\kappa = 2$, from Chebyshev's elementary result (1.1.4.1) on the distribution of primes.

28   1. The Structure of Sifting Arguments

It is however not entirely convenient to work with (5.1) in all situations of interest. We saw in (1.2.3.4), in the context of Goldbach's problem, that it is possible for $\rho(p)$, and hence the error term $O(1)$ in (5.1), to depend on the parameter X appearing in (1.1), in a way which would therefore have to be quantified. It is possible to handle the subject in a way based on the notion of *sifting density* as formulated in Definition 7 below, which will avoid having to deal with such questions in this and similar situations. Lemma 3 will make it clear that this particular ρ has sifting density 2.

In this connection the following observation is relevant: it ought to be sufficient to use a hypothesis on ρ which bounds $\rho(p)$ only from above. Consider first the context of a lower bound sieve given by an inequality of the type on the left of (1.5). If we decreased $\rho(p)$ then the number of excluded residue classes e_i described in (4.1) is decreased, so the available lower bound ought to become sharper (i.e. larger). Thus a decrease in $\rho(p)$ would be allowable in that it would not spoil the validity of a lower bound derived from (1.5). The situation is actually more subtle: we will see in Chapters 2 and 3 that it is the estimates for the quantities labelled $c^{\pm}(D, z)$ in (2.2) that will sharpen with decreasing $\rho(p)$, a remark that will apply to upper and lower bound sieves alike.

It appears already from our discussion of Legendre's sieve that the expression $V(P)$ defined in (1.8) plays a natural rôle in sieve arguments. In this volume we will normally express an important property of ρ in terms of the product V, as follows.

Definition 7. Say that κ is a *sifting density* for the function ρ if there exists a constant $L > 1$ (depending on κ) such that

$$\frac{V(P(w))}{V(P(z))} = \prod_{p | P(z)/P(w)} \left(1 - \frac{\rho(p)}{p}\right)^{-1} \leq K_w \left(\frac{\log z}{\log w}\right)^{\kappa} \quad \text{if} \quad 2 \leq w < z, \tag{5.2}$$

with

$$K_w \leq 1 + \frac{L}{\log w}. \tag{5.3}$$

In many situations it will be convenient to use somewhat less, namely that there exists a constant $K > 1$ such that

$$K_w \leq K \quad \text{when} \quad w \geq 2. \tag{5.4}$$

It will be clear from the occurrence of the constants K or L whether (5.3) or only the weaker version (5.4) is being assumed.

Note that the choice $w = p$, $z = p + \varepsilon$ shows that (5.2) implies

$$\left(1 - \frac{\rho(p)}{p}\right)^{-1} \leq K_p, \tag{5.5}$$

so that $\rho(p)$ is bounded away from p: in fact $\rho(p) \leq p(1 - 1/K_p)$.

1.3 A General Formulation of a Sifting Situation

In (5.2) we have explicitly written the consequence $p|P(z)$ of Hypothesis 2 for the sake of emphasis. It can be of some significance. For example, if $\mathcal{A}$ and P are as in Sect. 1.2.5 then we would have

$$\rho(p) = \begin{cases} 1 & \text{if } p \equiv 3, \mod 4 \\ 0 & \text{otherwise,} \end{cases}$$

from which it follows by a simple argument appearing in Sect. 4.5.2 that (5.2) holds with $\kappa = \frac{1}{2}$.

There are a few other remarks that should be made about Definition 7. In the first place it does not define a sifting density κ uniquely; if (5.2) holds for $\kappa = \kappa_0$ then it holds when $\kappa \geq \kappa_0$. This property will be useful. However, in connection with the sieve with weights considered in Chapter 5 it will sometimes be necessary to make the following assumption bounding $\rho(p)$ from both sides, in which case κ, if it exists, will be specified uniquely by ρ.

Definition 8. Say that κ is a *two-sided sifting density* for ρ if it satisfies Definition 7 and, additionally, there exists $L' \geq L$ such that

$$\frac{V(P(w))}{V(P(z))} \geq \left(1 - \frac{L'}{\log w}\right)\left(\frac{\log z}{\log w}\right)^{\kappa} \quad \text{if } 2 \leq w < z . \tag{5.6}$$

The "constant" L' is, of course, independent of w and z but depends on ρ, which in practice will have been chosen with other parameters of the current situation in mind. For example, in Sect. 1.2.3 the function ρ depends on $N = X$, and it will be found that (5.6) cannot be satisfied unless L' is also allowed to depend on X. In fact in that particular situation a choice $L' = c \log \log X$ can be found that satisfies (5.6).

A technicality that will arise is that in some applications it will be important that (5.2) is known with K_w sufficiently close to 1. This does not follow from (5.3) unless w is sufficiently large. This raises a problem, because in some situations we will prefer to avoid the technical complications involved in working with the more elaborate hypothesis (5.3) as opposed to the simpler (5.4) in which K is constant.

One way of dealing with this situation involves using an adjustment of the internal parameters of the sifting mechanism. The simplest version of this idea adjusts the sifting density κ; if (5.3) holds with $\kappa = \kappa_0$ then for any $\varepsilon > 0$ we obtain

$$\frac{V(P(w))}{V(P(z))} \leq K'\left(\frac{\log z}{\log w}\right)^{\kappa_0 + \varepsilon} \quad \text{if } 2 \leq w < z , \tag{5.7}$$

with $K' = K(\log w / \log z)^{\varepsilon}$. Thus (5.2) and (5.4) apply with κ replaced by $\kappa + \varepsilon$ and

$$K \leq 1 + O\left(\frac{1}{\log^{\varepsilon} z}\right) \quad \text{if } 2 \leq w < z . \tag{5.8}$$

30   1. The Structure of Sifting Arguments

If z is large throughout the argument, as will normally be the case, we can proceed using now a sifting density $\kappa_0 + \varepsilon$. Here ε can be taken to be an arbitrary constant, or a suitable decreasing function such as $1/\log\log z$. On the other hand if we choose $\varepsilon = 1$ then we obtain $K = 1 + O(1/\log z)$ in (5.8). There is an example where this type of reasoning is used in Sect. 3.3.5.

It will not always be convenient to vary the assumed sifting density κ in this way. For example in the discussions in Chapter 4 where we use information specific to $\kappa = 1$ a move to $\kappa = 1 + \varepsilon$ would be highly inconvenient. In some arguments we avoid using (5.3) with small w by arranging to apply the sifting argument only to a sequence $\mathcal{A}$ whose members are not divisible by very small primes,

$$a \in \mathcal{A},\ p|a \implies p \geq z_1 , \tag{5.9}$$

say. Thus we may take $\rho(p) = 0$ when $p < z_1$ in (1.1). Then in (5.3) we may take $w \geq z_1$, so that (5.4) holds with $K = 1 + O(1/\log z_1)$. To secure (5.9) we use a separate preliminary sifting using a method more suitable for dealing with small sifting primes. This type of argument is used in Sect. 4.4 and elsewhere.

Partial Summation. We have already referred to this well-known technique, in connection with (1.1.4.1), but it is appropriate to add a few words about it here.

The discussion is eased if we use the notation of Stieltjes integrals (but nothing of their theory). For our purposes we may define them by the formula of integration by parts,

$$\int_w^z f(t)\, dc(t) = c(z)f(z) - c(w)f(w) - \int_w^z f'(t)c(t)\, dt , \tag{5.10}$$

when f' is (say) piecewise continuous on the interval $[w, z]$ and c is of bounded variation. Then partial summation (actually a combination of summation by parts and estimating sums by integrals) may be formulated as in Lemma 1(i). Part (ii) is more specialised, but covers a situation of frequent occurrence in the sequel. Part (iii) will be useful at one point in Chap. 5.

Lemma 1. *Denote*

$$C(x, y) = \sum_{x \leq n < y} c_n$$

whenever the indicated numbers c_n are defined.

(i) *If $w \leq z$ and f' is piecewise continuous on $[w, z]$ then*

$$\sum_{w \leq n < z} c_n f(n) = \int_w^z f(t)\, dC(w, t) = -\int_w^z f(t)\, dC(t, z) , \tag{5.11}$$

where the integrals with respect to t are given by (5.10).

1.3 A General Formulation of a Sifting Situation

(ii) *Suppose, in addition, that f is positive and monotone in $[w, z]$ and that $C(x, y)$ is given by an expression*
$$C(x, y) = \gamma(y) - \gamma(x) + e(x, y),$$
where $e(x, y) \leq E$ whenever $w \leq x \leq y \leq z$. Then
$$\sum_{w \leq n < z} c_n f(n) \leq \int_w^z f(t)\, d\gamma(t) + E \max\{f(w), f(z)\}.$$

(iii) *If f and g both satisfy the conditions required of f in parts (i) and (ii) then*
$$\sum_{w \leq n < z} c_n f(n) g(n) \leq \int f(t) g(t)\, d\gamma(t) + E \max_{w \leq u \leq z} f(u) \max_{w \leq v \leq z} g(v).$$

In part (i), the formulation in terms of $dC(w, t)$ is perhaps more familiar but we will see that it may be more convenient to use the other. Since $C(w, w) = 0$, (5.10) gives
$$\int_w^z f(t)\, dC(w, t) = C(w, z) f(z) - \int \sum_{w \leq n < t < z} c_n f'(t)\, dt,$$
and we have only to perform the integration over t. Similarly
$$-\int_w^z f(t)\, dC(t, z) = C(w, z) f(w) + \int \sum_{w \leq t < n < z} c_n f'(t)\, dt,$$
and part (i) follows.

When f decreases in part (ii) the first equation in (5.11) gives
$$\sum_{w \leq n < z} c_n f(n) = \int_w^z f(t)\, d\gamma(t) + \int_w^z f(t)\, de(w, t).$$
Since $e(w, w) = 0$ the last integral satisfies
$$\int_w^z f(t)\, de(w, t) = e(w, z) f(z) - \int_w^z e(w, t) f'(t)\, dt$$
$$\leq E f(z) + E \int_w^z (-f'(t))\, dt = E f(w),$$
as required. When f increases the second equation in (5.11) leads similarly to
$$\sum_{w \leq n < z} c_n f(n) = -\int_w^z f(t)\, d(-\gamma(t)) - \int_w^z f(t)\, de(t, z)$$
$$= \int_w^z f(t)\, d\gamma(t) + e(w, z) f(w) + \int_w^z e(t, z) f'(t)\, dt,$$
and the stated result follows as before.

The purpose of part (iii) is that it covers the situation where one of f, g is increasing and the other decreasing. Part (ii) gives

$$\sum_{w \le n < z} c_n g(n) = \int_x^y g(t)\, d\gamma(t) + E(x,y),$$

where $E(x,y) \le E \max_{x \le t \le y} g(t)$. Apply part (ii) with c_n replaced by $c_n g(n)$, $\gamma(t)$ replaced by

$$\Gamma(t) = \int_x^t g(u)\, d\gamma(u),$$

and $e(x,y)$ replaced by $E(x,y)$. Thus

$$\sum_{x \le n < y} c_n f(n) g(n) \le \int_x^y f(t)\, d\Gamma(t) + E \max_{w \le u \le z} f(u) \max_{w \le v \le z} g(v),$$

and part (iii) of the lemma follows.

The following situation, which arises in Sects. 2.2 and 7.1, is related to those considered in Lemma 1, so is considered here.

Lemma 2. *Suppose that, when t and n are in an interval $[x,y)$, $f'(t)$ is continuous and $f(t)$ is positive and decreasing, $g(n) \ge 0$, and $E(t,n) \le A$. Then*

$$\int_{x < t < y} f(t)\, d \sum_{x \le n < t} g(n) E(t,n) \le A \sum_{x \le n < y} f(n) g(n).$$

Since $f'(t) \le 0$ the integral on the left is

$$f(y) \sum_{x \le n < y} g(n) E(y,n) + \int_{x < t < y} \sum_{x \le n < t} g(n) E(t,n) |f'(t)|\, dt$$

$$\le A f(y) \sum_{x \le n < y} g(n) + A \int_{x < t < y} \sum_{x \le n < t} g(n) |f'(t)|\, dt$$

$$= A \int_{x < t < y} f(t)\, d \sum_{x \le n < y} g(n) = A \sum_{x \le n < y} f(n) g(n),$$

the last step being an instance of Lemma 1(i).

Alternative Sifting Density Hypotheses. In connection with the λ^2 method of Selberg discussed in Chapters 2 and 7 it will be more convenient, although not essential, to replace the hypothesis embodied in Definition 7 by a slightly stronger one. Let $g(d)$ be the multiplicative function defined for squarefree d by specifying

$$g(p) = \frac{\rho(p)}{p - \rho(p)} \tag{5.12}$$

1.3 A General Formulation of a Sifting Situation

for primes p. Then in place of (5.2) we may specify the existence of a constant $A > 1$ such that

$$\sum_{w \leq p < z} g(p) \log p \leq \kappa \log \frac{z}{w} + A \quad \text{when} \quad 2 \leq w < z. \tag{5.13}$$

The case $z \to w = p$ gives $g(p) \leq A/\log p$, so that $\rho(p) < p/(1 + A^{-1} \log p)$ and $\rho(p)$ is again bounded away from p, as in (5.5).

We will show in Lemma 5 that (5.13) implies the usual sifting density hypothesis. A suitable example will show that the direct converse is false, but Lemma 4 shows that (5.3) does, however, imply a relation only slightly weaker than (5.13). From the point of view of those interested in applications rather than theory these implications are perhaps not of first-rate importance, since frequently a function ρ arising in practice can be directly seen to satisfy (5.3) and (5.13). A situation that often arises is described in the next lemma.

Lemma 3. *Suppose $\rho(p) < p$ when $p \leq \kappa$ and $\rho(p) \leq \kappa$ for all p. Then ρ has sifting density κ, in the stronger sense expressed by (5.3), and also satisfies (5.13).*

The inequality $\rho(p) < p$ guarantees $\rho(p) \leq p(1 - 1/K)$ when $p \leq \kappa$, for a certain $K > 1$ depending only on κ, so that $(1 - \rho(p)/p)^{-1} < K$ for these p.
Mertens' product (1.1.4.2) shows that

$$\prod_{w \leq p < z} \left(1 - \frac{1}{p}\right)^\kappa = \left(\frac{\log z}{\log w}\right)^\kappa \left(1 + O\left(\frac{1}{\log w}\right)\right),$$

where the O-constant may depend on κ. When $w \geq \kappa$ this gives

$$\prod_{w \leq p < z} \left(1 - \frac{\rho(p)}{p}\right)^{-1} \leq \left(\frac{\log z}{\log w}\right)^\kappa \left(1 + O\left(\frac{1}{\log w}\right)\right) \prod_{w \leq p < z} \frac{\left(1 - \frac{1}{p}\right)^\kappa}{\left(1 - \frac{\kappa}{p}\right)}.$$

The last product over primes p is

$$\exp\left(\sum_{w \leq p < z} \kappa \log\left(1 - \frac{1}{p}\right) - \log\left(1 - \frac{\kappa}{p}\right)\right) = \exp\left(\sum_{w \leq p < z} O\left(\frac{1}{p^2}\right)\right)$$

$$= \exp\left(O\left(\frac{1}{w}\right)\right) = 1 + O\left(\frac{1}{w}\right), \tag{5.14}$$

so (5.3) follows, with some value of the constant L. It also holds when $w < \kappa$, possibly with a larger value of L, because the contribution from primes $p < \kappa$ to the product in (5.2) is bounded.

In a similar style observe when $w > \kappa$

$$\sum_{w \leq p < z} \frac{\rho(p) \log p}{p - \rho(p)} \leq \sum_{w \leq p < z} \frac{\kappa \log p}{p - \kappa} \leq \kappa \log \frac{z}{w} + O(1),$$

where we used (1.1.4.1). Thus (5.13) follows for these w, and remains valid for smaller w because $g(p)$ is bounded when $p \leq \kappa$.

In one direction the relationship between (5.3) and (5.13) is straightforward.

Lemma 4. *Suppose ρ is such that the function g satisfies (5.13). Then the sifting density hypothesis follows in the form (5.3), with*

$$K \leq e^{A/\log w} = 1 + \frac{L}{\log w}.$$

Apply partial summation to (5.13), using Lemma 1(ii) with $c_p = g(p) \log p$ and $c_n = 0$ if n is not prime. This gives

$$\sum_{w \leq p < z} g(p) \leq \kappa \int_w^z \frac{1}{\log t} \, d\log t + \frac{A}{\log w}.$$

Now observe that

$$\left(1 - \frac{\rho(p)}{p}\right)^{-1} = 1 + \frac{\rho(p)}{p - \rho(p)} = 1 + g(p). \tag{5.15}$$

Hence

$$\prod_{w \leq p < z} \left(1 - \frac{\rho(p)}{p}\right)^{-1} = \exp \sum_{w \leq p < z} \log(1 + g(p))$$

$$\leq \exp \sum_{w \leq p < z} g(p) \leq \left(\frac{\log z}{\log w}\right)^{\kappa} e^{A/\log w}.$$

This gives (5.2), with $K_w \leq e^{A/\log w} \leq 1 + L/\log w$ for a certain L depending on A, so that (5.3) follows as required.

The A-dependence of the number L in this argument is rather poor. While $L = O(A)$ would follow if $\log w > A$ were given, in general L might be as large as e^A.

If we start from the sifting density hypothesis (5.3), then the resulting estimate for the sum in (5.13) is the somewhat untidy (5.20). For this reason we will prefer to adopt the slightly stronger property (5.13) when this is appropriate. There will be no need to use (5.20) in this book, but (5.16)–(5.19) will be referred to later on.

Lemma 5. *Suppose that ρ has sifting density κ, and that $2 \leq w < z$.*
  (i) *When (5.4) holds we find*

$$\sum_{w \leq p < z} \frac{\rho(p) \log p}{p} \leq (\kappa + \log K) \log z. \tag{5.16}$$

1.3 A General Formulation of a Sifting Situation

(ii) *If (5.3) holds then*

$$\sum_{t \leq p < z} \frac{\rho(p)}{p} \leq \kappa \log \frac{\log z}{\log t} + \frac{L}{\log t}, \qquad (5.17)$$

$$\sum_{w \leq p < z} \frac{\rho(p) \log p}{p} \leq \kappa \log \frac{z}{w} + L \log\left(e \frac{\log z}{\log w}\right). \qquad (5.18)$$

Furthermore the function $g(p)$ defined in (5.12) satisfies

$$g(p) \leq \frac{L}{\log p}, \qquad (5.19)$$

$$\sum_{w \leq p < z} g(p) \log p \leq \kappa \log \frac{z}{w} + L(1 + \kappa) \log\left(e \frac{\log z}{\log w}\right) + \frac{L^2}{\log w}. \qquad (5.20)$$

From (5.2) we obtain

$$\sum_{t \leq p < z} \frac{\rho(p)}{p} \leq -\sum_{t \leq p < z} \log\left(1 - \frac{\rho(p)}{p}\right) = \kappa \log \frac{\log z}{\log t} + E(t), \qquad (5.21)$$

where $E(t) \leq \log K_t$, and $E(z) = 0$. Lemma 1 now gives

$$\sum_{w \leq p < z} \frac{\rho(p) \log p}{p} = -\int_w^z \log t \, d\left(\kappa \log \frac{\log z}{\log t} + E(t)\right) = \kappa \log \frac{z}{w} + R(w, z),$$

with

$$R(w, z) = E(w) \log w + \int_w^z E(t) \frac{dt}{t}.$$

If $K_t \leq K$ as in (5.4) then $R(w, z) \leq \log K \log z$, and (5.16) follows. If (5.3) holds then $E(t) \leq \log K_t \leq L/\log t$, and (5.18) follows. Also (5.21) now gives (5.17).

Use of the choice $w = p$, $z = p + \varepsilon$ in (5.2) gives

$$1 + \frac{\rho(p)}{p - \rho(p)} = \left(1 - \frac{\rho(p)}{p}\right)^{-1} \leq K_p \leq 1 + \frac{L}{\log p},$$

since K_p satisfies (5.5) and (5.3). Now (5.19) follows.

Lastly, note from (5.12) that

$$g(p) = \frac{\rho(p)}{p} + \frac{\rho(p) g(p)}{p}.$$

Hence

$$\sum_{w \leq p < z} g(p) \log p = \sum_{w \leq p < z} \frac{\rho(p) \log p}{p} + \sum_{w \leq p < z} \frac{\rho(p) g(p) \log p}{p}.$$

36   1. The Structure of Sifting Arguments

Since (5.19) gives $g(p)\log p \le L$, the inequality (5.20) now follows using (5.17) and (5.18).

The following example shows that the occurrence of the entries $\log\log z$ in (5.18) and (5.20) is essential. Take $\rho(p) = \kappa(1 + 1/\log p)$. Then (for w large enough to ensure $\rho(p) < p$ when $p \ge w$)

$$\sum_{w\le p<z} g(p)\log p \ge \sum_{w\le p<z} \frac{\rho(p)\log p}{p} = \kappa \sum_{w\le p<z} \frac{\log p + 1}{p}$$

$$= \kappa \log \frac{z}{w} + \kappa \log \frac{\log z}{\log w} + O(1),$$

so that in particular (5.13) does not hold. On the other hand

$$\prod_{w\le p<z}\left(1 - \frac{\rho(p)}{p}\right)^{-1} = \prod_{w\le p<z}\left(1 - \frac{\kappa}{p} - \frac{\kappa}{p\log p}\right)^{-1}$$

$$\le \exp\left(\sum_{w\le p<z} \frac{\kappa}{p} + O\left(\frac{1}{p\log p}\right)\right) = \exp\left(\kappa \log \frac{\log z}{\log w} + O\left(\frac{1}{\log w}\right)\right)$$

$$= \left(\frac{\log z}{\log w}\right)^{\kappa}\left(1 + O\left(\frac{1}{\log w}\right)\right),$$

so that ρ has sifting density κ as described by (5.3). Thus (5.3) does not imply (5.13).

The following consequence of the earlier results of the section will frequently be useful.

Lemma 6. *Suppose that ρ satisfies (5.3) and that $2 \le D^{1/v} < D^{1/s}$.*

(i) *If a function F is monotone and $F(x) \le B$ on an interval $[1/v, 1/s]$, then*

$$\sum_{D^{1/v}\le p<D^{1/s}} F\left(\frac{\log p}{\log D}\right)\frac{\rho(p)}{p} \le \kappa \int_{1/v}^{1/s} F(x)\frac{dx}{x} + \frac{BL}{\log w}.$$

(ii) *If G is also monotone and $G(x) \le C$ on the same interval then*

$$\sum_{D^{1/v}\le p<D^{1/s}} F\left(\frac{\log p}{\log D}\right)G\left(\frac{\log p}{\log D}\right)\frac{\rho(p)}{p} \le \kappa \int_{1/v}^{1/s} F(x)\frac{dx}{x} + \frac{BCL}{\log w}.$$

The notation $w = D^{1/v}$, $z = D^{1/s}$ and the apparently arbitrary base $D > 1$ are inserted for future convenience. Let $c_n = \rho(p)/p$ if n is a prime p, and $c_n = 0$ otherwise. Then in Lemma 1(ii) we may take

$$f(t) = F\left(\frac{\log n}{\log D}\right), \qquad \gamma(t) = \kappa \log\log t, \qquad e(x,y) \le E = L/\log w,$$

because of (5.17) and (5.3). This yields part (i).

Part (ii) follows in a similar way, but using Lemma 1(iii).

1.3 A General Formulation of a Sifting Situation

The procedure in Lemma 6 is not precise enough for all our purposes, because it rests on an estimate

$$\int E(t)|f'(t)|\,dt \le E\int |f'(t)|\,dt$$

when $E(t) \le E$, and on occasions, as when deriving (5.18), we need more detailed information about $E(t)$.

1.3.6 The Sifting Limit $\beta(\kappa)$

In our proto-theorem (1.5) take $P = P(z)$, as in (1.9). Suppose that the function ρ has a finite sifting density, as in (5.2) and (5.4). We wish to characterise those pairs D, z for which $V^-(D, P(z)) > 0$, so that the lower bound in (1.5) is non-trivial. In the efficient sieve methods discussed later such information is available if $z^s \le D$ for certain bounded s, depending on κ and the constant K in (5.4). For an example of this situation one may examine Theorem 3.3.1 and its corollaries.

We are interested in the infimum of those s for which the desired inequality

$$V^-\left(z^s, P(z)\right) > 0 \tag{6.1}$$

holds. For this purpose assume that the constant K in (5.4) takes some fixed value, such as 1. The infimum in question now depends only on κ. Consider the class $\mathcal{C}$ of those ρ which satisfy (5.3) with $K = 1$.

Definition 9. The *sifting limit* $\beta(\kappa)$ is the infimum over $\mathcal{C}$ of those s for which (6.1) holds (for some function λ_D^-, depending on ρ, that obeys (1.2)).

One may also restrict the choice of λ_D in Definition 9 to some specified class, and then refer to the sifting limit for some specific sifting method, determined by the class in question. Thus we may speak of "the sifting limit for Brun's method", for example.

1.3.7 Composition of Sieves

The following method of constructing a sieve from two others is useful, for example when the products P_1, P_2 are composed respectively of small primes $< w$ and of large primes $\ge w$, and differing constructions are required in the two cases.

Suppose, for $i = 1, 2$, that functions $\lambda_{D_i}^\pm$ are supported on squarefree products P_i of disjoint sets of sifting primes, so that the products P_1, P_2 have only the integer 1 in common. Thus each divisor d of $P_1 P_2$ is uniquely expressible as $d = d_1 d_2$ with $d_i \mid P_i$. The principle of composing two sieves is then as follows.

38   1. The Structure of Sifting Arguments

Lemma 7. *Suppose $\lambda_{D_i}^{\pm}$ has level D_i, in the sense (1.3), and set $D = D_1 D_2$. Define*

$$\lambda_D^+(d) = \lambda_{D_1}^+(d_1)\lambda_{D_2}^+(d_2) \, , \tag{7.1}$$

$$\lambda_D^-(d) = \lambda_{D_1}^-(d_1)\lambda_{D_2}^+(d_2) + \lambda_{D_1}^+(d_1)\lambda_{D_2}^-(d_2) - \lambda_{D_1}^+(d_1)\lambda_{D_2}^+(d_2) \, . \tag{7.2}$$

If $\lambda_{D_i}^{\pm}$ are sifting functions for the products P_i, as in (1.2), then λ_D^+, λ_D^- are respectively upper and lower sifting functions of level D for the product $P_1 P_2$, in that they satisfy (1.2) and (1.3).

More generally, suppose the functions $\lambda_{D_i}^{\pm}$ satisfy not (1.2) but (1.4), in which the "weight" function ψ is now to satisfy

$$\psi(d_1) = 1, \ \psi(d_1 d_2) = \psi(d_2) \quad \text{when} \quad d \mid P_1 \, . \tag{7.3}$$

Then λ_D^+, λ_D^- also satisfy (1.4).

The corresponding sums $V^{\pm}$ defined by (1.6) satisfy

$$V^+(D, P) = V^+(D_1, P_1) V^+(D_2, P_2)$$
$$V^-(D, P) = V^-(D_1, P_1) V^+(D_2, P_2) + V^+(D_1, P_1) V^-(D_2, P_2) \tag{7.4}$$
$$- V^+(D_1, P_1) V^+(D_2, P_2) \, .$$

Furthermore, if

$$\lambda_{D_i}^{\pm}(d_i) = \mu(d_i) \chi^{\pm}(d_i) \, , \quad \text{where} \quad 0 \leq \chi^{\pm}(d_i) \leq 1 \, , \tag{7.5}$$

then $|\lambda_D^{\pm}(d)| \leq 1$.

It is clear that $\lambda_D^{\pm}$ have level D because $d = d_1 d_2 \leq D$ when $d_i \leq D_i$. For brevity denote

$$\Lambda_i^{\pm} = \sum_{d_i \mid (A, P_i)} \lambda_{D_i}^{\pm}(d_i), \quad \Lambda^{\pm} = \sum_{d \mid (A, P_1 P_2)} \lambda_D^{\pm}(d) \, .$$

Then on writing $d = d_1 d_2$ we obtain from (1.2)

$$\Lambda^+ = \Lambda_1^+ \Lambda_2^+ \geq S(A, P_1) S(A, P_2) = S(A, P_1 P_2) \, , \tag{7.6}$$

so that λ^+ satisfies the second inequality in (1.2), this being straightforward since Λ_i^+ and $S(A, P_i)$ are non-negative.

For Λ^- a directly similar argument would fail because it is not guaranteed that $\Lambda_i^- \geq 0$. However, from (7.2) we obtain

$$\Lambda^- = \Lambda_1^- \Lambda_2^+ + \Lambda_1^+ \Lambda_2^- - \Lambda_1^+ \Lambda_2^+$$
$$= \Lambda_1^- \Lambda_2^- - (\Lambda_1^+ - \Lambda_1^-)(\Lambda_2^+ - \Lambda_2^-) \, . \tag{7.7}$$

Apart from reducing to $S(A, P_1)S(A, P_2)$ in the extreme case when (1.2) holds with equality, this expression increases with Λ_1^- and Λ_2^- since $\Lambda_i^+ \geq 0$. Since $\Lambda_i^- \leq S(A, P_i)$, (7.7) now gives

$$\Lambda^- \leq S(A, P_1)S(A, P_2) - \left(\Lambda_1^+ - S(A, P_1)\right)\left(\Lambda_2^+ - S(A, P_2)\right) \\ \leq S(A, P_1)S(A, P_2) = S(A, P) , \tag{7.8}$$

as required, because $S(A, P_i) \leq \Lambda_i^+$.

A condition such as (7.3) is required for the generalisation involving the "weight" function ψ because the choices of ψ required in Chapter 5 will not be multiplicative. However, when (7.3) holds we still find

$$S(A, P_1, \psi)S(A, P_2, \psi) = \sum_{d_1|(A,P_1)} \mu(d_1)\psi(d_1) \sum_{d_2|(A,P_2)} \mu(d_2)\psi(d_2) \\ = \sum_{d|(A,P)} \mu(d)\psi(d) = S(A, P, \psi) ,$$

analogous to the equalities in (7.6) and (7.8), so that the inequalities (1.4) follow in the same way as did (7.6) and (7.8).

For the identities (7.4), multiply (7.1) and (7.2) by $\rho(d_1)\rho(d_2)/d_1 d_2$ and sum over d_1 and d_2.

Lastly, note that when (7.5) holds the inequality $\left|\lambda_D^+(d)\right| \leq 1$ follows immediately from (7.1), the sign of $\mu(d_i)$ not being important. For $\lambda_D^-(d)$ observe that (7.5) gives

$$\mu(d_1)\mu(d_2)\lambda_D^-(d) = \chi^-(d_1)\chi^+(d_2) + \chi^+(d_1)\chi^-(d_2) - \chi^+(d_1)\chi^+(d_2) .$$

This expression increases with $\chi^-(d_1)$ and $\chi^-(d_2)$. Consequently it lies between $-\chi^+(d_1)\chi^+(d_2)$ and

$$\chi^+(d_1) + \chi^+(d_2) - \chi^+(d_1)\chi^+(d_2) = 1 - \left(1 - \chi^+(d_1)\right)\left(1 - \chi^+(d_2)\right) ,$$

both of which lie in the interval $[-1, 1]$. Hence $\left|\lambda_D^-(d)\right| \leq 1$, as asserted.

1.4 Notes on Chapter 1

Sect. 1.1.1. A record of the work of Eratosthenes on the sieve appears in the work of Nicomachus from the first century A.D. [Dickson (1920)] lists four printed editions of Nicomachus's works, all in Greek. The account in the text is based upon an English translation [Nicomachus (1926)].

Sect. 1.1.2. Legendre's account is in [Legendre (1808)]. As is now usual, our account refers to the Möbius function μ, but this did not appear until [Möbius (1832)], and then not in this context. The underlying idea of the Inclusion-Exclusion principle is older. In number theory it occurs in [Euler (1775)], and in connection with probability goes back to the writings of D. Bernoulli and others in the early 18th century.

Sect. 1.3.4. The extensive material presented in "Lectures on Sieves" in [Selberg (1991)] starts with a set of axioms of this type.

Sect. 1.3.5. The one-sided form of the specification of sifting density, as in (5.2), appeared in [Halberstam and Richert (1974)] (also in [Halberstam and Richert (1971)], actually in a form more closely related to (5.13). It was not, however, very extensively used by these authors. For the most part they worked with a two-sided condition on ρ related more closely to (5.1). With the difficulty mentioned in Sect. 1.2.3 in mind, they dealt with the entry $O(1)$ in (5.1) in an unsymmetrical way, in that they placed it in an interval $(-L, A)$ and made the (relatively mild) L-dependence of their results explicit.

In this volume we work almost entirely on the basis of (5.2) or (5.13). An exception will arise in Sect. 5.3. Results derived from (5.2) or (5.13) are, of course, independent of any such parameter L.

Sect. 1.3.7. The principle of composing two sieves described here appears in [Selberg (1972)]. More recently, it was used in the "vector sieve" [Brüdern and Fouvry (1994), (1996)]. It will be used fairly extensively in this volume, on grounds of convenience rather than necessity. This point is amplified somewhat in our note on Sect. 4.4.1.

2. Selberg's Upper Bound Method

The sieve method developed by Brun around the year 1920 was, with some reason, generally perceived as being rather complicated. We will discuss Brun's ideas in Chap. 3, in a way which we hope may appear less complicated that some of the earliest accounts. It must be agreed, however, that to follow the details of Brun's methods through to even the simplest of its applications in number theory is still some way from being a trivial matter.

It was thus a welcome event when Selberg described his sieve method in 1947. As he pointed out, his method was simpler than Brun's and (at any rate in the situations more likely to be of direct interest) also led to better results. In many situations this remark remains true today, notwithstanding the subsequent developments of combinatorial sieve methods which we will describe later, more particularly in Chap. 4.

Selberg's method, in its basic form, is an upper bound sieve method: in the first place it delivers only an inequality of the type appearing on the right of our prototype inequality (1.3.1.2), involving an upper sifting function λ_D^+. It is this basic upper bound construction that we discuss in this chapter. The question of how Selberg's ideas should be turned to the lower bound situation is a much more difficult one, which has still not been given a complete solution. There is a discussion of this matter in Chap. 7.

The problem of constructing a satisfactory upper bound sieve is that of satisfying the right-hand inequality in (1.3.1.2):

$$S(\mathcal{A}, P(z)) \leq \sum_{d \mid (a, P(z))} \lambda_D^+(d) ,$$

while keeping the level of support $d < D$ on which $\lambda(d) \neq 0$ satisfactorily small, so that the effect of the remainder terms $r_\mathcal{A}(d)$ in (1.3.1.5) does not become excessively large. In Selberg's method this requirement is guaranteed in a very simple way, by arranging that the sum over d is a square:

$$\sum_{d \mid A} \lambda_D^+(d) = \left(\sum_{d_1 < \sqrt{D}} \lambda(d_1) \right)^2 ,$$

with $\lambda(1) = 1$. This will allow us to operate effectively with $z = \sqrt{D}$, very substantially larger than the value $\log D$ arising in Legendre's method, for example.

2. Selberg's Upper Bound Method

Selberg's result, as described in Sect. 2.1, is very general, requiring only the universal hypothesis $\rho(p) < p$ and the expression (1.3.1.1) of $\mathcal{A}_d$ in terms of the multiplicative function ρ, which may be regarded as a mere definition of the remainder term $r(\mathcal{A}, d)$. The result involves a sum

$$G(x) = \sum_{d<x} \frac{\mu^2(n)\rho(n)}{\rho^*(n)} ,$$

for a certain related multiplicative function ρ^* which we will define.

There then arises the question of estimating the sum $G(x)$, which requires some information about the function ρ. Initially we will assume essentially that ρ is precisely that which occurs in the application under consideration, but these *ad hoc* assumptions will be replaced in Sect. 2.2 by consequences of the condition that ρ has sifting density κ, as described in Sect. 1.3.5. These will lead to estimations of $G(x)$ valid under a moderate degree of generality. This matter will be taken further in Chap. 7.

Some selected applications of the results in Sects. 2.1 and 2.2 to situations in Number Theory are given in Sect. 2.3.

2.1 The Sifting Apparatus

Selberg's basic theorem on the upper bound sieve is Theorem 1 below. This rests on the inequality (1.14), which follows since squares of real numbers are non-negative. This will raise the question of optimising a certain quadratic form $V^+(\lambda)$ in a large number of variables $\lambda(d)$, subject to a single normalising constraint $\lambda(1) = 1$. The optimisation process uses the simple identity for $V^+(\lambda)$ recorded in Lemma 1.

Theorem 1 raises the question of estimating a main term (depending on a sum G of multiplicative functions) and an associated error term E. The estimation of E will be helped by the convenient bound on the optimised $\lambda(d)$ recorded in Lemma 2. In certain circumstances a sharper bound on E can be obtained as in Theorem 3.

The sum G arising in the main term is discussed in Sect. 2.2. However, in certain situations (essentially when $\rho(p) = 1$ apart from finitely many "bad" primes p) the simple procedure described in Theorem 4 and extended in Sect. 2.3.1 suffices to give a result of excellent quality.

2.1.1 Selberg's Theorem

We assume familiarity with some of the notations and definitions from Chap. 1. In particular we need the expression

$$|\mathcal{A}_d| = X\frac{\rho(d)}{d} + r_\mathcal{A}(d) \quad \text{if} \quad d|P, \ d < D \tag{1.1}$$

2.1 The Sifting Apparatus

of the number $\mathcal{A}_d$ of a in $\mathcal{A}$ divisible by a squarefree number d, where ρ is multiplicative. Also, we frequently refer to the expression $S(\mathcal{A}, P)$ for the number of a in $\mathcal{A}$ divisible by no prime dividing P, and to the associated expression $S(a, P)$ that counts such numbers a. The symbol $[d_1, d_2]$ denotes a least common multiple.

A function ρ^* will be the multiplicative function "conjugate" to ρ in that its values at primes p are given as

$$\rho^*(p) = p - \rho(p) . \tag{1.2}$$

This defines $\rho^*(d)$ for squarefree d, the only d that will be relevant in the sequel. Then the function from (1.3.5.12) is

$$g(n) = \begin{cases} \rho(n)/\rho^*(n) & \text{if } n \text{ is squarefree} \\ 0 & \text{otherwise.} \end{cases} \tag{1.3}$$

In Theorem 1 and elsewhere we shall denote

$$G(x) = \sum_{n < x; n | P} g(n) , \tag{1.4}$$

where P is a specified product of primes.

Theorem 1. *Suppose (1.1) holds. Then*

$$S(\mathcal{A}, P) \leq \frac{X}{G(\sqrt{D})} + E(D, P) , \tag{1.5}$$

where G is as in (1.4),

$$E(D, P) = \frac{1}{\lambda^2(1)} \sum_{d_i < \sqrt{D}, d_i | P} \lambda(d_1) \lambda(d_2) r_\mathcal{A}\big([d_1, d_2]\big) , \tag{1.6}$$

and the real numbers $\lambda(d)$ are given, for some $C \neq 0$, by

$$\frac{\lambda(d)\rho(d)}{d} = C \mu(d) \sum_{\substack{h \equiv 0, \bmod d \\ h < \sqrt{D}, h | P}} g(h) \quad \text{if} \quad \rho(d) \neq 0 , \tag{1.7}$$

with $\lambda(d) = 0$ if $\rho(d) = 0$.

Taking $d = 1$ in (1.7) shows

$$C = \frac{\lambda(1)}{G(\sqrt{D})} . \tag{1.8}$$

The usual normalisation in Theorem 1 is to make $\lambda(1) = 1$.

2. Selberg's Upper Bound Method

The value of $\lambda(d)$ when $\rho(d) = 0$ is irrelevant to the coefficient of X in Theorem 1. The choice $\lambda(d) = 0$ for these d is the most efficient one when the error term $E(D,P)$ is considered.

Since g is as in (1.3), substituting $h = dk$ in (1.7) gives

$$\lambda(d) = C\,\mu(d)\,\frac{d}{\rho^*(d)} \sum_{\substack{(k,d)=1 \\ k<\sqrt{D}/d; k|P}} g(k). \tag{1.9}$$

The proof of Theorem 1 embodies the more general statement (1.15) in which $\lambda(d)$ need not be as stated. The form (1.7) as given is the one in which $\lambda(d)$ has been chosen so as to optimise the coefficient of X in (1.5).

We will need the following identity, in which the numbers $\lambda(d)$ may be arbitrary, subject only to a restriction that they are supported on an interval $1 \leq d < \sqrt{D}$, so that all the sums appearing in Lemma 1 are finite. The restriction to $\rho(h) \neq 0$ in (1.11) is a natural one, because for other h the summand $x^2(h)/g(h)$ is a multiple of $\rho(h)$.

Lemma 1. *Denote*

$$V^+(\lambda) = \sum_{d_1|P} \sum_{d_2|P} \frac{\lambda(d_1)\lambda(d_2)\rho([d_1,d_2])}{[d_1,d_2]}. \tag{1.10}$$

Then

$$V^+(\lambda) = \sum_{h<\sqrt{D};\rho(h)\neq 0} \frac{x^2(h)}{g(h)}, \tag{1.11}$$

where g is as in (1.3) and

$$x(h) = \sum_{d\equiv 0,\,\mathrm{mod}\,h} \frac{\lambda(d)\rho(d)}{d}. \tag{1.12}$$

Moreover

$$\frac{\lambda(d)\rho(d)}{d} = \sum_k \mu(k)x(kd). \tag{1.13}$$

This lemma is almost immediate. Using $[d_1,d_2] = d_1d_2/(d_1,d_2)$ in the equation (1.10) gives

$$V^+(\lambda) = \sum_{\substack{d_1\ d_2 \\ \rho((d_1,d_2))\neq 0}} \frac{\lambda(d_1)\rho(d_1)}{d_1}\,\frac{\lambda(d_2)\rho(d_2)}{d_2}\,\frac{(d_1,d_2)}{\rho((d_1,d_2))}.$$

2.1 The Sifting Apparatus

Since $f = (d_1, d_2)$ is squarefree the last fraction may be expressed as

$$\frac{f}{\rho(f)} = \prod_{p|f} \frac{p}{\rho(p)} = \prod_{p|f}\left(1 + \frac{\rho^*(p)}{\rho(p)}\right) = \sum_{h|f}\frac{1}{g(h)}.$$

This gives (1.11), where $x(h)$ is as in (1.12). Lastly, (1.12) inverts to

$$\sum_k \mu(k)x(kN) = \sum_{kN|d}\mu(k)\frac{\lambda(d)\rho(d)}{d} = \sum_{kNl=d} = \sum_{mN=d}\frac{\lambda(d)\rho(d)}{d}\sum_{k|m}\mu(k)$$

$$= \frac{\lambda(N)\rho(N)}{N},$$

which is (1.13). Here we used the characteristic property (1.1.2.8) of the Möbius function.

Proof of Theorem 1. Suppose, for the moment, that $\lambda(d)$ are arbitrary real numbers supported on squarefree $d < \sqrt{D}$. The starting point is the inequality

$$\lambda^2(1)S(a,P) \le \left(\sum_{d|(a,P)}\lambda(d)\right)^2. \tag{1.14}$$

This holds because if $(a, P) = 1$ then $S(a, P) = 1$ and both sides of (1.14) take the value $\lambda^2(1)$, but if $(a, P) > 1$ then $S(a, P) = 0$ and the right side is non-negative.

The right side of (1.14) can be written

$$\sum_{d_1|(a,P)}\sum_{d_2|(a,P)}\lambda(d_1)\lambda(d_2) = \sum_{d_1|P}\sum_{d_2|P}\lambda(d_1)\lambda(d_2)\sum_{[d_1,d_2]|a}1.$$

On summing over a in $\mathcal{A}$ and expressing the result in terms of ρ from (1.1) this gives

$$\lambda^2(1)S(\mathcal{A}, P) \le X V^+(\lambda) + \lambda^2(1)E(D, P), \tag{1.15}$$

with $V^+(\lambda)$ as in (1.10) and $E(D, P)$ as stated in (1.6).

In discussing $V^+(\lambda)$ we may streamline the notation by using the convention that $\rho(d) = 0$ if $d \nmid P$, together with a natural one that terms with $\rho(d) = 0$ are not to be included in summations over d. This will save a certain amount of repetition of the conditions $d|P$ or $\rho(d) \ne 0$, where these are implied. Similarly, the conditions $d < \sqrt{D}$, $\mu^2(d) = 1$ need not be explicit in a sum involving $\lambda(d)$.

Theorem 1 will follow by choosing $x(k)$ so as to minimise, for given $\lambda(1)$, the expression (1.11) for the quantity $V^+(\lambda)$. The constraint

$$\sum_{k<\sqrt{D}}\mu(k)x(k) = \lambda(1) \tag{1.16}$$

is now required by (1.13).

46   2. Selberg's Upper Bound Method

In this connection recall Cauchy's inequality

$$\left(\sum_{h<H} a_h b_h\right)^2 \le \sum_{h<H} a_h^2 \sum_{h<H} b_h^2,$$

with holds with equality when there is a constant C with $a_h = Cb_h$ whenever $h < H$. This is an immediate consequence of the identity

$$\tfrac{1}{2}\left(\sum_{h<H, k<H} a_h b_k - a_k b_h\right)^2 = \sum_{h<H} a_h^2 \sum_{k<H} b_k^2 - \left(\sum_{h<H} a_h b_h\right)^2.$$

Apply Cauchy's inequality to the condition (1.16) in a way that relates to (1.11):

$$\lambda^2(1) = \left(\sum_{k<\sqrt{D}} \frac{\mu(k)x(k)}{\sqrt{g(k)}} \cdot \sqrt{g(k)}\right)^2$$

$$\le \left(\sum_{k<\sqrt{D}} \frac{x^2(k)}{g(k)}\right)\left(\sum_{k<\sqrt{D}} g(k)\right) \le V^+(\lambda) G(\sqrt{D}), \quad (1.17)$$

where $G(\sqrt{D})$ is as stated in (1.4). Equality occurs in (1.17) if

$$x(k) = C\mu(k)g(k) \quad \text{when} \quad k < \sqrt{D}, \quad (1.18)$$

in which case (1.15) gives the conclusion (1.5) required in Theorem 1. This situation is attained if we make $\lambda(d)$ satisfy (1.13) with these $x(k)$, so that

$$\frac{\lambda(d)\rho(d)}{d} = C \sum_{k<\sqrt{D}/d} \mu(k)\mu(kd)g(kd) = C\mu(d) \sum_{\substack{h \equiv 0, \bmod d \\ h < \sqrt{D}}} g(h).$$

This is the value stated in (1.7), in which the implied condition $h \mid P$ was written explicitly. This completes the proof of Theorem 1.

2.1.2 The Numbers $\lambda(d)$

Theorem 2 below will give a convenient property of the numbers $\lambda(d)$ appearing in Theorem 1. This is related to the following remark, which will be useful in the discussion of some of the applications of the theorem in Sect. 2.2. For given d, a squarefree number h may be written uniquely as $h = fn$, where $f \mid d$ and $(n,d) = 1$. Hence

$$\sum_{h<\sqrt{D}} g(h) \le \sum_{f \mid d} g(f) \sum_{(n,d)=1, n<\sqrt{D}} g(n). \quad (2.1)$$

2.1 The Sifting Apparatus

Observe from (1.2) that for squarefree d the sum over f in (2.1) is

$$\sum_{f|d} g(f) = \prod_{p|d}\left(1 + \frac{\rho(p)}{p - \rho(p)}\right) = \frac{d}{\rho^*(d)}. \tag{2.2}$$

As in Sect. 2.1.1, the fact that f, g and h divide P may be taken as implicit in the definition of the function ρ.

Theorem 2. *The numbers $\lambda(d)$ described in Theorem 1 satisfy the inequality*

$$|\lambda(d)| \leq |\lambda(1)|.$$

Corollary 2.1. *In Theorem 1, the error term satisfies*

$$E(D,P) \leq \sum_{d_i < \sqrt{D}, d_i | P} |r_A([d_1, d_2])|.$$

Corollary 2.1 follows directly from (1.6) and Theorem 2. Since each prime has three representations as $[d_1, d_2]$ it implies

$$E(D,P) \leq \sum_{d<D} \mu^2(d) 3^{\nu(d)} |r_A(d)|,$$

but this estimate is not sharp with respect to those d exceeding $\sqrt{D}$.

Proof of Theorem 2. Let C be the normalising constant (1.8), so that

$$\lambda(1) = C \sum_{h < \sqrt{D}} g(h).$$

We need an inequality in the opposite sense to (2.1):

$$|\lambda(1)| \geq |C| \sum_{f|d} g(f) \sum_{\substack{(n,d)=1 \\ n < \sqrt{D}/d}} g(n). \tag{2.3}$$

With the identity (2.2) this gives

$$|\lambda(1)| \geq |C| \frac{d}{\rho^*(d)} \sum_{\substack{(n,d)=1 \\ n < \sqrt{D}/d}} g(n) = |C| \frac{d}{\rho(d)} \sum_{\substack{h \equiv 0, \bmod d \\ h < \sqrt{D}}} g(h) = |\lambda(d)|,$$

by the expression (1.9) for $\lambda(d)$. This proves Theorem 2.

The inequality (2.3) is not at all sharp as soon as d is significantly larger than 1, so the question arises whether one can improve upon the bound in

48   2. Selberg's Upper Bound Method

Corollary 2.1. This requires some further input relating to the quantity $r_A(d)$. One context where this is available is that considered in Sect. 1.2.3. There the function $\rho(p)$ arose as the number of roots of a polynomial congruence, so that $\rho(p) \geq 1$ when $p|P$.

Theorem 3 discusses a rather special case of this situation. There is a slight refinement, and an extension to a more general context, in Sect. 2.2.3.

Theorem 3. *Suppose that in* (1.1) *we have*

$$r_A(d) \leq \rho(d) , \quad \rho(d) = 1 \quad \text{when} \quad d|P . \tag{2.4}$$

Then in Theorem 1 we obtain

$$E(D,P) \leq \frac{A^2 D}{G^2(\sqrt{D})},$$

where A is a certain absolute constant.

In the current situation $g(h) = 1/\phi(h)$, where ϕ is Euler's function. Now (1.7) gives

$$\sum_{\substack{d<\sqrt{D} \\ d|P}} |\lambda(d)| = |C| \sum_{d<\sqrt{D}} d \sum_{\substack{h \equiv 0, \bmod d \\ h<\sqrt{D}, h|P}} \frac{\mu^2(h)}{\phi(h)} = |C| \sum_{\substack{h<\sqrt{D} \\ h|P}} \frac{\mu^2(h)\sigma(h)}{\phi(h)} ,$$

where σ is the sum-of-divisors function. Here

$$\frac{\mu^2(h)\sigma(h)}{\phi(h)} = \prod_{p|h}\left(\frac{p+1}{p-1}\right) = \prod_{p|h}\left(1 + \frac{2}{p-1}\right) = \sum_{f|h} \frac{\mu^2(f)2^{\nu(f)}}{\phi(f)} ,$$

where $\nu(f)$ is the number of (distinct) prime factors of f. Thus

$$\sum_{\substack{d<\sqrt{D} \\ d|P}} |\lambda(d)| = |C| \sum_{\substack{h<\sqrt{D} \\ h|P}} \sum_{f|h} \frac{\mu^2(f)2^{\nu(f)}}{\phi(f)} \leq |C| \sum_{f<\sqrt{D}} \frac{\mu^2(f)2^{\nu(f)}}{\phi(f)} \sum_{\substack{h \equiv 0, \bmod f \\ h<\sqrt{D}, h|P}} 1 .$$

Here the inner sum does not exceed $\sqrt{D}/f$, whence

$$\sum_{d<\sqrt{D}} |\lambda(d)| \leq |C|\sqrt{D} \sum_{f<\sqrt{D}} \frac{\mu^2(f)2^{\nu(f)}}{f\phi(f)} \leq |C|\sqrt{D} \prod_p \left(1 + \frac{2}{p(p-1)}\right) . \tag{2.5}$$

When (2.4) holds (1.6) gives

$$E(D,P) \leq \frac{1}{\lambda^2(1)}\left(\sum_{d<D, d|P} |\lambda(d)|\right)^2 \leq \frac{C^2}{\lambda^2(1)} A^2 D ,$$

with A given by the convergent product over primes p appearing in (2.5). This establishes Theorem 3, since $\lambda(1)$ and C are related as in (1.8).

2.1.3 A Simple Application

We will see that Theorem 1 leads easily to the following bound for the number of primes in an interval $(Y - X, Y]$.

Theorem 4. *If $Y \geq X \geq 2$, then*

$$\pi(Y) - \pi(Y - X) \leq \frac{2X}{\log X} + O\left(\frac{X}{\log^2 X}\right),$$

where the implied constant is absolute.

As first mentioned in Sect. 1.1.4, this bound has the significant property of being uniform in Y, no matter how large Y may be in terms of X.

Let $A = \{a \in \mathbb{N} : Y - X < a \leq Y\}$, as discussed in Sect. 1.2.2. Thus (1.1) holds with

$$\rho(p) = 1 \text{ for all } p, \qquad |r(A, d)| \leq 1 \text{ for all } d, \tag{3.1}$$

so that the conditions of Theorem 3 are satisfied. We use (3.1) when $d < D$, where D is to be specified.

As in all our simpler applications we will use Theorem 1 with $z = \sqrt{D}$. Let $P(z)$ consist of all the primes less than z. Then the conditions $h|P$, $d_i|P$ in Theorem 1 become redundant. Arguing as first seen in Sect. 1.1.2, observe

$$\pi(Y) - \pi(Y - X) \leq S(A, P(z)) + \pi(z), \tag{3.2}$$

where the entry $\pi(z)$ may be omitted if $Y - X > z$.

Estimate the right side of (3.2) from above using Theorem 1. In the situation (3.1) the equation (1.2) reduces to $\rho^*(p) = p - 1$, so that in Theorem 1 we now find

$$G(\sqrt{D}) = \sum_{h < \sqrt{D}} \frac{\mu^2(h)}{\phi(h)}, \tag{3.3}$$

where ϕ is Euler's function. This sum can be estimated asymptotically by standard techniques, but for our purposes it is better to use the following explicit bound from below:

$$\sum_{h < \sqrt{D}} \frac{\mu^2(h)}{\phi(h)} = \sum_{h < \sqrt{D}} \frac{\mu^2(h)}{h} \prod_{p|h} \left(1 - \frac{1}{p}\right)^{-1} = \sum_{h < \sqrt{D}} \frac{\mu^2(h)}{h} \sum_{p|m \Rightarrow p|h} \frac{1}{m}$$

$$\geq \sum_{n < \sqrt{D}} \frac{1}{n} > \log \sqrt{D}, \tag{3.4}$$

the last inequality following by comparing the sum with the corresponding integral in the elementary way already used in Sect. 1.1.3.

For the error term in Theorem 1 we may use Corollary 2.1. Then (3.1) gives

$$|E(D, P(\sqrt{D}))| \leq \left(\sum_{d < \sqrt{D}} 1\right)^2 \leq D. \tag{3.5}$$

Because of (3.3) and (3.4), Theorem 1 now gives

$$S(\mathcal{A}, P(\sqrt{D})) \leq \frac{2X}{\log D} + D. \tag{3.6}$$

This leads to the variant of Theorem 4 incorporating a slightly weaker error term $O(X \log \log X / \log^2 X)$. We may choose $D = X/\log^2 X$, for example.

The better result recorded in Theorem 4 follows on using Theorem 3 instead of Theorem 2. It gives

$$S(\mathcal{A}, P(\sqrt{D})) \leq \frac{2X}{\log D} + \frac{A^2 D}{\log^2 D}$$

in place of (3.5). This permits the choice $D = X$, thereby obtaining Theorem 4 as stated.

There are other applications that could be considered in the same way, in particular to questions involving primes in arithmetic progressions, but these are postponed until after discussion of some general technical questions in Sect. 2.2.

2.2 General Estimates of $G(x)$ and $E(D, P)$

The estimation of $G(\sqrt{D})$ made in Sect. 2.1.3 used specialised methods suitable in the particular context in hand. In this section we proceed under versions of a more general hypothesis that the function ρ has a sifting density κ. A little care is required to obtain treatments that proceed under these one-sided hypotheses, which in this section are stated in terms of the function g.

It is sufficient for all the theorems in this section if we adopt the stronger form of the sifting density hypothesis

$$\sum_{w \leq p < z} g(p) \log p \leq \kappa \log \frac{z}{w} + A \quad \text{when} \quad 2 \leq w < z.$$

first given in (1.3.5.13). It will be seen, however, that for each of these theorems a significantly weaker assumption of this type is sufficient. In particular the sifting density hypothesis from Sect. 1.3.5 makes an assumption about the behaviour of $\rho(p)$ over each interval, no matter how short, and no such assumption is required for Theorems 1 and 2.

2.2 General Estimates of $G(x)$ and $E(D, P)$

The principal result (Theorem 2) provides an asymptotic inequality for $G(x)$. A weaker estimate, given in Theorem 1, holds under less prescriptive conditions on ρ. In situations where Theorem 2 also applies, Theorem 1 would typically give an estimate of $G(x)$ from below which would be weaker to the extent of a constant factor. In situations where an O-estimate is all that is required from Theorem 2.1.1 such a result would be entirely satisfactory. In particular Theorem 1 would suffice in most of those situations where the older literature called upon the methods of V. Brun that will be discussed in Chap. 3.

In Theorem 3 we return to the question of providing better estimations, in certain circumstances, of the error term $E(D, P)$ in Theorem 2.1.1 than those that follow from Theorem 2.1.2.

2.2.1 An Estimate by Rankin's Device

In connection with the expression $G(x)$ introduced in Theorem 2.1.1 consider the so-called "incomplete" sum

$$G_z(x) = \sum_{d \leq x; d \mid P(z)} g(d) . \tag{1.1}$$

For future reference recall that the equation $g(p) = \rho(p)/(p - \rho(p))$ first introduced in Sect. 1.3.5 can be rewritten as

$$\left(1 - \frac{\rho(p)}{p}\right) g(p) = \frac{\rho(p)}{p}, \quad 1 + g(p) = \left(1 - \frac{\rho(p)}{p}\right)^{-1} . \tag{1.2}$$

Then note that $G_z(x)$ increases with z, so that in particular

$$G(x) = G_x(x) \geq G_z(x) , \tag{1.3}$$

when $z \leq x$. In connection with Theorem 2.1.1 a lower bound for $G(x)$ is required (with $x = \sqrt{D}$), so a lower bound for $G_z(x)$ would be relevant.

In a natural notation, denote

$$G_z(\infty) = \sum_{d \mid P(z)} g(d) ,$$

so that $G_z(x) = G_z(\infty)$ whenever $x \geq P(z)$. Observe, referring to (1.2) where necessary,

$$G_z(x) \leq G_z(\infty) = \prod_{p \mid P(z)} (1 + g(p)) = \prod_{p \mid P(z)} \left(1 - \frac{\rho(p)}{p}\right)^{-1} = \frac{1}{V(P(z))} . \tag{1.4}$$

This gives a connection between the sum $G_z(x)$ arising in Selberg's method and the product $V(P)$ that we might expect to arise in a sieve argument, following its occurrence in Legendre's sieve in Theorem 1.3.2. The inequality in (1.4) is in the wrong sense for our purposes, but one in the useful direction is recorded in Theorem 1.

For Theorem 1 the only assumptions about ρ that we need are the universal hypothesis $0 \leq \rho(p) < p$ from Sect. 1.3.1 and an upper bound

$$B(z) \leq B \; ; \quad \text{where} \quad B(z) = \frac{1}{\log z} \sum_{p<z} \frac{\rho(p) \log p}{p} \; . \tag{1.5}$$

Lemma 1.3.5 showed that such a bound, for all $z > 2$, is a (rather weak) consequence of the standard sifting density hypothesis: (1.5) followed with $B = \kappa + \log K$. In fact Theorem 1 uses (1.5) only for that z that appears in the theorem, so that the sharpest choice of B would be the number $B(z)$ itself.

Theorem 1. *Suppose that (1.5) holds, where $z \geq 2$, and write $z = D^{1/s}$. Then $G_z(\sqrt{D})$, as defined by (1.1), satisfies*

$$\frac{1 - \exp\bigl(-\psi_B(\tfrac{1}{2}s)\bigr)}{V(P(z))} \leq G_z(\sqrt{D}) \leq \frac{1}{V(P(z))} \; , \tag{1.6}$$

where, for each $v > 0$,

$$\psi_B(v) = \max\Bigl\{0, \, v \log \frac{v}{B} - v + B\Bigr\} = \int_{B<t<v} \log \frac{t}{B} \, dt \; . \tag{1.7}$$

In particular

$$\psi_B(\tfrac{1}{2}s) \sim \tfrac{1}{2} s \log s \quad \text{as } s \to \infty \; ,$$
$$\exp\bigl(-\psi_B(\tfrac{1}{2}s)\bigr) \leq e^{7B-s} \quad \text{for all } s > 0 \; .$$

Theorem 1 estimates $G_z(\sqrt{D})$ quite well when s is large. Inspection of the ensuing proof will show that the occurrence of the factors $\frac{1}{2}$ in Theorem 1 arises from the quadratic nature of Selberg's method and the associated occurrence of $\sqrt{D}$ in the statement of Theorem 1. This may be compared with the corresponding Corollary 3.3.1.2 derived in Sect. 3.3 by Brun's method, which is superior when s is large in that the factor $\frac{1}{2}$ is absent.

Possibly the most frequently applicable result in this book is the following, which is expressed using the notations from Sects. 1.2 and 1.3.

Corollary 1.1. *Suppose that there is a B, independent of z, such that (1.5) holds for all $z \geq 2$. Let E be as in Corollary 2.1.2.1. Suppose $w \geq z = D^{1/s}$, with $s > 7B$. Then*

$$S(\mathcal{A}, P(w)) \leq X\, V(P(z))\bigl(1 + O(e^{-s})\bigr) + E(D, P(z)) \; .$$

Observe that $S(\mathcal{A}, P(w)) \leq S(\mathcal{A}, P(z))$ when $w \geq z$, all numbers counted in the former being also counted in the latter. It therefore suffices to establish

2.2 General Estimates of $G(x)$ and $E(D,P)$

the result when $w = z$. Choose z so that $s > 7B$. Corollary 1.1 now follows from Theorem 1. The reader can check that the O-constant can, for example, be taken as 2 whenever $s > 7B + 1$.

An Application. Corollary 1.1 is entirely satisfactory in situations where the value of a constant multiplying $X V(P(z))$ is not considered to be important. When applied to the situation considered in Theorem 2.2.1, for example, Corollary 1.1 gives

$$\pi(Y) - \pi(Y - X) = O\left(\frac{X}{\log X}\right),$$

on choosing $D = X/\log X$, taking a suitable value of $s > 2$, and using Mertens' formula (1.1.4.2) for the product $V(P)$.

Theorem 1 will follow from Lemma 1, which will be used again in Chap. 7.

Lemma 1. *Suppose that (1.5) holds, and denote*

$$I_z(x) = G_z(\infty) - G_z(x) = \sum_{d \geq x; d | P(z)} g(d), \tag{1.8}$$

and write $x = z^v$. Then for each $c \geq 0$

$$I_z(x) \leq \frac{1}{V(P(z))} \exp(-cv + B(e^c - 1)). \tag{1.9}$$

In particular

$$I_z(x) \leq \frac{\exp(-\psi_B(v))}{V(P(z))}, \tag{1.10}$$

where ψ_B is as in Theorem 1.

Lemma 1 follows using Rankin's device. Take $\varepsilon \geq 0$. Then

$$I_z(x) \leq \frac{1}{x^\varepsilon} \sum_{d | P(z)} g(d) d^\varepsilon = \frac{1}{x^\varepsilon} \prod_{p | P(z)} (1 + p^\varepsilon g(p)).$$

Hence

$$I_z(x) V(P(z)) \leq \frac{1}{x^\varepsilon} \prod_{p | P(z)} \left(\frac{p - \rho(p)}{p}\right) \left(1 + \frac{p^\varepsilon \rho(p)}{p - \rho(p)}\right)$$

$$\leq \frac{1}{x^\varepsilon} \prod_{p | P(z)} \left(1 + \frac{\rho(p)}{p}(p^\varepsilon - 1)\right) \leq \frac{1}{x^\varepsilon} \exp\left(\sum_{p | P(z)} \frac{\rho(p)}{p}(p^\varepsilon - 1)\right).$$

We may choose $\varepsilon = c/\log z$, provided $c \geq 0$. Then $z^\varepsilon = e^c$. Observe (e.g. from its Maclaurin series) that $(e^t - 1)/t$ increases when $t > 0$. Hence when $p < z$

$$\frac{p^\varepsilon - 1}{\varepsilon \log p} = \frac{e^{c \log p/\log z} - 1}{c \log p/\log z} \leq \frac{e^c - 1}{c}.$$

When $z = x^{1/v}$ use of (1.5) now gives

$$I_z(x) V(P(z)) \leq \frac{1}{e^{cv}} \exp\left(\frac{e^c - 1}{\log z} \sum_{p < z} \frac{\rho(p)}{p} \log p\right)$$

$$\leq \exp(-cv + B(e^c - 1)),$$

as required in (1.9).

The optimal choice of c satisfies $v = Be^c$, i.e. $c = \log v - \log B$, provided $v > B$. If $v \leq B$ then the best permissible choice is $c = 0$. This gives (1.10), as required by Lemma 1.

Proof of Theorem 1. The second inequality in (1.6) is just (1.3). The proof of (1.6) is completed by noting that (1.8), (1.4) and (1.10) give

$$G_z(x) = G_z(\infty) - I_z(x) \geq \frac{1 - \exp(-\psi_B(v))}{V(P(z))},$$

and taking $x = \sqrt{D}$, so that the notations $z^v = x$ and $z = D^{1/s}$ require $v = \frac{1}{2}s$. The comparatively simple consequence $\exp(\psi_B(\frac{1}{2}s)) \leq e^{7B-s}$ is now most easily seen by making the sub-optimal choice $c = 2$ in Lemma 1 and noting that $e^2 - 1 < 7$.

2.2.2 Asymptotic Formulas

In situations where one wishes the sharpest possible result one might seek an asymptotic formula for $G(x)$. Theorem 2 is an asymptotic inequality, rather than a formula, in that it estimates $G(x)$ only from the side that we need, on the basis for an assumed average bound for $g(p)$ only from the side that is essential. It would be rather easier to show that a two-sided assumption such as (2.16) would lead to asymptotic equality between the two sides of (2.3), so that the estimate (2.3) is in fact a sharp one.

Theorem 2 estimates $G_z(x)$ in the case $z = x$ which, as will be illustrated in Sect. 2.3, is that arising in straightforward applications of Theorem 2.1.1. The general case in which $z < x$ is allowed will be dealt with in Sect. 7.1.

Denote

$$\sum_{p < v} g(p) \log p = \kappa \log v + \eta(v). \tag{2.1}$$

In Theorem 2 we require that $\eta(t)$ is bounded above. This is stronger than (2.2), and also slightly stronger than can be inferred from Lemma 1.3.5.

2.2 General Estimates of $G(x)$ and $E(D, P)$

It would be possible to proceed on the basis of the standard form of the sifting density hypothesis and its consequence Lemma 1.3.5, but it appears that we would then inherit some error terms involving $\log \log z$ that it will be tidier to avoid.

Theorem 2. *Let $G(z) = G_z(z)$ and $\eta(t)$ be as specified in (1.3), (1.1) and (2.1). Assume*
$$\eta(t) \leq A \quad \text{when} \quad 2 \leq t < z \tag{2.2}$$
where $A > 1$. Then
$$G(z) \geq \frac{e^{-\gamma\kappa}}{\Gamma(\kappa+1)V(P(z))}\left(1 + O\left(\frac{A}{\log z}\right)\right), \tag{2.3}$$
where $V(P)$ is as in Sect. 1.3. The implied constant depends only on κ.
In (2.3) we may use
$$\frac{e^{-\gamma\kappa}}{V(P(x))} = \log^\kappa x \left(1 + O\left(\frac{1}{\log x}\right)\right) \prod_{p<x}\left(1 - \frac{1}{p}\right)^\kappa\left(1 - \frac{\rho(p)}{p}\right)^{-1}. \tag{2.4}$$

The relation (2.4) follows from Mertens' product (1.1.4.2). It is in the form on the right (with $x = z$) that the quantity in (2.4) will arise in the proof of Theorem 2.

The Integral Equation. Theorem 2 will rest on the approximate integral equation for the expression $G_z(x)$, to be established in Lemma 2. For this purpose write
$$g_z(n) = \begin{cases} g(n) & \text{if } n|P(z) \\ 0 & \text{otherwise,} \end{cases}$$
so that the expression (1.1) is now
$$G_z(x) = \sum_{\substack{n<x \\ n|P(z)}} g(n) = \sum_{n<x} g_z(n). \tag{2.5}$$

Theorem 2 refers only to the case when $z = x$, but Lemma 2 will also be used in the discussion in Chap. 7 of the general case in which $z \leq x$.

Lemma 2 is independent of any hypothesis about the quantity η in (2.1).

Lemma 2. *When $z \leq x$ the expression $G_z(x)$ given in (2.5) satisfies*
$$G_z(x)\log x = \int_1^x G_z(t)\frac{dt}{t} + \kappa\int_{x/z}^x G_z(t)\frac{dt}{t} + \delta_z(x), \tag{2.6}$$
where $\delta_z(x)$ is expressible as
$$\delta_z(x) = \sum_{n<x} g_z(n)E_{z,n}\left(\frac{x}{n}\right) \quad \text{with} \quad E_{z,n}(t) \leq \eta(\min\{t,z\}). \tag{2.7}$$

Since m is squarefree when $g_z(m) \neq 0$ we have

$$\sum_{m<x} g_z(m) \log m = \sum_{np<x} g_z(np) \log p$$

$$= \sum_{n<x} g_z(n) \sum_{\substack{p<x/n \\ p<z}} g(p) \log p - \sum_{n<x} g_z(n) \sum_{\substack{p<x/n, p<z \\ p|n}} g(p) \log p \,.$$

Write $n = pk$ in the last sum. This gives

$$\sum_{m<x} g_z(m) \log m = \sum_{n<x} g_z(n) \sum_{\substack{p<x/n \\ p<z}} g(p) \log p - \sum_{k<x} g_z(k) \sum_{\substack{p<\sqrt{x/k} \\ p<z, p\nmid k}} g^2(p) \log p$$

$$= \sum_{n<x} g_z(n) F_{z,n}\left(\frac{x}{n}\right),$$

where use of (2.1) shows

$$F_{z,n}(y) = \sum_{\substack{p<y \\ p<z}} g(p) \log p - \sum_{\substack{p<\sqrt{y}, p<z \\ (p,k)=1}} g^2(p) \log p$$

$$\leq \sum_{\substack{p<y \\ p<z}} g(p) \log p = \kappa \log \min\{y, z\} + \eta(\min\{y, z\}) \,.$$

Hence

$$\sum_{n<x} g_z(n) \log n = \kappa \sum_{n<x} g_z(n) \log \min\left\{\frac{x}{n}, z\right\} + \delta_z(x)$$

$$= \kappa \sum_{n<x} g_z(n) \log \frac{x}{n} - \kappa \sum_{n<x/z} g_z(n) \log \frac{x/z}{n} + \delta_z(x) \,, \qquad (2.8)$$

where, setting $E_{z,n}(y) = F_{z,n}(y) - \kappa \log y$, we have

$$\delta_z(x) = \sum_{n<x} g_z(n) E_{z,n}\left(\frac{x}{n}\right) \quad \text{with} \quad E_{z,n}(t) \leq \eta(\min\{t, z\}) \,, \qquad (2.9)$$

the expression for δ_z stated in Lemma 2.

Next, note the relation, valid for $x \geq 1$,

$$\sum_{m<x} g(m) \log \frac{x}{m} = \sum_{m<x} g(m) \int_m^x \frac{dt}{t} = \int_1^x \sum_{m<t} g(m) \frac{dt}{t} \,.$$

Using this we can rewrite (2.8) as

$$G_z(x) \log x - (\kappa+1) \int_1^x G_z(t) \frac{dt}{t} + \kappa \int_1^{x/z} G_z(t) \frac{dt}{t} = \delta_z(x) \,.$$

This is equivalent to (2.6), so the proof of Lemma 2 is complete.

2.2 General Estimates of $G(x)$ and $E(D, P)$ 57

Proof of Theorem 2. Theorem 2 refers to $G_z(x)$ only in the case $z = x$, when Lemma 2 reduces to

$$G(x)\log x = (\kappa + 1)\int_1^x G(t)\frac{dt}{t} + \delta(x), \qquad (2.10)$$

where

$$\delta(x) = \sum_{n<x} g(n) E_n\left(\frac{x}{n}\right). \qquad (2.11)$$

The primes p exceeding z are irrelevant to Theorem 2. Therefore for present purposes we may re-define $\rho(p)$ for these p by specifying

$$1 - \frac{\rho(p)}{p} = \left(1 - \frac{1}{p}\right)^\kappa \quad \text{if} \quad p \geq z. \qquad (2.12)$$

This will make questions relating to the convergence (as $x \to \infty$) of the product in (2.4) become trivial. Also, when $p > z$ the function g in (1.2) now satisfies

$$g(p) = \frac{\rho(p)/p}{1 - \rho(p)/p} = -1 + \left(1 - \frac{1}{p}\right)^{-\kappa} = \frac{\kappa}{p} + O\left(\frac{1}{p^2}\right). \qquad (2.13)$$

Now (2.2) extends to an inequality valid also when $t \geq z$,

$$\eta(t) \leq A_1 \ll A \quad \text{when} \quad 2 \leq t, \qquad (2.14)$$

where the constant A_1 may be somewhat larger than A.

A familiar procedure for determining an integrating factor in (2.10) leads to

$$\int_2^x \frac{d\delta(t)}{\log^{\kappa+1} t} = \int_2^x \frac{1}{\log^{\kappa+1} t}\left(\log t\, dG(t) - \kappa\frac{G(t)}{t}dt\right) = \int_2^x d\frac{G(t)}{\log^\kappa t}. \qquad (2.15)$$

Here, as usual, the Stieltjes integrals could have been avoided, at greater length, by expressing them by an integration by parts formula.

If we were to assume a two-sided hypothesis on ρ in place of (2.2), say

$$|\eta(t)| < A, \qquad (2.16)$$

then we could infer a corresponding bound for $|\delta(t)|$. Then writing (2.15) in the form

$$\frac{G(x)}{\log^\kappa x} - \frac{G(2)}{\log^\kappa 2} = \left(\int_2^\infty - \int_x^\infty\right)\frac{d\delta(t)}{\log^{\kappa+1} t}$$

would show $G(x) \sim J\log^\kappa x$, for a certain constant J, the infinite integrals being convergent. Since we do not wish to assume (2.16) we use the more delicate argument leading to (2.20) below.

There follows a certain one-sided form of Cauchy's convergence criterion. Denote

$$H(x) = \frac{G(x)}{\log^\kappa x} \quad \text{if} \quad x \geq 2. \qquad (2.17)$$

2. Selberg's Upper Bound Method

Then, for $2 \leq x < y$, (2.15) and (2.11) give

$$H(y) - H(x) = \int_x^y \frac{d\delta(t)}{\log^{\kappa+1} t} = \int_x^y \frac{1}{\log^{\kappa+1} t} d \sum_{x \leq n < t} g(n) E_n\left(\frac{t}{n}\right).$$

Here $E_n(t/n) \leq A_1$ because of (2.7) and (2.14), so Lemma 1.3.2 applies, to give

$$H(y) - H(x) \leq A_1 \sum_{x \leq n < y} \frac{g(n)}{\log^{\kappa+1} n}. \tag{2.18}$$

The sum in (2.18) can be expressed in terms of the function H by a "condensation" procedure. Choose $c > 1$. When $U \geq 2$ we find

$$\sum_{U \leq n < U^c} \frac{g(n)}{\log^{\kappa+1} n} \leq \frac{1}{\log U} \frac{G(U^c)}{\log^\kappa U} \leq c^\kappa \frac{1}{\log U} H(U^c).$$

Hence

$$\sum_{x \leq n < y} \frac{g(n)}{\log^{\kappa+1} n} = \sum_{r \geq 0} \sum_{\substack{x^{c^r} \leq n < x^{c^{r+1}} \\ n < y}} \frac{g(n)}{\log^{\kappa+1} n} \leq c^\kappa \sum_{r \geq 0} \frac{1}{c^r \log x} \sup_{x \leq t \leq y} H(t).$$

Now (2.18) gives

$$H(y) - H(x) \leq \frac{A_2}{\log x} \sup_{x \leq t \leq y} H(t), \tag{2.19}$$

with $A_2 = CA_1$. Here, the constant C arising from the sum over r takes the value $c^{\kappa+1}/(c-1) \leq e(\kappa+1)$, if we choose $c = 1 + 1/(\kappa+1)$.

Let $0 < \varepsilon < \frac{1}{2}$, and let $J = \limsup_{t \to \infty} H(t)$. Then J is finite, for otherwise there would exist a sequence y_n such that

$$H(y_n) \to \infty \text{ as } n \to \infty, \qquad H(y_n) > (1-\varepsilon) \sup_{t \leq y_n} H(t).$$

Then (2.19) would give $1 - \varepsilon \leq A_2/\log x$, which is false for large x.

If x is sufficiently large, then $\sup_{x \leq t} H(t) < J + \varepsilon$. We can choose $y > x$ so that $H(y) > J - \varepsilon$. Now (2.19) gives

$$H(x) \geq J\left(1 - \frac{A_2}{\log x}\right) - \varepsilon\left(1 + \frac{A_2}{\log x}\right).$$

Now take $0 < \varepsilon < J/\log x$, so that with (2.17) this gives

$$\frac{G(x)}{\log^\kappa x} = H(x) \geq J\left(1 + O\left(\frac{A}{\log x}\right)\right), \tag{2.20}$$

initially for sufficiently large x, and thence for $x \geq 2$.

2.2 General Estimates of $G(x)$ and $E(D,P)$

We need a partial identification of the constant J. For this purpose introduce the Dirichlet series

$$Z_g(s) = \sum_{n=1}^{\infty} \frac{g(n)}{n^s} = \prod_p \left(1 + \frac{g(p)}{p^s}\right).$$

When $s > 0$ we find

$$Z_g(s) = \int_1^{\infty} \frac{1}{t^s} dG(t) = -\int_1^{\infty} G(t) \, dt^{-s}$$

$$\leq -\int_1^{\infty} (J + o(1)) \log^{\kappa} t \, dt^{-s} = s \int_0^{\infty} (J + o(1)) u^{\kappa} e^{-su} \, du$$

$$\sim \frac{J \, \Gamma(\kappa + 1)}{s^{\kappa}} \quad \text{as} \quad s \to 0+,$$

in which the quantities $o(1)$ are as $t \to \infty$ and $u \to \infty$. On multiplication by the appropriate power of the Riemann ζ-function we infer

$$\limsup_{s \to 0+} \prod_p \left(1 - \frac{1}{p^{s+1}}\right)^{\kappa} \left(1 + \frac{g(p)}{p^s}\right) = \limsup_{s \to 0+} \frac{Z_g(s)}{\zeta^{\kappa}(s+1)} \leq J\Gamma(\kappa+1). \quad (2.21)$$

Our choice (2.12) of $\rho(p)$ when $p \geq z$ ensures that the product on the left converges when $s = 0$, being in fact finite for this s. It is easy to infer

$$J\Gamma(\kappa + 1) = \prod_{p < z} \left(1 - \frac{1}{p}\right)^{\kappa} \left(1 - \frac{\rho(p)}{p}\right)^{-1};$$

we supply details below. Theorem 2 now follows in the form given by (2.4) on taking $x = z$ in (2.20).

This argument needs that the product in (2.21) is continuous in $s \geq 0$. When $z \leq x < y$ the contribution to (2.21) from large p is

$$\exp \sum_{x \leq p < y} \left(\kappa \log\left(1 - \frac{1}{p^{s+1}}\right) + \log\left(1 + \frac{g(p)}{p^s}\right)\right)$$

$$= \exp \sum_{x \leq p < y} \left(\frac{g(p) - \kappa/p}{p^s} + O\left(g^2(p) + \frac{1}{p^2}\right)\right).$$

Because of (2.13) the summand is $O(1/p^2)$ when $s \geq 0$, so the sum converges uniformly to a sum continuous in $s \geq 0$ as required.

2.2.3 The Error Term

We resume the discussion of the error term $E(D, P(\sqrt{D}))$ arising in Theorem 2.1.1. As already observed in Theorem 2.1.3, some progress follows from the condition on the remainder term $r_A(d)$ arising in the applications to polynomial sequences discussed in Sect. 1.2.3. Theorem 3 also assumes

$$\sum_{y \leq p < x} \frac{\rho(p) \log p}{p} \ll 1 + \log \frac{x}{y}, \tag{3.1}$$

a weak consequence of (2.2).

Theorem 3. *Suppose that* (2.1.1.1) *holds with*

$$|r_A(d)| \leq \rho(d), \qquad \rho(p) \geq 1 \quad \text{when} \quad p | P, \tag{3.2}$$

and assume that ρ satisfies (3.1). *Then the error term in Theorem 1 satisfies*

$$E(D, P(z)) \ll \frac{D}{\log^2 D},$$

where the implied constant depends only on that in (3.1).

Theorem 3 depends on Lemma 3 below. First note that the following weaker estimate, which is almost as useful for many applications, can be obtained in a simpler way.

Theorem 4. *Assume* (3.2), *and that ρ has sifting density κ. Then*

$$E(D, P(z)) \ll D \log^{2\kappa} z.$$

Corollary 2.1.2.1 and (3.2) give

$$E(D, P) \leq \sum_{d_i < \sqrt{D}, d_i | P} \frac{\rho(d_1)\rho(d_2)}{\rho((d_1, d_2))} \leq \left(\sum_{d < \sqrt{D}, d | P} \rho(d) \right)^2. \tag{3.3}$$

Then

$$\sum_{d < \sqrt{D}, d | P(z)} \rho(d) \leq \sqrt{D} \sum_{d | P(z)} \frac{\rho(d)}{d} \leq \sqrt{D} \prod_{p < z} \left(1 + \frac{\rho(p)}{p}\right) \leq \frac{\sqrt{D}}{V(P(z))}.$$

Here $V(P(z))^{-1} \ll K \log^{\kappa} z$ because ρ has sifting density κ, so Theorem 4 follows.

For Theorem 3 we need the following lemma on sums of multiplicative functions, simpler than the theorem of the previous section. In the case $f(p) = \rho(p)$ the hypothesis of Lemma 3 is a weak corollary of (2.2), and hence of Lemma 1.3.2.

2.2 General Estimates of $G(x)$ and $E(D, P)$

Lemma 3. *Suppose $f(p)$ is a positive multiplicative function supported on the squarefree numbers for which*

$$\sum_{y \leq p < x} \frac{f(p) \log p}{p} \leq a \log \frac{x}{y} + b,$$

for all x, y with $y < x$ and for some constants b and $a > 0$. Then

$$\sum_{m < x} f(m) \ll (a + b) \frac{x}{\log x} \sum_{m < x} \frac{f(m)}{m},$$

where the implied constant is absolute.

By partial summation as in Lemma 1.3.1 we find

$$\sum_{p < x} f(p) \log p \leq \int_1^x t \, d\left(a \log \frac{x}{t}\right) + bx \leq (a + b)x.$$

With use of Chebyshev's idea on the squarefree number m, as in Lemma 2, this gives

$$\sum_{m < x} f(m) \log m = \sum_{np < x} f(np) \log p \leq \sum_{n < x} f(n) \sum_{p < x/n} f(p) \log p$$

$$\leq (a + b) x \sum_{n < x} \frac{f(n)}{n}. \tag{3.4}$$

A further partial summation gives

$$\sum_{m < x} f(m) = 1 + \int_2^x \frac{1}{\log t} d \sum_{2 \leq m < t} f(m) \log m$$

$$= 1 + \frac{1}{\log x} \sum_{2 \leq m < x} f(m) \log m + \int_2^x \sum_{2 \leq m < t} f(m) \log m \, \frac{dt}{t \log^2 t}.$$

Hence use of (3.4) gives

$$\sum_{m < x} f(m) \leq 1 + (a + b) \left(\sum_{n < x} \frac{f(n)}{n}\right) \left(\frac{x}{\log x} - \int_2^x t \, d\frac{1}{\log t}\right)$$

$$= 1 + (a + b) \left(\sum_{n < x} \frac{f(n)}{n}\right) \left(\frac{2}{\log 2} + \int_2^x \frac{dt}{\log t}\right)$$

$$\ll (a + b) \frac{x}{\log x} \sum_{n < x} \frac{f(n)}{n},$$

as stated in Lemma 3.

2. Selberg's Upper Bound Method

Proof of Theorem 3. As elsewhere, adopt the convention that $\rho(p) = 0$ (whence $g(p) = 0$) when $p \nmid P$. Theorem 2.1.1 and (3.2) imply

$$E(D, P(z)) \leq \frac{1}{\lambda^2(1)} \sum_{d_i < \sqrt{D}} |\lambda(d_1)||\lambda(d_2)| \frac{\rho(d_1)\rho(d_2)}{\rho((d_1, d_2))}$$

$$\leq \frac{1}{\lambda^2(1)} \left(\sum_{d | P(z), d < \sqrt{D}} |\lambda(d)| \rho(d) \right)^2.$$

The specification in Theorem 2.1.1 of the numbers $\lambda(d)$ gives

$$\sum_{d < \sqrt{D}} |\lambda(d)\rho(d)| = |C| \sum_{d < \sqrt{D}} d \sum_{h \equiv 0 \bmod d} g(h) = |C| \sum_{h < \sqrt{D}} g(h)\sigma(h),$$

where $\sigma(h)$ denotes the sum of the divisors of h. Here $\lambda(1) = C\, G(\sqrt{D})$, so that

$$E(D, P(z)) \leq \left(\frac{1}{G(\sqrt{D})} \sum_{h < \sqrt{D}} g(h)\sigma(h) \right)^2. \tag{3.5}$$

From (3.1) with $y = p$, $x = p + \varepsilon$ we infer $\rho(p) \ll p/\log p$. Hence

$$g(p) \frac{\sigma(p)}{p} \ll g(p) = \frac{\rho(p)}{p - \rho(p)} \ll \frac{\rho(p)}{p},$$

so $f(p) = g(p)\sigma(p)$ inherits the property (3.1) from $\rho(p)$. Thus Lemma 3 gives

$$\sum_{h < \sqrt{D}} g(h)\sigma(h) \ll \frac{\sqrt{D}}{\log D} \sum_{m < \sqrt{D}} \frac{g(m)\sigma(m)}{m}. \tag{3.6}$$

Here

$$\sum_{m < \sqrt{D}} \frac{g(m)\sigma(m)}{m} = \sum_{ln < \sqrt{D}} \frac{g(ln)n}{ln} \ll G(\sqrt{D}), \tag{3.7}$$

the implied constant depending only on that in (3.1) because

$$\sum_l \frac{g(l)}{l} \ll \prod_p \left(1 + \frac{g(p)}{p} \right) \ll \prod_p \left(1 + \frac{1}{p \log p} \right) \ll 1.$$

Theorem 3 now follows from (3.5), (3.6) and (3.7).

2.3 Applications

Most of the ensuing discussions will refer to the estimate of $G(D, P)$ obtained in Sect. 2.2. In this respect our first application needs only the ideas used already in Sect. 2.1.3. It has not been given earlier only because it is convenient to refer to the estimate of the error term given in Theorem 2.2.3.

2.3.1 Arithmetic Progressions

Let $\pi(x; l, k)$ denote the number of primes p satisfying $p \leq x$ in the arithmetic progression $p \equiv l$, mod k. A small extension of the discussion in Sect. 2.1.3 leads to a bound for this number, or more generally for the number of primes in this progression lying in the interval $y - x < p \leq y$. We may suppose $(k, l) = 1$, since otherwise the primes in question all divide (k, l).

As before, our result will be uniform in y and (a remark that remains significant when $y = 1$) will be explicit in its dependence on the modulus k, which may be nearly as large as the length x of the interval. In contrast, the prime number theorem for arithmetic progressions applies only in a much more restricted region such as $k \leq \log^B x$.

Any result of the following general type may be called the "Brun-Titchmarsh Theorem", the weaker original version having been obtained by Titchmarsh using Brun's method.

Theorem 1. *If $k < x \leq y$ then*

$$\pi(y; l, k) - \pi(y - x; l, k) \leq \frac{2x}{\phi(k) \log(x/k)} + O\left(\frac{x}{k \log^2(x/k)}\right),$$

the implied constant being absolute.

We require to bound the number of primes $p = kn + l$ with $y - x < p \leq y$. Thus n is confined to an interval of length $X = x/k$, actually $(Y - X, Y]$ with $Y = (y - l)/k$. Take $\mathcal{A} = \{kn + l : Y - X < n \leq Y\}$, as in Sect. 1.2.4, and let $P(z)$ be the product of those primes $\varpi < z$ not dividing k. Then

$$\pi(y; l, k) - \pi(y - x; l, k) \leq S(\mathcal{A}, P(z)) + \pi(z; l, k).$$

As in Sect. 1.2.4, $|\mathcal{A}_d|$ is given by (2.1.1.1) with $\rho(p) = 1$ if $p \nmid k$, $\rho(p) = 0$ otherwise, and $|r_\mathcal{A}(d)| \leq 1$. We bound $S(\mathcal{A}, P(z))$ by Theorem 2.1.1.

It is important that the estimate for $G(\sqrt{D})$ should be explicit in its dependence on k. This is easily achieved using (2.1.2.1):

$$\sum_{f|k} \frac{\mu^2(f)}{\phi(f)} \sum_{\substack{(n,k)=1 \\ n < \sqrt{D}}} \frac{\mu^2(n)}{\phi(n)} \geq \sum_{h < \sqrt{D}} \frac{\mu^2(h)}{\phi(h)}.$$

The sum over f is just $k/\phi(k)$, as in (2.1.2.2), and the sum over h was estimated in (2.1.3.4). Thus

$$G(\sqrt{D}) = \sum_{\substack{(g,k)=1 \\ g<\sqrt{D}}} \frac{\mu^2(g)}{\phi(g)} \geq \frac{\phi(k)}{k} \log \sqrt{D}.$$

If we used Theorem 2.1.2 then the error term in Theorem 2.1.1 could be estimated as

$$E(D, P(\sqrt{D})) \leq \left(\sum_{d<\sqrt{D};(d,k)=1} 1 \right)^2 \leq D, \tag{1.1}$$

although the k-dependence of this estimate can be improved somewhat. Then Theorem 2.1.1 would give

$$S(\mathcal{A}, P(\sqrt{D})) \leq \frac{k}{\phi(k)} \frac{X}{\log \sqrt{D}} + D.$$

As in Sect. 2.1.3, this would give a result weaker than Theorem 1 by a factor $\log \log X = \log \log(x/k)$.

The result stated follows by estimating the error term using Theorem 2.2.3 in place of (1.1) (a form that is very slightly weaker, by the occurrence of factors $k/\phi(k)$ in the error term, follows using Theorem 2.1.3 instead). This gives

$$S(\mathcal{A}, P(\sqrt{D})) \leq \frac{k}{\phi(k)} \frac{X}{\log \sqrt{D}} + O\left(\frac{D}{\log^2 D} \right).$$

Choose $D = X = x/k$ to obtain Theorem 1.

2.3.2 Prime Twins and Goldbach's Problem

The discussion here will be completely elementary, and will illustrate the utility of a one-sided sifting density hypothesis in dealing with an X-dependence of the function ρ first pointed out in Sect. 1.2.3.

For these particular problems, sharper results can be obtained by sifting the sequences $\{p - 2 : p < x\}$ and $\{N - p : p < N\}$, in which p is prime. Some details are supplied in the notes.

We proceed by sifting certain sequences

$$\mathcal{A} = \left\{ f(n) : 1 \leq n < X \right\}, \tag{2.1}$$

where f is an integer polynomial. The estimate from Sect. 1.2.3 is

$$|\mathcal{A}_d| = X \frac{\rho(d)}{d} + O(\rho(d)), \tag{2.2}$$

where $\rho(d)$ is the number of roots of the congruence $f(n) \equiv 0$, mod d.

2.3 Applications

For Theorem 2 we sift the numbers $n(n+2)$. Let $\pi_2(X)$ denote the number of n with $1 \le n < X$ such that both n and $n+2$ are prime.

Theorem 2. *Let*

$$C = \prod_{p>2} \frac{p(p-2)}{(p-1)^2}, \qquad (2.3)$$

where the product is over all odd primes p. Then

$$\pi_2(X) \le \frac{16CX}{\log^2 X}\left(1 + O\left(\frac{\log \log X}{\log X}\right)\right).$$

The upper bound associated with Goldbach's problem may be approached in a similar way, by sifting the set of numbers $n(N-n)$.

Theorem 3. *Let $F_2(n)$ denote the number of representations of an even number N as a sum of two primes. Then*

$$F_2(N) \le \frac{16CN}{\log^2 N}\left(\prod_{2<p|N} \frac{p-1}{p-2}\right)\left(1 + O\left(\frac{\log \log N}{\log N}\right)\right),$$

where C is the constant in Theorem 2.

Proof of Theorem 2. Take $f(n) = n(n+2)$. Then the function ρ in (2.2) is given by

$$\rho(p) = 2 \text{ if } p > 2, \qquad \rho(2) = 1.$$

Thus $\rho(p) \le 2$ for all p, so (as in Lemma 1.3.3) ρ has sifting density 2, and (2.2.2.2) holds with $\eta(z,w) = O(1)$. Then Theorem 2.3.2 gives

$$G(\sqrt{D}) \ge \frac{1}{2!}A_D \log^2 \sqrt{D}\left(1 + O\left(\frac{1}{\log D}\right)\right),$$

where

$$A_D = \tfrac{1}{2} \prod_{2<p<\sqrt{D}} \left(1 - \frac{1}{p}\right)^2 \left(1 - \frac{2}{p}\right)^{-1}. \qquad (2.4)$$

Because of Theorems 2.1.1 and 2.2.4 this gives

$$\pi_2(X) \le S(\mathcal{A}, P(\sqrt{D})) + \pi(\sqrt{D}) \le \frac{2X}{A_D \log^2 \sqrt{D}} + O(D \log^{2\kappa} D). \qquad (2.5)$$

Theorem 2 follows as stated on choosing $D = X/(\log X)^{2\kappa+3}$ and observing

$$A_D = \frac{1}{2C}\left(1 + O\left(\frac{1}{\sqrt{D}}\right)\right),$$

C being the constant in (2.3). Here the term $O(1/\sqrt{D})$ arises as in (1.3.5.14).

Proof of Theorem 3. Take $\mathcal{A} = \{n(N-n) : 1 \leq n < N\}$, so that

$$F_2(N) \leq S(\mathcal{A}, P(z)) + \pi(z).$$

Now (2.2) holds with $X = N$ and

$$\rho(p) = 2 \quad \text{if } p \nmid N, \qquad \rho(p) = 1 \quad \text{if } p \mid N.$$

The N-dependence of ρ does not create a problem if Theorem 2.2.2 is used. Rather as with Theorem 2 it leads to

$$G(D, P(\sqrt{D})) \leq \frac{1}{2!} B_D(N) \log^2 \sqrt{D} \left(1 + O\left(\frac{\log \log N}{\log N}\right)\right),$$

where

$$B_D(N) = \prod_{\substack{p \mid N \\ p < \sqrt{D}}} \left(1 - \frac{1}{p}\right) \prod_{\substack{p \nmid N \\ p < \sqrt{D}}} \left(1 - \frac{1}{p}\right)^2 \left(1 - \frac{2}{p}\right)^{-1} = A_D \prod_{\substack{2 < p \mid N \\ p < \sqrt{D}}} \frac{p-2}{p-1},$$

with A_D the product in (2.4). Estimate the associated error term $E(D, P)$ and choose D as in Theorem 2. This gives Theorem 3, in which the bounded number of primes exceeding $\sqrt{D}$ and dividing N contribute a negligible factor $1 + O(1/\sqrt{D})$.

If the better but more difficult Theorem 2.2.3 were invoked to deal with the error terms $E(D, P)$ it would give an error term $O(D/\log^2 D)$ in (2.5). Then the choice $D = X/\log X$ would again lead to Theorems 2 and 3, with different O-constants.

2.3.3 Polynomial Sequences

The situation considered in Theorem 2 can be extended to deal with more or less arbitrary polynomials, as follows. On the other hand Theorem 3 does not follow as a special case of Theorem 4 because of the dependence of the O-constant on the coefficients of the polynomial f. Let $\pi(f, X)$ denote the number of integers n for which $1 \leq n \leq X$ and $f(n)$ is prime.

Theorem 4. *Let $f(n)$ be a polynomial having k irreducible factors over the integers and having no constant factor. Then*

$$\pi(f, X) \leq 2^k k! C \frac{X}{\log^k X} \left(1 + O\left(\frac{\log \log X}{\log X}\right)\right),$$

where

$$C = \prod_p \left(1 - \frac{\rho(p)}{p}\right) \left(1 - \frac{1}{p}\right)^{-k},$$

ρ as in (2.2), and the implied constant may depend on f.

2.3 Applications

The hypothesis that $f(n)$ has no constant factor excludes cases such as $n^2 + n + 2$, which is always even. Then $\rho(p) < p$ for all p, so that we may take $P(z)$ to be the product of all primes p with $p < z$ without violating Hypothesis 1.3.1.

The bound in Theorem 4 exceeds the generally conjectured asymptotic value by the factor $2^k k!$.

Take $\mathcal{A}$ as in (2.1). Then

$$\pi(f, z) \leq S(\mathcal{A}, P(z)) + \pi(z),$$

by the usual argument. Write

$$f(n) = f_1(n) f_2(n) \ldots f_k(n),$$

where each f_i is irreducible.

The situation is simpler when each factor f_i is linear:

$$f(n) = (n - a_1)(n - a_2) \ldots (n - a_k).$$

Then in (2.2)

$$\rho(p) \begin{cases} \leq k & \text{for all } p \\ = k & \text{if } p \nmid \Delta_f, \end{cases}$$

where Δ_f is certain constant depending on f. Then (because of Lemma 1.3.1) Theorem 2.2.2 applies with $\kappa = k$, to give

$$G(\sqrt{D}) \geq \frac{1}{k!} \log^k \sqrt{D} \left(1 + O\left(\frac{1}{\log D}\right)\right) \prod_{p < \sqrt{D}} \frac{\left(1 - \frac{1}{p}\right)^\kappa}{\left(1 - \frac{\rho(p)}{p}\right)} \quad (3.1)$$

$$= \frac{\log^k D}{2^k k! C} \left(1 + O\left(\frac{1}{\log D}\right)\right), \quad (3.2)$$

where C is as in Theorem 4. The stated convergence of this product follows via Cauchy's criterion using a calculation of the type first given in (1.3.5.14).

In the general case let $\rho_i(p)$ be the number of roots of $f_i(n) \equiv 0$, mod p. Then

$$\rho(p) = \sum_{i=1}^{k} \rho_i(p)$$

for all except finitely many primes p. It is then a consequence of the Prime Ideal Theorem that

$$\sum_{p < z} \frac{\rho_i(p) \log p}{p} = \log z + O(1). \quad (3.3)$$

Consequently

$$\sum_{w\leq p<z} \frac{\rho(p)\log p}{p} = k\log\frac{z}{w} + O(1),\qquad(3.4)$$

so that (3.1) and (3.2) follow as before.

Theorem 4 now follows after dealing with the error term $E(D,P)$ in the manner used in the proofs of Theorems 2 and 3.

2.4 Notes on Chapter 2

Sect. 2.1.1. Theorem 1 appeared in [Selberg (1947)], though the proof was not precisely as given here. The differences, if there are any, relate to the technique used in determining the conditional minimum of the quadratic form. It has been stated that Selberg used the method of Lagrange's multipliers, but there is no evidence for this statement in Selberg's paper: he simply writes "... we easily find ...". Of course one method is simply to resort to the process of completion of the square. The reader who embarks on the details of such an exercise may well experience a sense of *déja vu*, and it may have been such an experience that led the late P. Turán to an argument like given in the text: see [Fluch (1959)], or [Levin (1965)].

Sect. 2.1.2. The inequality in Theorem 2 appears in [van Lint and Richert (1965)]. The author is uncertain of the origin of Theorem 3, which he saw in Iwaniec's Rutgers lecture notes.

Sect. 2.1.3. Again, the reasoning in this section is as least as old as the paper [van Lint and Richert (1965)].

Sect. 2.2.1. This use of Rankin's trick appears in [Halberstam and Richert (1974)] and in [Selberg (1991)].

Sect. 2.2.2. The integral equation in Lemma 2 appears in [Halberstam and Richert (1974)], where a theorem like our Theorem 2 is obtained, but using a two-sided assumption of the type (2.16). As these authors indicate, the approach is derived from [Wirsing (1961)].

The first proof in the literature of a theorem like our Theorem 2 is in [Rawsthorne (1982)], by a method distinctly different from that given here. Rawsthorne started from the corresponding theorem in [Halberstam and Richert (1974)], for which a two-sided hypothesis on the function g (or equivalently ρ) was required. Following a process suggested by W. B. Jurkat in a lecture in Urbana in 1973 (see some remarks in [Halberstam and Richert (1988)]), a certain "topping-up" procedure was used, in which the values of $\rho(p)$ were selectively increased until the lower bound for $G(z)$ assumed the form stated in our theorem. It is important to this process that the product $G(z)V(P(z))$ decreases as this topping-up proceeds. It should perhaps be mentioned that Rawsthorne's paper operates in a more general context that we have deferred to Chap. 7, and that some of the details of his paper arise from this fact.

The direct proof of Theorem 2 in the text is developed from a manuscript by M. N. Huxley and the author, dated October 1979, which remained unpublished at the time. A revised version of this ms. is to appear in the proceedings of a conference held in Turku in 1999, and may well be in print by the time this book is published.

Sect. 2.2.3. This section is derived almost entirely from Iwaniec's Rutgers lecture notes.

Selberg's Method as a "Large" Sieve. Selberg's method (unlike those described later in this book) retains its effectiveness when the function $\rho(p)$ does not have a bounded sifting density κ but may become large as p increases. Accordingly we will add a few words about the estimation of the sum $G(x)$ in this situation. The same sum arises in the arithmetical form of the large sieve (for which see, for example, [Montgomery (1971)], and [Gallagher (1973)] for a discussion of a multidimensional version.

Also extremely useful, when it is applicable, is the "larger" sieve of [Gallagher (1971)], which appears, however, to have a limitation in that it does not appear to generalise to the multi-dimensional situation.

For a specimen application where the methods just mentioned were used, see [Greaves (1992)]. In this and other applications the function ρ satisfies $\rho(p) > Cp$ for each p, possibly apart from those dividing a product m of "bad" primes. Suppose that this condition is satisfied.

First consider the case $m = 1$. In this case little is lost (actually only a factor $\log x$) by estimating the sum $G(x)$ as

$$G(x) \geq \sum_{p<x} \frac{\rho(p)}{p - \rho(p)} \geq \sum_{p<x} \frac{C}{1-C} \gg \frac{x}{\log x},$$

the implied constant depending on C. In the case $m > 1$ we may adapt a device used in Theorem 2.3.1. Define $g(p) = C$ when $p|m$. Then

$$\sum_{\substack{n<x \\ (n,m)=1}} g(n) \prod_{p|m}(1 + g(p)) = \sum_{n<x} g(n) > \sum_{p<x} g(p) \gg \frac{x}{\log x}.$$

The product over primes dividing m may be estimated as

$$\prod_{p|m} \frac{1}{1-C} = \prod_{p|m} 2^\alpha = d^\alpha(m),$$

where $\alpha = -\log(1-C)/\log 2$.

2. Selberg's Upper Bound Method

An estimate that is useful for smaller values of $g(p)$, say of the order of $\log p$, is given in [Vaughan (1973)].

Sect. 2.3.1. The prototype version of Theorem 1 appeared in [Titchmarsh (1930)]. For the result given here, cf. [van Lint and Richert (1965)] and [Halberstam and Richert (1974)].

A considerable body of work has been devoted to improvements on Theorem 1 for various ranges of the modulus k, for the most part using ideas to be discussed in Chap. 6 and the notes thereon. The first significant improvement was in [Motohashi (1975)]. For a recent reference, see [Friedlander and Iwaniec (1997)]. A characteristic of these results is that, if we write $k = x^\theta$, then any improvement on the constant 2 in Theorem 1 tends to zero as $\theta \to 0$ or 1.

Sect. 2.3.2. The "prime twins" application was given in [Selberg (1947)], but without details of how the sum $G(z)$ had been estimated. We have given an elementary treatment of the prime twin and Goldbach upper-bound problems here, because the latter gives a simple illustration of the evasion of difficulties obtained by working with a one-sided sifting density hypothesis. The difficulty in question is that alluded to in Sect. 1.2.3, that an unpleasant X-dependence of the function ρ will arise if a two-sided hypothesis on ρ were needed.

For the alternative procedure in which one sifts sequences of numbers $p - 2$ and $N - p$, involving an appeal to the Bombieri-Vinogradov theorem as in Sect. 1.2.6, see [Halberstam and Richert (1974)]. This approach results in a halving of the constant 16 appearing in Theorems 2 and 3. In the case of Theorem 2 a further reduction of the constant to $7 + \varepsilon$, for any $\varepsilon > 0$, is possible after an appeal to extensions of the Bombieri-Vinogradov theorem, for which see [Bombieri, Friedlander and Iwaniec (1986)] and [Fouvry (1987)].

3. Combinatorial Methods

We shall not attempt a definition of the term "combinatorial sieve" here, save to say that there is a class of sieve methods which are by common consent called combinatorial, and that methods such as those described in this chapter in which the functions $\lambda_D^\pm(d)$ appearing in (1.3.1.2) take only the values $\mu(d)$ or 0 are certainly combinatorial. In the context of the Weighted Sieve of Chapter 5 other combinatorial methods will appear that do not exactly answer this description. Selberg's λ^2 method from Chapter 2 is however not combinatorial. We will describe the general principles underlying combinatorial sieves in Sect. 3.1.

The first really significant sieve method to appear after Legendre was V. Brun's "Pure Sieve" in 1915. This led to the interesting application that the sum of the reciprocals of the prime twins

$$\sum_{p,\, p+2 \text{ primes}} \frac{1}{p}$$

is convergent (or possibly finite). We describe this method and application in Sect. 3.2. This section may be read independently of the more general discussion appearing in Sect. 3.1.

A much more powerful form of Brun's ideas appeared in 1920, leading in particular to the applications mentioned in the introduction to this book. We will describe a development of Brun's method in Sect. 3.3. Our account draws also on some more modern ideas due to Rosser, Iwaniec and others, which we develop more fully in Chapter 4. A briefer account of what is essentially Brun's version of his method is given in Sect. 3.4.

The defect of Legendre's method, as used in (1.1.3.2), for example, is that the number of integers d that are involved is an excessively large function $2^{\pi(z)}$ of the size z of the largest sifting prime employed. Thus in dealing with a question involving sifting integers of size up to X we were forced to choose z relatively small ($z = \log X$) in deriving (1.1.3.3). In a more general framework we noted at (1.3.2.1) that if we are to write $d \leq D$ in Theorem 1.3.1 then, with this construction, the severe restriction $D = P(z)$ would apply.

Brun's contribution greatly improved this aspect of the matter, by a suitable modification of the ideas that led via (1.1.2.4) to (1.1.2.5). We shall see in Sect. 3.3 that his ideas led to significant results in which the restriction on D is of the type $z^s \leq D$ for bounded s. For example, in the problem initially discussed in Sect. 1.1.3 we will be able to choose $z = X^{1/s}$ rather than $z = \log X$. In this respect the Pure Sieve of Sect. 3.2 was already a great advance on Legendre's method.

This feature is of course also present in the later λ^2 method of Selberg discussed in Chapter 2. One significant difference is that Brun's methods apply in a parallel fashion to both the upper bound and lower bound aspects of the sifting problem.

3.1 The Construction of Combinatorial Sieves

We will begin in Sect. 3.1.1 with an overview aimed at providing a gentle introduction to the general ideas of Brun's methods. The reader may proceed directly to Sect. 3.1.2, if this is preferred. There, the general features of his construction are described in Lemma 1. We also lay down the general procedures used in making the necessary estimations of the resulting expressions of the type appearing in (1.3.1.4). These are embodied in the identities given in Lemma 4. The prototype of all identities of this type is the relatively well-known identity of Buchstab which we mention in passing in Sect. 3.1.3.

The objective in this section is the description, in Sect. 3.1.4, of a general combinatorial construction leading to a sieve theorem of the type first postulated in Theorem 1.3.1.

3.1.1 Preliminary Discussion of Brun's Ideas

We conduct a re-examination of Legendre's construction from Sect. 1.1.2.

The discussion leading to the expression (1.1.2.4) indicates that already it gives an upper bound for $S(\mathcal{A}, P)$, because the weight

$$1 - \binom{r}{1} + \binom{r}{2}$$

that it attaches to numbers n having r prime factors from P is easily verified to be non-negative. A similar observation can be shown to apply to

$$\sum_{d|n; \nu(d) \leq r} \mu(d)$$

whenever r is even; here $\nu(d)$ is the number of prime factors of d. A similar inequality in the opposite sense applies when r is odd. This observation alone leads to the Pure Sieve of Sect. 3.2, which is thus constructed by a simple truncation of the process of Legendre.

But Legendre's process may be modified in a much more far-reaching way, in which multiple but partial truncations are introduced as follows. Suppose

3.1 The Construction of Combinatorial Sieves

that our purpose is to bound $S(\mathcal{A}, P)$ from below. In (1.1.2.4) the suggested rôle of the terms $[X/p_1p_2]$ was to compensate for the negative count

$$\sum_{n \in \mathcal{A}} \left(1 - \sum_{p_1 | n; \, p_1 < z} 1\right) = X - \sum_{p_1 < z} \left[\frac{X}{p_1}\right]$$

that would be achieved without them. The new idea is that this compensation may be performed in a partial way; in place of (1.1.2.4) we may consider the smaller quantity

$$\sum_{n \in \mathcal{A}} \left(1 - \sum_{\substack{p_1 | n \\ p_1 < z}} 1 + \sum_{\substack{p_1 p_2 | n \\ p_2 < p_1 < z; \, p_2 < B_2}} 1\right),$$

where B_2 is arbitrary, possibly depending on p_1 as well as on z and X. In this expression, those n divisible by three primes p_1, p_2, p_3 with $p_3 < p_2$ have been counted with weight 1, so we should add a correcting term

$$-\sum_{n \in \mathcal{A}} \sum_{\substack{p_1 p_2 p_3 | n \\ p_3 < p_2 < p_1 < z; \, p_2 < B_2}} 1 \,,$$

but at the next step we make a partial correction by adding only

$$\sum_{n \in \mathcal{A}} \sum_{\substack{p_1 p_2 p_3 p_4 | n \\ p_4 < p_3 < p_2 < p_1 < z; \, p_2 < B_2, \, p_4 < B_4}} 1 \,,$$

for some B_4. Continue in this style, introducing B_6, B_8 ... in turn. In this construction, the rôle of B_2, B_4 will be to bound the size of $d = p_1 p_2 \ldots$ by means other than Legendre's device of (severely) restricting the size z of the largest prime involved.

3.1.2 Fundamental Inequalities and Identities

Following the discussion of Sect. 3.1.1 we formulate a definitive version of a "Fundamental Inequality" in Lemma 1 below. In this and subsequent lemmas we use the following notation for a squarefree number $d > 1$: write

$$d = p_1 p_2 \ldots p_r \,, \quad p_1 > p_2 > \ldots > p_r \,, \tag{2.1}$$

where $p_1, p_2, \ldots$ are prime.

Lemma 1. *Write d as in (2.1). Let $B_i = B_i(p_1, \ldots, p_{i-1})$ be arbitrary real numbers. Define*

$$\chi^-(d) = \begin{cases} 1 & \text{if } p_{2i} < B_{2i} \text{ when } 2 \leq 2i \leq r \\ 0 & \text{otherwise,} \end{cases} \tag{2.2}$$

$$\chi^+(d) = \begin{cases} 1 & \text{if } p_{2i+1} < B_{2i+1} \text{ when } 1 \leq 2i+1 \leq r \\ 0 & \text{otherwise,} \end{cases} \tag{2.3}$$

3. Combinatorial Methods

so that in particular $\chi^+(1) = \chi^-(1) = 1$. Then

$$\sum_{d|A} \mu(d)\chi^-(d) \leq \sum_{d|A} \mu(d) \leq \sum_{d|A} \mu(d)\chi^+(d) . \qquad (2.4)$$

This lemma, which provides a construction of the type defined by (1.3.1.2), will follow using the identities in Lemma 2 below. In formulating these identities, view $\mu(d)$ as a first approximation to $\mu(d)\chi^-(d)$. The first respect in which it is inaccurate is that it includes terms $d = p_1 p_2 \ldots$ when $p_2 \geq B_2$, so we should remove these, to obtain a second approximation

$$\mu(d)\chi_2^-(d) = \mu(d) - \sum_{\substack{d=p_1 p_2 d_2 \\ B_2 \leq p_2}} \mu(d_2) ,$$

where d is written as in (2.1). Thus $\chi_2^-(d) = 1$ only for those d for which $r \leq 1$ or $p_2 < B_2$. The next correction that should be made is to take account of B_4: we should form

$$\mu(d)\chi_4^-(d) = \mu(d)\chi_2^-(d) - \sum_{\substack{d=p_1 p_2 p_3 p_4 d_4 \\ p_2 < B_2,\, B_4 \leq p_4}} \mu(d_4) ,$$

so that $\chi_4^-(d) = 1$ only for those d with $p_2 < B_2$, $p_4 < B_4$ or alternatively $r \leq 3$ and $\chi_2(d) = 1$. Continue in this style.

The "Fundamental Identity" in Lemma 2 provides a definitive version of this discussion.

Lemma 2. *Express d as in (2.1). Define*

$$\bar{\chi}^-(p_1 \ldots p_{2n}) = 1 \quad \text{if} \quad \begin{cases} B_{2n} \leq p_{2n} \\ p_{2i} < B_{2i} \text{ when } i < n , \end{cases} \qquad (2.5)$$

$$\bar{\chi}^+(p_1 \ldots p_{2n+1}) = 1 \quad \text{if} \quad \begin{cases} B_{2n+1} \leq p_{2n+1} \\ p_{2i+1} < B_{2i+1} \text{ when } i < n . \end{cases} \qquad (2.6)$$

In all other cases define $\bar{\chi}^-(d) = 0$, $\bar{\chi}^+(d) = 0$. Then the functions $\chi^-(d)$, $\chi^+(d)$ defined in (2.2), (2.3) satisfy

$$\chi^-(d) = 1 - \sum_{j \geq 1} \sum_{\substack{d=p_1\ldots p_{2j} d_{2j} \\ d_{2j} | P(p_{2j})}} \bar{\chi}^-(p_1 \ldots p_{2j}) , \qquad (2.7)$$

$$\chi^+(d) = 1 - \sum_{j \geq 0} \sum_{\substack{d=p_1\ldots p_{2j+1} d_{2j+1} \\ d_{2j+1} | P(p_{2j+1})}} \bar{\chi}^+(p_1 \ldots p_{2j+1}) . \qquad (2.8)$$

3.1 The Construction of Combinatorial Sieves

The condition
$$d_j \,|\, P(p_j), \qquad (2.9)$$
repeated in (2.7) and (2.8), is already implied by the notation (2.1).

Proof of Lemma 2. To establish (2.7), distinguish two cases. The number d may be such that all the inequalities $p_{2i} < B_{2i}$ are satisfied for $2i \leq r$. In this case $\chi^-(d) = 1$ and all the entries $\bar{\chi}$ in (2.5) are 0, so (2.7) holds. In the contrary case there is a least J with $2J \leq r$ for which $p_{2J} \geq B_{2J}$. Then
$$\chi^-(d) = 0, \quad \bar{\chi}^-(p_1 \ldots p_{2J}) = 1,$$
and all other entries $\bar{\chi}$ in (2.7) are 0. This proves (2.7), and (2.8) follows in a similar way.

Proof of Lemma 1. To deduce Lemma 1 from Lemma 2, multiply (2.7) and (2.8) by $\mu(d)$ and sum over d dividing A. Note that when $d\,|\,A$ the condition on d_{2j} in (2.7) is $d_{2j}\,|\,A_{2j}$, where
$$A_{2j} = \left(\frac{A}{p_1 \ldots p_{2j}}, P(p_{2j}) \right), \qquad (2.10)$$
because of (2.9). Thus from (2.7) we obtain
$$\sum_{d\,|\,A} \mu(d)\chi^-(d) = \sum_{d\,|\,A} \mu(d) - \sum_{j} \sum_{p_1 \ldots p_{2j}\,|\,A} \bar{\chi}^-(p_1 \ldots p_{2j}) \sum_{d_{2j}\,|\,A_{2j}} \mu(d_{2j}) \qquad (2.11)$$
$$\leq \sum_{d\,|\,A} \mu(d),$$
because the sum over d_{2j} takes only the values 1 or 0, so in particular is non-negative. This proves the first inequality in (2.4). The other follows similarly using
$$\sum_{d\,|\,A} \mu(d)\chi^+(d) = \sum_{d\,|\,A} \mu(d) + \sum_{j} \sum_{p_1 \ldots p_{2j+1}\,|\,A} \bar{\chi}^-(p_1 \ldots p_{2j+1}) \sum_{d_{2j+1}\,|\,A_{2j+1}} \mu(d_{2j+1})$$
in place of (2.11), the changed sign of the last sum arising from a factor $\mu(p_1 \ldots p_{2j+1})$.

Alternative formulations of Lemma 2. For use in later chapters it is worthwhile to re-express Lemma 2 using a more compact notation, and also to rewrite it in other equivalent ways. Here and elsewhere let $q(d)$ and $Q(d)$ denote, respectively, the least and the greatest prime factor of d. When necessary, adopt the conventions $q(1) = \infty$, $Q(1) = 1$.

If $d > 1$, then
$$\bar{\chi}^-(d) = \chi^-\big(d/q(d)\big) - \chi^-(d), \qquad (2.12)$$
$$\bar{\chi}^+(d) = \chi^+\big(d/q(d)\big) - \chi^+(d). \qquad (2.13)$$

76   3. Combinatorial Methods

To see that (2.5) is equivalent to (2.12) when $d > 1$, observe that (2.12) says the following. First, it gives $\bar{\chi}^-(d) = 1$ or 0, because it is impossible that $\chi^-(d) = 1$ if $\chi^-(d/q(d)) = 0$. Second, (2.12) gives $\bar{\chi}^-(d) = 1$ if and only if $\chi^-(d) = 0$ with $\chi^-(d/q(d)) = 1$, which is precisely what was specified in (2.5). Similarly (2.13) is a reformulation of (2.6).

The definitions (2.2) and (2.3) are both instances of

$$\chi(1) = 1, \qquad \chi(d) = 1 \quad \text{if} \quad p_j < B_j \quad \text{when} \quad 1 \leq j \leq n,$$

where d is again expressed as in (2.1). One has only to take B_j vacuously large ($= +\infty$, say) for all j of specified parity in order to recover (2.2) and (2.3). Then (2.12) and (2.13) say

$$\bar{\chi}(1) = 0, \qquad \bar{\chi}(d) = \chi(d/q(d)) - \chi(d),$$

in which χ may be χ^+ or χ^-.

Lemma 2 then assumes the form

$$\chi(d) = 1 - \sum_{\substack{d=tf \\ Q(f)<q(t)}} \bar{\chi}(t), \qquad (2.14)$$

in which the condition $t > 1$ is to be understood because $\bar{\chi}(1) = 0$, and the term with $f = 1$ is included in the sum in accordance with the convention $Q(1) = 1$. In this form the identity can be verified directly as follows. Again write d as in (2.1). Then the numbers t in (2.14) are precisely those of the form $p_1 p_2 \ldots p_j$ with $j \geq 1$. The right side of (2.14) is

$$1 - \sum_{\substack{d=tf \\ Q(f)<q(t)}} \left(\chi(t/q(t)) - \chi(t)\right)$$

$$= 1 + (\chi(p_1) - \chi(1)) + \cdots + (\chi(p_1 \ldots p_n) - \chi(p_1 \ldots p_{n-1}))$$

$$= \chi(p_1 \ldots p_n) = \chi(d),$$

as a result of a wholesale cancellation of terms.

Another expression of Lemma 2 is now as follows. The consequence (2.11) of the identity (2.7) reads

$$\sum_{d|A} \mu(d)\chi(d) = \sum_{d|A} \mu(d) - \sum_{t|A} \mu(t)\bar{\chi}(t) \sum_{\substack{f|A/t \\ f|P(q(t))}} \mu(f)$$

$$= S(A) - \sum_{t|A} \mu(t)\bar{\chi}(t) S\left(\frac{A}{t}, P(q(t))\right), \qquad (2.15)$$

in which the condition $Q(f) < q(g)$ has been re-expressed, and the notation S from Definition 1.3.1 has been used.

3.1 The Construction of Combinatorial Sieves

In connection with the sieve with weights discussed in Chapter 5 we will use a generalisation of (2.15), obtained by multiplying (2.14) by an arbitrary arithmetic function $\mu(d)\psi(d)$ rather than by $\mu(d)$. Thus we obtain a superficially more formidable "Fundamental Identity"

$$\sum_{d|A} \mu(d)\chi(d)\psi(d) = \sum_{d|A} \mu(d)\psi(d) - \sum_{t|A} \mu(t)\bar{\chi}(t) \sum_{\substack{f|A/t \\ f|P(q(t))}} \mu(f)\psi(ft). \quad (2.16)$$

This generalisation is equivalent to (2.14), and thus to the identities in Lemma 2, since we could take

$$\psi(d) = \mu(d) \quad \text{if} \quad d = d_1, \qquad \psi(d) = 0 \quad \text{otherwise},$$

to recover (2.14) at $d = d_1$.

3.1.3 Buchstab's Identity

The simplest special case of the situation discussed in Sect. 3.1.2 arises when we take $B_1 = 0$ in (2.3), so that $\chi^+(d) = 1$ only for $d = 1$ and $\bar{\chi}^+(d) = 1$ only for $d = p_1$. Then (2.15) gives

$$\sum_{d|A} \mu(d) = 1 - \sum_{p_1|A} \sum_{d_1|(A/p_1, P(p_1))} \mu(d_1). \quad (3.1)$$

This, or equivalently (3.2) below, is one form of "Buchstab's Identity". It is immediate from the observation that if $d \neq 1$ is squarefree then it is uniquely expressible as $d = p_1 d_1$ with $d_1 \mid P(p_1)$, so that p_1 is (as in (2.1)) the largest prime factor of d.

In the case $A = (a, P)$, (3.1) says

$$S(a, P) = 1 - \sum_{p_1|P} S(a/p_1, P(p_1)), \quad (3.2)$$

It is possible to view the identities and inequalities of Lemmas 1 and 2 as following via multiple application of (3.1). This was the procedure used by Brun, although the name "Buchstab's Identity" would not appear until more than a decade later.

There are some consequences of (3.2) which are also referred to as "Buchstab's Identity". Two applications of (3.2) show

$$S(a, P(z)) = S(a, P(w)) - \sum_{\substack{w \leq p_1 < z \\ p_1|P}} S(a/p_1, P(p_1)) \quad \text{if} \quad w \leq z.$$

Then summing this identity over all $a \in \mathcal{A}$ gives, when $w \leq z$,

$$S(\mathcal{A}, P(z)) = S(\mathcal{A}, P(w)) - \sum_{\substack{w \leq p_1 < z \\ p_1|P}} S(\mathcal{A}_{p_1}, P(p_1)). \quad (3.3)$$

We shall say a little more about this contribution of Buchstab in Sect. 7.2.

3.1.4 The Combinatorial Sieve Lemma

The inequality we give below is not yet directly useful in applications because we have not specified a choice of the parameters B_i in the definitions of χ in Lemma 1, a matter we defer until Sect. 3.3.

Lemma 3. *Let χ^-, χ^+ be as specified in Lemma 1. Then*

$$X V^-(P) + R^-(P) \le S(\mathcal{A}, P) \le X V^+(P) + R^+(P), \qquad (4.1)$$

where the expressions $V^-(P)$, $V^+(P)$, $R^-(P)$, $R^+(P)$ are given by

$$V^\pm(P) = \sum_{d|P} \frac{\mu(d)\chi^\pm(d)\rho(d)}{d}, \quad R^\pm(P) = \sum_{d|P} \mu(d)\chi^\pm(d) r_\mathcal{A}(d).$$

Lemma 3 is a restatement of the inequalities of Theorem 1.3.1 in the current context. Lemma 1 shows that we may take

$$\lambda^-(d) = \mu(d)\chi^-(d), \quad \lambda^+(d) = \mu(d)\chi^+(d),$$

in Theorem 1.3.1, any suffices D being redundant at this stage. Lemma 3 follows.

The Main Term. We will need a procedure for estimating the expressions $V^\pm(P)$ appearing in (4.1). This is initiated by the same Lemma 2 that has been used already in (2.16). Take $\psi(d) = \rho(d)/d$ and sum over all d dividing P. The first term on the right in (2.16) gives

$$\sum_{d|P} \mu(d)\psi(d) = V(P),$$

in the notation defined in (1.3.1.8), and similarly in the second terms we encounter (since ψ is currently multiplicative)

$$\sum_{f|P((q(t))} \mu(f)\psi(f) = V\big(P(q(t))\big),$$

the condition on f being as in (2.9). This gives the following identities for $V^\pm(P)$.

Lemma 4. *In Lemma 3, we obtain*

$$V^-(P) = V(P) - T^-(P), \quad V^+(P) = V(P) + T^+(P),$$

where

$$T^-(P) = \sum_{j=1}^{\infty} T_{2j}(P), \quad T^+(P) = \sum_{j=0}^{\infty} T_{2j+1}(P),$$

with T_n given in terms of $\bar{\chi}^-$, $\bar{\chi}^+$ (as defined in Lemma 2) by

$$T_n(P) = \sum_{\nu(t)=n;\, t|P} \frac{\bar{\chi}^{(-)^{n+1}}(t)\rho(t)}{t} V\big(P(q(t))\big). \qquad (4.2)$$

Here

$$\bar{\chi}^{(-)^{n+1}} = \begin{cases} \bar{\chi}^- & \text{if } n \text{ is even} \\ \bar{\chi}^+ & \text{if } n \text{ is odd.} \end{cases}$$

No questions of convergence arise in Lemma 4 because the sums over j are finite.

The identities in Lemma 4 will be used after we make a specific choice of χ in Sect. 3.3. A simple case of them will be used in a similar way in our discussion of Brun's Pure Sieve in Sect. 3.2.

3.2 Brun's Pure Sieve

As was indicated in Sect. 3.1.1, the Pure Sieve is based on a truncation

$$\sum_{d|a;\, \nu(d)<r} \mu(d)$$

of Legendre's construction. As usual, $\nu(d)$ denotes the number of prime factors of d.

This simple idea is surprisingly effective. We shall see in Corollary 1.1 that if (1.3.1.11) and (1.3.5.2) hold (for some κ) then

$$S(A, P(z)) \sim X V(P(z)) \text{ as } x \to \infty \text{ uniformly in } z \leq X^{1/(c\kappa \log \log X)},$$

where $c = 3 \cdot 591\ldots$ is a certain constant. Here, the admissible range for z is very much better than those $z = O(\log X)$ allowed by Legendre's method.

For the application to prime twins p, $p+2$ with $p \leq X$ we sift the sequence $f(n) = n(n+2)$ to obtain an upper bound for their number. In (1.2.3.2) we now have $\rho(p) = 2$ when $p > 2$, so for some constant A we will find

$$V(P(z)) \sim \frac{A}{\log^2 z} \text{ as } z \to \infty.$$

This leads to bounds for $S(A, P(z))$ of the order $X(\log \log X)^2/\log^2 X$, as appears in (4.3). This is weaker only by the factor $(\log \log X)^2$ than the bound given by the full-blown version of Brun's method, or the method of Selberg from Chapter 2.

3.2.1 Inequalities and Identities

The following instance of an inequality of the type (1.3.1.4) should be plausible following the discussion in Sect. 3.1.1.

Lemma 1. *For each integer $j \geq 1$ we obtain*

$$\sum_{d|a;\, \nu(d)<2j} \mu(d) \leq \sum_{d|A} \mu(d) \leq \sum_{d|a;\, \nu(d)<2j-1} \mu(d) \,.$$

The reader who has studied Lemma 3.1.1 may take B_i vacuously large ($B_i = +\infty$, say) if $i < k$, and $B_i = 0$ otherwise. Then we have

$$\chi^{\pm}(d) = 1 \quad \text{if and only if} \quad \nu(d) < k \,, \tag{1.1}$$

where we take $k = 2j$ for χ^- and $k = 2j - 1$ for χ^+. Then Lemma 1 follows.

A self-contained treatment of Lemma 1 is as follows. As in Sect. 3.1.2 we deduce (1.1) from a corresponding identity. As in Sect. 3.1, $q(k)$ denotes the least prime factor of k, with the convention $q(1) = \infty$.

Lemma 2. *For any multiplicative function ψ we obtain*

$$\sum_{d|A;\, \nu(d)<r} \mu(d)\psi(d) = \sum_{d|A} \mu(d)\psi(d)$$
$$+ (-1)^{r+1} \sum_{k|A;\, \nu(k)=r} \psi(k) \sum_{l|(A,P(q(k)))} \mu(l)\psi(l) \,, \tag{1.2}$$

where $P(\cdot)$ is as in (1.3.1.10).

This is a small extension of Buchstab's identity (3.1.3.1) and may be proved in the same way. When $\nu(d) \geq r$ in the first sum on the right, express d uniquely as $d = kl$, where k has r prime factors and $l \mid P(q(k))$ (so that k is the product of the r largest prime factors of d). Then

$$\mu(d)\psi(d) = (-1)^r \psi(k)\mu(l)\psi(l) \,,$$

so (1.2) follows.

Proof of Lemma 1. Take $\psi(k) = 1$, so that the sum over l is the nonnegative quantity $S(A, P(q(k)))$, as in (1.2.1.2).

Lemma 3. *For each $j \geq 1$, the quantity $S(A, P)$ satisfies*

$$X M_{2j}(P) - R_{2j}(A, P) \leq S(A, P) \leq X M_{2j-1}(P) + R_{2j-1}(A, P) \,, \tag{1.3}$$

where

$$M_r(P) = \sum_{d|P;\, \nu(d)<r} \frac{\mu(d)\rho(d)}{d}, \quad R_r(\mathcal{A},P) = \sum_{d|P;\, \nu(d)<r} |r_\mathcal{A}(d)|. \tag{1.4}$$

Lemma 1 provides an instance of the basic inequalities (1.3.1.2), where now $|\lambda(d)| \leq 1$ because $\lambda(d)$ is either $\mu(d)$ or 0. The inequalities in Lemma 3 are now an instance of those in Corollary 1.3.1.1.

Lemma 4. *The sum $M_r(P)$ appearing in (1.4) is expressible as*

$$M_r(P) = V(P) + (-1)^{r+1} G_r(P), \tag{1.5}$$

where $V(P)$ is as in (1.3.1.8) and

$$G_r(P) = \sum_{k|P;\, \nu(k)=r} \frac{\rho(k)}{k} V\bigl(P(q(k))\bigr). \tag{1.6}$$

Apply Lemma 2 with $A = P$ and $\psi(d) = \rho(d)/d$. The first term on the right of (1.2) is now $V(P)$ and the sum over l reduces to $V\bigl(P(q(k))\bigr)$. This gives the identity in Lemma 4.

Lemma 5. *The quantity $S(\mathcal{A},P)$ satisfies*

$$-X G_{2j}(P) - R_{2j}(\mathcal{A},P) \leq S(\mathcal{A},P) - X V(P)$$
$$\leq X G_{2j-1}(P) + R_{2j-1}(\mathcal{A},P), \tag{1.7}$$

where R_r is as in Lemma 3.

This follows from (1.5) and (1.3).

3.2.2 The "Pure Sieve" Theorem

To make the results of Sect. 3.2.1 into an effective instrument we need to estimate the quantity $G_r(P)$ appearing in (1.6). We will obtain Theorem 1, in which we meet the number c for which

$$\left(\frac{c}{e}\right)^c = e. \tag{2.1}$$

Its numerical value may be verified to be $3 \cdot 591\ldots$, but it is immediate from (2.1) that $c > e$, which is all that we use below. The number c will also appear in other contexts later in this chapter.

In the following theorem choose P and z so that z exceeds all the prime factors of P, so that we may write $P = P(z)$. Bear in mind from Definition 1.3.6 that $V(P) < 1$, so that $\log V(P) < 0$. We shall see later that

3. Combinatorial Methods

Theorem 1 is weaker than that obtained in the full version of Brun's method only by the factor V^{-c} in the θ_1 term.

Theorem 1. *When $z \geq 2$ let $s = \log D / \log z$, so that $D = z^s$, and suppose $s \geq 1 + c|\log V(P)|$, where c is as in (2.1) and $P = P(z)$. Under the conditions of Lemma 3 we obtain*

$$S(A, P) = X V(P)\left(1 + \theta_1 (V(P))^{-c} e^{-s}\right) + \theta_2 R(A, P) ,$$

where $|\theta_i| \leq 1$ and

$$R(A, P) = \sum_{d \leq D;\, d|P} |r_A(d)| .$$

In the treatment below we abbreviate $V(P)$ to V. The inequality required for s is not a serious restriction because it asks only $V^{-c} e^{-s} \leq 1/e$. In the applications of this theorem we would want better than this in any case, and will choose s somewhat larger than has been specified.

It is sufficient to use the trivial estimate $V(P(q(k))) \leq 1$ in (1.6). Since there are $r!$ permutations of the r prime factors of d, we obtain

$$G_r(P) \leq \sum_{d|P;\, \nu(d)=r} \frac{\rho(d)}{d} \leq \frac{1}{r!} \left(\sum_{p|P} \frac{\rho(p)}{p} \right)^r = \frac{G^r}{r!} , \tag{2.2}$$

say. Here

$$G \leq \sum_{p|P} -\log\left(1 - \frac{\rho(p)}{p}\right) = -\log V = |\log V| . \tag{2.3}$$

The well-known inequality $e^r \geq r^r/r!$ is easily obtained by selecting a term from the Maclaurin series for e^r. Actually $u_r = r! e^r / r^r$ increases with r, because

$$\frac{u_{r+1}}{u_r} = (r+1) e \frac{r^r}{(r+1)^{r+1}} = \frac{e}{(1 + 1/r)^r} \geq 1 .$$

Thus

$$e^r \geq e \frac{r^r}{r!} \quad \text{if} \quad r \geq 1 , \quad e^r \geq \frac{e^2}{2} \frac{r^r}{r!} \quad \text{if} \quad r \geq 2 , \tag{2.4}$$

and so on.

On using the first of the inequalities (2.4), we obtain from (2.2) and (2.3)

$$G_r(P) \leq \frac{1}{e} \left(\frac{eG}{r}\right)^r \leq \frac{1}{e} \left(\frac{e}{r} |\log V|\right)^r .$$

For (1.7) to be of good quality we require to choose r so that $G_r(P)$ is somewhat smaller than V. Write $r = b|\log V| = -b \log V$, so that

$$G_r(P) \leq \frac{1}{e} \left(\frac{e}{b}\right)^{b|\log V|} = \exp\Big(-|\log V|(b \log b - b) - 1\Big) . \tag{2.5}$$

3.2 Brun's Pure Sieve

The equation (2.1) can be rewritten as $c \log c - c = 1$. To secure the required estimate we take b slightly larger than c.

We will make
$$r \geq s - 1, \tag{2.6}$$
so that $r \geq c |\log V|$. Hence $b \geq c$, and the Mean Value Theorem gives
$$(b \log b - b) - 1 \geq (b - c) \log \xi \geq b - c,$$
actually for some $\xi > c$. Thus (2.5) and (2.6) give
$$G_r(P) \leq \exp\bigl(-|\log V|(b - c + 1) - 1\bigr) = \exp\bigl(-r - |\log V|(-c + 1) - 1\bigr)$$
$$\leq \exp(-s - (1 - c)|\log V|) = V^{1-c} e^{-s}. \tag{2.7}$$

To use (2.7) on both sides of (1.7) we need to obtain (2.6) with r being of specified parity. This can be done with
$$s - 1 \leq r \leq s + 1.$$

In the expression (1.4) we may rewrite the condition on ν as $\nu(d) \leq r - 1$, so that $\nu(d) \leq s$. Since $d | P$ the prime factors of d do not exceed z, whence $d \leq z^s$. Thus the terms $R_r(\mathcal{A}, P)$ in Lemma 3 are now as stated in Theorem 1. The terms $M_r(P)$ are also as required for Theorem 1 because of (2.7) and Lemma 4.

3.2.3 A Corollary

Next, we add some standard assumptions about the quantities $r_\mathcal{A}(d)$ and $\rho(d)$. These will be satisfied in the subsequent application to the prime-twin problem. In Corollary 1.1, the sifting primes in the product $P(z)$ do not exceed z, as in Sect. 3.2.2.

Corollary 1.1. *Suppose that* $|r_\mathcal{A}(d)| \leq \rho(d)$, *as in* (1.3.1.11), *and that ρ has sifting density* $\kappa > 0$. *Then*
$$S(\mathcal{A}, P(z)) \sim X V(P(z)) \text{ as } X \to \infty, \text{ uniformly in } z \leq X^{1/c\kappa \log \log X},$$
but not necessarily uniformly in κ. *Here* $c = 3 \cdot 591 \ldots$ *is as in* (2.1).

We can use (1.3.1.12), which gives
$$R(\mathcal{A}, P) \leq D/V = z^s/V,$$
where $V = V(P) = V(P(z))$. In Theorem 1, make z and s be related by the equation $z^s = X e^{-s}$, so that $s = \log X / \log(ez)$. Then $R(\mathcal{A}, P) \leq X V^{-1} e^{-s}$. Now Theorem 1 gives
$$S(\mathcal{A}, P) = XV(1 + 2\theta V^{-c} e^{-s}),$$

provided its condition $V^{-c}e^{-s} \leq 1/e$ is satisfied, as it will be below.

From (1.3.5.2) we have $V^{-1} < K(2\log z)^\kappa$ for $z \geq 2$. Also, with z as stated, $\log ez/\log X \leq 1/(c\kappa \log \log X) + 1/\log X$, so for sufficiently large X

$$e^{-s} = e^{-\log X/\log ez} \ll 1/(\log X)^{c\kappa}, \quad V^{-1} < 2^\kappa K \left(\frac{\log X}{c\kappa \log\log X}\right)^\kappa.$$

Then we obtain

$$S(\mathcal{A}, P(z)) = XV(P(z))\left(1 + O\left(\frac{1}{(\log\log X)^{c\kappa}}\right)\right),$$

and Corollary 1.1 follows.

3.2.4 Prime Twins

Prime twins are by definition pairs of numbers p, $p+2$ both of which are prime. Brun's earliest result on this topic was as follows.

Theorem 2. *The series*

$$\sum_{p, p+2 \text{ prime}} \frac{1}{p}$$

of reciprocals of prime twins is convergent (or perhaps finite).

Let

$$\pi_2(X) = \sum_{\substack{p \leq X \\ p, p+2 \text{ prime}}} 1 \qquad (4.1)$$

denote the number of prime twins $p, p+2$ with $p \leq X$. The essential content of Theorem 2 is the upper bound (4.3) for $\pi_2(X)$. As elsewhere, take

$$\mathcal{A} = \{n(n+2) : 1 \leq n \leq X\},$$

so that (1.2.3.2) gives that (1.3.1.1) applies in the form (1.3.1.11). Here, $\rho(p) = 2$ for primes $p > 2$, and $\rho(2) = 1$. In (1.3.1.8) take $P = P(z)$ to be the product of all primes $p < z$. Then when $w \geq 2$ we obtain

$$\frac{V(P(w))}{V(P(z))} = \exp\left\{-\sum_{w \leq p < z} \log\left(1 - \frac{2}{p}\right)\right\} = \exp\left\{2\sum_{w \leq p < z} \frac{1}{p} + O\left(\sum_{p \geq w} \frac{1}{p^2}\right)\right\}$$

$$= \exp\left\{2\log\frac{\log z}{\log w} + O\left(\frac{1}{w}\right)\right\} = \left(\frac{\log z}{\log w}\right)^2 \left(1 + O\left(\frac{1}{w}\right)\right), \quad (4.2)$$

so that (1.3.5.2) and (1.3.5.4) hold with $\kappa = 2$ (with a great deal to spare in respect of the O-term).

Argue as in Sect. 1.2.1. Of the $\pi_2(X)$ prime twins counted in (4.1), those with $z \le p \le X$ are counted by $S(\mathcal{A}, P)$. Corollary 1.1 applies, and leads to

$$\pi_2(X) \le z + X V(P(z))(1 + o(1)) \quad \text{as} \quad X \to \infty,$$

where $z = X^{1/2c \log \log X}$. In particular it follows using (4.2) that

$$\pi_2(X) \ll \frac{X}{\log^2 z} \ll \frac{X(\log \log X)^2}{\log^2 X}. \tag{4.3}$$

Theorem 2 now follows, perhaps most elementarily by noting the consequence

$$\sum_{\substack{2^k < p \le 2^{k+1} \\ p+2 \text{ prime}}} \frac{1}{p} \ll 2 \left(\frac{\log \log 2^{k+1}}{\log 2^{k+1}} \right)^2, \quad \text{uniformly in } k,$$

whence the stated convergence follows from that of $\sum \log^2 k / k^2$.

3.3 A Modern Edition of Brun's Sieve

In the full version of Brun's Sieve the ideas of Sect. 3.1 are used in a much more comprehensive way than in the Pure Sieve of Sect. 3.2.

We have to specify a choice for the functions χ in Lemma 3.1.1. Our choice is not Brun's, but one arising in work of J. B. Rosser dating from the 1950's. We shall have more to say about Rosser's sieve, and will give a detailed account of the estimations it involves, in Chapter 4. Here we will deal with these estimations in a more approximate fashion derived from Brun's papers.

Brun's choice of χ will be discussed in Sect. 3.4.

Suppose that none of the primes in the product P exceed z (so that $P = P(z)$) and that ρ has sifting density κ. In this situation we will show (in Corollary 1.1) that

$$S(\mathcal{A}, P) = X V(P(z))(1 + 2\theta_1 K^{10} e^{9-s}) + \theta_2 \sum_{d \le D} |r_\mathcal{A}(d)|,$$

where $z = D^{1/s}$ and $|\theta_1| \le 1$. This improves on Theorem 3.2.1 on the Pure Sieve in that the term V^{1-c} is not now present.

The construction also yields results that are in some respects sharper than Corollary 1.1. In particular we will show how it leads to proofs of Brun's theorems enunciated in the introduction.

3.3.1 Rosser's Choice of χ

In Lemma 3.1.1, take the parameters B_r as given by the condition
$$p_r \leq B_r \iff p_1 p_2 \ldots p_{r-1} p_r^{\beta+1} \leq D , \tag{1.1}$$
where $p_1 > p_2 \ldots$ as in (3.1.2.1), and the real number $\beta \geq 1$ is to be specified. Note that provided $p_1 \leq D$ the condition
$$\chi^{\pm}(d) \neq 0 \Longrightarrow d \leq D$$
follows at once. Thus the symbol D has been used in a way consistent with our notational convention (1.3.1.3) for the function
$$\lambda_D^{\pm}(d) = \mu(d)\chi^{\pm}(d) .$$
When (1.1) holds we write $\chi = \chi_D$ to denote this fact.

The choice (1.1) has a fairly simple motivation which we briefly outline. For a given ρ (actually only the sifting density κ is important) our objective is to establish an inequality
$$\sum_{d \leq D;\, d | P(z)} \frac{\mu(d)\chi_D^-(d)\rho(d)}{d} > 0 \quad \text{if} \quad s > \beta , \tag{1.2}$$
where $z = D^{1/s}$, and β is to be as small as possible (the "sifting limit"). The notation has been chosen so that the use of β in (1.1) and (1.2) is consistent.

In the notations (3.1.2.1) and (3.1.2.5) the sum over d in (1.2) contains terms where
$$d = p_1 p_2 \ldots p_{2r} d_{2r} , \quad d_{2r} | P(p_{2r}) ,$$
which sum to
$$\frac{\rho(p_1 \ldots p_{2r})}{p_1 \ldots p_{2r}} \sum_{\substack{d_{2r} | P(p_{2r}) \\ d_{2r} \leq D/p_1 \ldots p_{2r}}} \frac{\mu(d_{2r})\chi_D^-(p_1 \ldots p_{2r} d_{2r})\rho(d_{2r})}{d_{2r}} . \tag{1.3}$$

In considering (1.3), we may view $\mathcal{A}_{p_1 \ldots p_{2r}}$ as the sequence to be sifted, in the light of the expression
$$|\mathcal{A}_{p_1 \ldots p_{2r} d}| = \frac{X \rho(p_1 \ldots p_{2r})}{p_1 \ldots p_{2r}} \frac{\rho(d)}{d} + r_{\mathcal{A}}(p_1 \ldots p_{2r} d) .$$
Assuming that the same sifting limit β applies, we expect a significant contribution from precisely those d_{2r} for which
$$d_{2r} | P\bigl(D/p_1 \ldots p_{2r}\bigr)^{1/\beta} .$$
This suggests that the condition
$$p_{2r}^{\beta} \leq D/p_1 \ldots p_{2r} ,$$
viz. (1.1), ought to be the one controlling the size of p_{2r} in the definition of $\chi_D^-(d)$.

3.3.2 A Technical Estimate

We deal next with a technical point which is independent of the particular choice (1.1) of the parameters B_r.

To make use of Lemma 3.1.3 we need to bound $V^+(P)$ from above and $V^-(P)$ from below, so that we need upper bounds for the quantities

$$T_r(P) = \sum_{\nu(d)=r} \frac{\bar{\chi}_D^{\pm}(d)\rho(d)}{d} V(P(q(d)))$$

appearing in Lemma 3.1.4. We use the notation (3.1.2.1), so that the least prime factor $q(d)$ of d is also written p_r. The condition $\bar{\chi}_D(d) = 0$ implies $p_r > B_r$. In Brun's work the parameter B_r was independent of the other prime factors of d. In the current context a corresponding inequality $p_r > z_r$ will be obtained in a less direct way in Lemma 4. In either situation the next estimate is of use.

Lemma 1. *Define*

$$\tau_r(z_r, z) = \sum_{z_r < q(d);\, \nu(d)=r} \frac{\rho(d)}{d} V(P(q(d))) , \qquad (2.1)$$

and suppose that ρ has sifting density κ. Then

$$\frac{\tau_r(z_r, z)}{V(P(z))} \leq \frac{1}{r!}\left\{\log\left(K\left(\frac{\log z}{\log z_r}\right)^{\kappa}\right)\right\}^r K\left(\frac{\log z}{\log z_r}\right)^{\kappa} .$$

In Lemma 1, the ρ-dependence of the estimate for τ_r is carried entirely by the factor $V(P(z))$ on the left.

In (2.1) we have $V(P(q(d))) \leq V(P(z_r))$. Proceeding as in Theorem 3.2.1 we obtain

$$\frac{\tau_r(z_r, z)}{V(P(z))} \leq \frac{1}{r!}\left(\sum_{z_r < p < z} \frac{\rho(p)}{p}\right)^r \frac{V(P(z_r))}{V(P(z))} .$$

Here the sum over p does not exceed

$$\sum_{z_r \leq p < z} -\log\left(1 - \frac{\rho(p)}{p}\right) = \log \frac{V(P(z_r))}{V(P(z))} .$$

Hence

$$\frac{\tau_r(z_r, z)}{V(P(z))} \leq \frac{1}{r!}\left(\log \frac{V(P(z_r))}{V(P(z))}\right)^r \frac{V(P(z_r))}{V(P(z))} .$$

On invoking (1.3.5.2) we obtain the result of Lemma 1.

3.3.3 A Simplifying Approximation

Take B_r as specified in (1.1). In this context we need to determine admissible numbers z_r with which Lemma 1 can be used. A certain simplification, with only an unimportant loss of precision, follows from the following observations. We will write d as in (3.1.2.1).

Lemma 2. *Suppose* $p_1 < z = D^{1/s}$ *with* $s \geq \beta$, *where* β *and* B_r *are as in* (1.1). *Then if either* $\chi_D^+(d) = 1$ *or* $\chi_D^-(d) = 1$ *we obtain*

$$p_1 p_2 \ldots p_{j-1} p_j^\beta \leq D \quad \text{when} \quad 1 \leq j \leq r, \tag{3.1}$$

these inequalities holding for j of either parity.

This follows for $r = 1$ because $s \geq \beta$ and for $r > 1$ because for each j the hypothesis about $\chi_D^\pm(d)$ gives one of the inequalities

$$p_1 p_2 \ldots p_{j-1}^{\beta+1} \leq D, \quad p_1 p_2 \ldots p_{j-1} p_j^{\beta+1} \leq D,$$

and Lemma 2 follows.

The inequalities in Lemma 2 have the following consequence.

Lemma 3. *Suppose the inequalities* (3.1) *hold for* $1 \leq r \leq N$. *Then*

$$p_1 p_2 \ldots p_N \leq D^{1-(1-1/\beta)^N}. \tag{3.2}$$

Raise the jth of the inequalities (3.1) to the power $(1 - 1/\beta)^{N-j}$ and multiply the resulting inequalities together. Because

$$1 + \left(1 - \frac{1}{\beta}\right) + \left(1 - \frac{1}{\beta}\right)^2 + \cdots + \left(1 - \frac{1}{\beta}\right)^{m-1} = \beta - \beta\left(1 - \frac{1}{\beta}\right)^m,$$

which we use with $m = N$ and $m = N - r$, this gives

$$(p_1 p_2 \ldots p_N)^\beta \leq \left(D^\beta\right)^{1-(1-1/\beta)^N},$$

as required.

In particular we obtain $d \leq D^{1-(1-1/\beta)^{\nu(d)}}$ whenever $\chi_D^\pm(d) = 1$. For the numbers characterised by $\bar{\chi}_D$ we infer the following.

Lemma 4. *Suppose that*

$$\bar{\chi}_D^+(d) = 1 \text{ if } \nu(d) \text{ is odd}, \quad \bar{\chi}_D^-(d) = 1 \text{ if } \nu(d) \text{ is even}.$$

Then the least prime factor $q(d) = p_n$ of the number $d = p_1 p_2 \ldots p_n$ satisfies

$$q(d) > D^{\beta^{-1}(1-\beta^{-1})^n} \,. \tag{3.3}$$

When $\bar{\chi}_D^{\pm}(d) = 1$ the definition of $\bar{\chi}$ in Lemma 3.1.2 gives first (via Lemma 2) that the inequalities (3.2) hold for $1 \leq N \leq n-1$. Then it gives, because of Lemma 3, that

$$p_n^{\beta+1} > \frac{D}{p_1 p_2 \ldots p_{n-1}} > D^{(1-\beta^{-1})^{n-1}} \,.$$

Thus

$$p_n > D^{\frac{1}{\beta}\frac{\beta}{\beta+1}\left(\frac{\beta-1}{\beta}\right)^{n-1}} > D^{\frac{1}{\beta}\left(\frac{\beta-1}{\beta}\right)^n} \,,$$

as asserted.

3.3.4 A Combinatorial Sieve Theorem

From this point onwards we rewrite $T_n(P)$ as $T_n(D, P)$ and $T^{\pm}(P)$ as $T^{\pm}(D, P)$, since dependence on D is now significant. The basic result in Sect. 3.3 is the following estimation of the quantities T_n given in (3.1.4.2), when χ is specified as in (1.1).

Lemma 5. *Suppose ρ has sifting limit κ, and*

$$z = D^{1/s} \quad \text{with} \quad s \geq \beta \,. \tag{4.1}$$

Let d be any number for which

$$\left(\frac{\beta}{\beta-1}\right)^{\kappa} \leq e^d \,. \tag{4.2}$$

Define $a = de^{d+1}$, $f = 1 + 1/d$. Then

(i) $\quad T_n(D, P(z)) \leq V(P(z)) a^n K^f / e \,,$

(ii) $\quad T_n(D, P(z)) = 0 \quad \text{when} \quad n \leq s - \beta \,,$

(iii) $\quad \sum_{n \geq 1} T_n(D, P(z)) \leq V(P(z)) \frac{a^{s-\beta}}{1-a} \frac{K^f}{e} \quad \text{when} \quad a < 1 \,.$

The expression (3.1.4.2) for T_n is

$$T_n(D, P(z)) = \sum_{p_n < \cdots < p_1 < z} \frac{\rho(p_1 \ldots p_n)}{p_1 \ldots p_n} \bar{\chi}_D(p_1 \ldots p_n) V(P(p_n)) \,,$$

where we have expanded the integer d in the form (3.1.2.1), and where

$$\bar\chi_D = \bar\chi_D^- \text{ if } n \text{ is even}, \quad \bar\chi_D = \bar\chi_D^+ \text{ if } n \text{ is odd}.$$

Of the conditions defining $\bar\chi$ in Lemma 3.1.2 we retain only

$$D < p_1 p_2 \ldots p_{n-1} p_n^{\beta+1}, \tag{4.3}$$

together with the consequence (3.3) of the other conditions. Here χ is as in (1.1).

In this way we obtain

$$T_n(D, P(z)) \leq \sum_{\substack{z_n < p_n < \ldots < p_1 < z \\ D < p_1 \ldots p_{n-1} p_n^{\beta+1}}} \frac{\rho(p_1 p_2 \ldots p_n)}{p_1 p_2 \ldots p_n} V(P(p_n)),$$

with

$$z_n = D^{\beta^{-1}(1-\beta^{-1})^n} > z^{(1-\beta^{-1})^n},$$

as in Lemma 4. Thus

$$\frac{\log z}{\log z_n} < \left(\frac{\beta}{\beta - 1}\right)^n.$$

On invoking Lemma 1 we obtain

$$T_n(D, P(z)) \leq \frac{V(P(z))}{n!} K\left(\frac{\beta}{\beta-1}\right)^{n\kappa} \left\{\log\left(K\left(\frac{\beta}{\beta-1}\right)^{n\kappa}\right)\right\}^n.$$

Because of (4.2) this gives

$$T_n(D, P(z)) \leq \frac{V(P(z))}{n!} K e^{nd} \left(\log(K e^{nd})\right)^n$$

$$= \frac{V(P(z))}{n!} K e^{nd} (nd)^n \left(1 + \frac{\log K}{nd}\right)^n$$

$$\leq V(P(z)) \frac{n^n}{n!} (d e^d)^n K^{1+1/d},$$

where we used the well-known inequality $(1+x/n)^n \leq e^x$. Since $e^{n-1} \geq n^n/n!$, as was noted at (3.2.2.4), this gives

$$T_n(D, P(z)) \leq V(P(z)) (d e^{d+1})^n K^{1+1/d}/e,$$

which is the result of part (i) of this lemma.

Part (ii) follows since (4.3) gives $D < p_1^{n+\beta}$, while $p_1 < z = D^{1/s}$, so that the sum defining $T_n(D, P(z))$ is empty unless $n + \beta > s$.

Part (iii) follows at once from parts (i) and (ii) and the hypothesis $s \geq \beta$.

Next, we show how to choose an exponent β in (1.1) so that the inequality $a < 1$ will follow, as is essential in part (iii) of Lemma 5. The number c appearing in Lemma 6 is that already encountered in Sect. 3.2.2.

Lemma 6. *Suppose* (4.1) *holds. Let* $c = 3.591\ldots$ *be the solution of*

$$e^{1/c} = c/e \,. \tag{4.4}$$

Suppose

$$\beta = 1 + b\kappa, \text{ where } b > c, \tag{4.5}$$

and that $z = D^{1/s}$, $s > \beta$ *are as in Lemma 5. Then under the conditions of Lemma 5 the sums* $T^{\pm}(D, P(z))$ *defined in Lemma 3.1.4 satisfy*

$$T^{\pm}(D, P(z)) \le V(P(z)) \frac{a^{s-\beta}}{1-a} \frac{K^{b+1}}{e},$$

where $a = b^{-1} e^{1+1/b} < 1$.

Since $a = de^{d+1}$ in Lemma 5 we need $d < 1/c$ to secure $a < 1$. We need to specify β so that such a choice of d is consistent with (4.2). This task is eased if we invoke the inequality

$$\left(\frac{\beta}{\beta-1}\right)^{\kappa} = \left(1 + \frac{1}{\beta-1}\right)^{\kappa} \le e^{\kappa/(\beta-1)}$$

(note, however, that this inequality is not particularly sharp unless β is large compared with 1). Thus $d = \kappa/(\beta - 1)$ is a legitimate choice in (4.2). Then $d < 1/c$ because of (4.5). In fact $d = 1/b$ and in Lemma 5 we now have

$$a = de^{d+1} = \frac{1}{b} e^{1+1/b} < 1 \,. \tag{4.6}$$

Also $1 + 1/d = 1 + b$, so now Lemma 5(i) gives

$$T_n(D, P(z)) \le V(P(z)) a^n K^{1+b}/e \,.$$

On summing this estimate over the range $n > s-\beta$ allowed by Lemma 5(ii) we obtain Lemma 6.

On assembling the results of Lemma 3.1.4 and Lemma 6 we obtain the following instance of a combinatorial sieve theorem, as typified by Lemma 3.1.3.

Theorem 1. *Let* $c = 3.591\ldots$ *be as in* (4.4). *Suppose* ρ *has sifting density* κ *and that* $\beta = 1 + b\kappa$ *with* $b > c$. *Also suppose* $z = D^{1/s}$ *with* $s > \beta$. *Then*

$$S(\mathcal{A}, P(z)) = X V^{\pm}(D, P(z)) + \theta_2 \sum_{d \le D;\, d | P(z)} |r_{\mathcal{A}}(d)| \,,$$

where

$$V^{\pm}(D, P(z)) = V(P(z)) \left(1 + \theta_1 \frac{a^{s-\beta}}{1-a} \frac{K^{1+b}}{e}\right),$$

for some θ_i *with* $|\theta_i| \le 1$, *and* $a = b^{-1} e^{1+1/b} < 1$ *is as in Lemma 6.*

In Theorem 1 the expressions for $V^-(D, P(z))$ and $V^+(D, P(z))$ have the same form, possibly with different numbers θ_i.

Note that $c < 4$ (this follows from (4.4) and the inequality $e^{5/4} < 4$, equivalent to $\log 2 > \frac{5}{8}$). In Theorem 1 it is essential to take $b > c$, but to obtain a reasonable bound for $V(D, P(z))$ we need to choose b slightly larger.

The Fundamental Lemma. We may take $b = 9$ in Theorem 1. Then we obtain $a < 1/e$ in (4.6), because $\log ae = 2 + \frac{1}{9} - 2\log 3$. This leads to Corollary 1.1 below.

The bounds just used on $\log 2$ and $\log 3$ are easily checked using, for example,
$$\log \frac{1+x}{1-x} \geq 2(x + \tfrac{1}{3}x^3 + \cdots) \quad \text{for} \quad 0 \leq x < 1.$$

Corollary 1.1. *Set $D = z^s$. Assume that ρ has sifting density κ, and suppose $s > 9\kappa + 1$. Then in Theorem 1 we may take*
$$V(D, P(z)) = X\, V(P(z))\bigl(1 + 2\theta K^{10} e^{9\kappa - s}\bigr),$$
where $|\theta| < 1$.

Corollary 1.1 may be compared with Theorem 3.2.1 on the Pure Sieve. Observe that the factors V^{-c} are now absent.

The description "Fundamental Lemma" refers to the fact that provided D is chosen to be suitably small, in terms of X, Corollary 1.1 provides an asymptotic expression for $S(\mathcal{A}, P(z))$ valid as $s \to \infty$. If this is our sole objective, however, then a sharper result can be obtained by choosing a somewhat larger value of b.

Corollary 1.2. *Set $D = z^s$. In Theorem 1 we may take*
$$V(D, P(z)) = X\, V(P(z))\Bigl(1 + O\bigl(K^6 e^{-s\log s + s\log\log Ks + O(s)}\bigr)\Bigr),$$
when s exceeds a certain $s_0(\kappa)$ depending only on κ.

The proviso $s > s_0(\kappa)$ is needed only to ensure that the following choice of b is consistent with the requirement $s > \beta$ in Theorem 1. It is sufficient if $s_0(\kappa) = 2 + 8\kappa + e^{2\kappa}$, for example.

In Theorem 1, take $b = 4 + s/\log Ks$. Since $a = b^{-1} e^{1+1/b}$ decreases with increasing b, $b > s/\log Ks$, and $\beta = 1 + b\kappa > 0$ we obtain
$$a^{s-\beta} < a^s < \exp\Bigl(-s\log s + s\log\log Ks + s + \log s + \log K\Bigr).$$
Here $a < \tfrac{1}{4}e^{5/4} < 1$. Also
$$K^{1+b} < \exp\!\left(\left(5 + \frac{s}{\log Ks}\right)\log K\right) < K^5 e^s.$$
Now Corollary 1.2 follows from Theorem 1.

3.3 A Modern Edition of Brun's Sieve

The Sifting Limit. By choosing b relatively close to c we can determine a good lower bound β_0 of those β for which we can obtain
$$V(D, P(z)) > 0$$
in Theorem 1, on the hypothesis that K in (1.3.5.2) is sufficiently close to 1. Then β_0 is an upper bound for the sifting limit $\beta(\kappa)$ introduced in Sect. 1.3.6. Following Theorem 1, we can obtain such a bound that is not substantially greater than $1 + c\kappa$.

Lemma 7. *Suppose*
$$a = \frac{1}{b}e^{1+1/b}, \quad b = c + \delta, \quad \eta = \frac{8}{\delta}\log\frac{8}{\delta},$$
where $0 < \delta < 1$ and $c = 3\cdot591\ldots$ is as in (4.4). Then $a^\eta/(1-a) < 1$.

Since $da/db < -e/b^2$ we obtain $(1-a)/\delta > e/\xi^2$ for some $\xi \in (c,b)$, so that $1 - a > \delta e/b^2 > \delta e/(c+1)^2 > \frac{1}{8}\delta$. Hence $\log a < \log(1 - \frac{1}{8}\delta) < -\frac{1}{8}\delta$. Thus
$$\frac{-\log(1-a)}{-\log a} < \frac{-\log(\delta/8)}{\delta/8} = \eta,$$
as required.

This leads to the following bound for the sifting limit.

Corollary 1.3. *The sifting limit $\beta(\kappa)$ for the sieve of density κ satisfies*
$$\beta(\kappa) \leq 1 + c\kappa + 4\sqrt{\kappa}\log\kappa + (1 + 8\log 8)\sqrt{\kappa} \quad \text{if} \quad \kappa \geq 1.$$
In particular
$$\limsup_{\kappa \to \infty} \frac{\beta(\kappa)}{\kappa} \leq c,$$
where $c = 3\cdot591\ldots$ is defined by (4.4).

In Theorem 1, $K^{b+1} < e$ when K is sufficiently close to 1 and $b < c+1$ as in Lemma 7. In this situation Lemma 7 shows that $V(D, P(z)) > 0$ in Theorem 1, provided that
$$s > \beta + \eta = 1 + (c+\delta)\kappa + \frac{8}{\delta}\log\frac{8}{\delta}.$$
If $\kappa \geq 1$ we may choose $\delta = 1/\sqrt{\kappa}$. Then Corollary 1.3 follows as stated.

3. Combinatorial Methods

An Improved Lower Bound. The lower bound aspect of Theorem 1 can be improved, because the expression for $V^-(P)$ given in Lemma 3.1.4 involves T_n only for even $n \geq 2$. Also for $n \geq 2$ we can use the better estimate $n^n/n! \leq 2e^{n-2}$ noted in (3.2.2.4). Proceeding as in Lemma 6, we obtain the following addendum to that result.

Lemma 8. *Under the hypotheses of Lemma 6 we obtain*

$$T^-(D, P(z)) \leq V(P(z)) \frac{a^{2(s-\beta)}}{1-a^2} \frac{K^{b+1}}{\frac{1}{2}e^2}.$$

By numerical work rather more precise than that employed earlier it can be verified that if we choose $b = 4$ then we find $a < \frac{7}{8}$ in (4.6), whence

$$\frac{a^2}{(1-a^2)\frac{1}{2}e^2} < \frac{8}{9}.$$

As in Lemma 6 we can choose any s with $s > \beta = 4\kappa + 1$. This shows that for any $\kappa > 0$ the sifting limit $\beta(\kappa)$ satisfies $\beta(\kappa) \leq 4\kappa + 1$. Moreover, by continuity considerations we obtain similar conclusions if the choice $b = 4$ is replaced by a slightly smaller number, as follows.

Corollary 1.4. *There exists a constant $b_0 < 4$ such that the quantity V^- in Theorem 1 satisfies*

$$V^-(D, P(z)) \geq \left(1 - \tfrac{9}{10}K^5\right) V(P(z))$$

if $z = D^{1/s}$ with $s > 1 + b_0\kappa$. Hence the lower bound in Theorem 1 may be replaced by

$$S(\mathcal{A}, P(z)) \geq X V(P(z)) \left(1 - \tfrac{9}{10}K^5\right) - \sum_{d \leq D;\, d | P(z)} |r_\mathcal{A}(d)|.$$

A calculation shows that this corollary holds with $b_0 = 3.99$, for example.

3.3.5 Applications

The results in Sect. 3.3.4 are sufficiently strong to lead directly to slightly improved versions of some of Brun's results. We will discuss one of them.

Theorem 2. *There are infinitely many n such that both of n and $n + 2$ have at most eight prime factors.*

The "conjugate" result that every sufficiently large even number can be represented as $m+n$, where both of m and n have at most eight prime factors, can be treated in a similar fashion.

Take
$$\mathcal{A} = \{n(n+2) : 1 \leq n \leq X\}. \tag{5.1}$$

The discussion in Sect. 1.2.3 applies with $f(n) = n(n+2)$ and shows

$$\mathcal{A}_d = x\frac{\rho(d)}{d} + r_\mathcal{A}(d) \quad \text{with} \quad |r_\mathcal{A}(d)| \leq \rho(d), \tag{5.2}$$

where the multiplicative function ρ satisfies

$$\rho(2) = 1, \quad \rho(p) = 2 \text{ for primes } p > 2.$$

Take P to be the product of all primes not exceeding z. Then, because of Lemma 1.3.3, we obtain

$$\frac{V(P(w))}{V(P(z))} \leq K\left(\frac{\log z}{\log w}\right)^2, \tag{5.3}$$

i.e. (1.3.5.2), with $\kappa = 2$ and for some K as in (1.3.5.4).

If w is small, and in particular if $w = 2$, we can obtain something more explicit. Using Mertens' product (1.1.4.2) we find

$$\frac{1}{V(P(z))} = 2 \prod_{2<p<z}\left(1-\frac{2}{p}\right)^{-1} = 2\prod_{2<p<z}\left(\frac{(1-1/p)^2}{1-2/p}\left(1-\frac{1}{p}\right)^{-2}\right)$$

$$\leq 2e^{2\gamma}(\log^2 z + O(\log z))\prod_{p>2}\left(1+\frac{1}{p(p-2)}\right).$$

This shows, however, that for small w the number K in (1.3.5.2) exceeds 1 by an amount large enough for Corollary 1.4 not to be directly applicable.

We may adopt the device suggested in Sect. 1.3.5 of taking κ slightly larger, $\kappa = 2 + \varepsilon$, say. Then, as in (1.3.5.8),

$$\frac{V(P(w))}{V(P(z))} \leq K\left(\frac{\log z}{\log w}\right)^2 \leq K'\left(\frac{\log z}{\log w}\right)^{2+\varepsilon},$$

with $K' = 1 + O(1/\log^\varepsilon z)$.

Apply Corollary 1.4 with $\kappa = 2 + \varepsilon$. We obtain

$$V^-(D, z) \geq C V(P(z)), \tag{5.4}$$

for any constant $C < \frac{1}{10}$, provided $z = z(\varepsilon)$ is large enough and provided

$$s = \log D / \log z > b_0\kappa + 1 = b_0(2+\varepsilon) + 1.$$

In particular (5.4) holds with $z = D^{1/s_0}$ for a certain $s_0 < 9$. One can check that the value $s_0 = 8 \cdot 99$ is admissible, provided ε is taken to be suitably small.

96   3. Combinatorial Methods

Since (5.2) holds we may use the estimate (1.3.1.12) in Corollary 1.4 to obtain

$$S(\mathcal{A}, P(z)) \geq CXV(P(z)) - \frac{D}{V(P(z))} \quad \text{if} \quad z < D^{1/s_0},$$

where $\mathcal{A}$ is as in (5.1). Here $V(P(z)) \gg 1/\log^2 z$ from (5.3). Accordingly we may take

$$D = X/\log^3 X, \quad z = (X+2)^{1/9},$$

while retaining the result $S(\mathcal{A}, P(z)) > CXV(P(z))$ for each $C < \frac{1}{10}$, provided X is large enough. Since $\mathcal{A}$ is as specified in (5.1) this is sufficient to establish Theorem 2.

3.4 Brun's Version of his Method

In Brun's work the parameters B_r of Lemma 3.1.1 did not depend on the prime factors of the number d, as was the case in the development discussed in Sect. 3.3. The details of his work are perhaps now of mainly historical interest. We will however give an account here, without striving towards the best results that could possibly be obtained within the limitations of his methods. We will see that the results are of somewhat lower quality than those obtained in Sect. 3.3. In particular the asymptotic value obtained for the sifting limit β_κ as the sifting density κ becomes large is worse by a factor e than that found in Corollary 3.3.1.3.

Our discussion will assume $\kappa \geq 1$. The method is valid for $\kappa > 0$ but some adjustment of detail would become necessary. Sharper results valid when $\kappa > 0$ were recorded in Sect. 3.3.

3.4.1 Brun's Choice of χ

In Brun's work a central idea is that the parameters B_r may be chosen so that their logarithms lie in a geometric progression.

Lemma 1. *Choose $z > 2$ and denote $z^s = D$. Specify*

$$B_n = z^{e^{1-n/\beta}} \quad \text{for} \quad n \geq 1. \tag{1.1}$$

Then in Lemma 3.1.1 we obtain

$$d | P(z), \ \chi^\pm(d) \neq 0 \implies d \leq D \tag{1.2}$$

provided

$$2 \leq \beta \leq s/e. \tag{1.3}$$

The principle of this lemma is to choose B_n in such a way that for each r we have
$$B_1 \geq B_2 \geq \ldots \geq B_r, \quad zB_1B_2\ldots B_r \leq D. \tag{1.4}$$
Then in the notation (3.1.2.1) we will obtain from (3.1.2.2), (3.1.2.3) that
$$\chi^-(d) = 1 \Longrightarrow d = p_1 p_2 \ldots p_r \leq zB_2^2 B_4^2 \ldots \leq D,$$
$$\chi^+(d) = 1 \Longrightarrow d \leq zB_1 B_3^2 \ldots \leq D,$$
since $p_1 < z$ when $d|P(z)$. Then (1.2) will follow.

The first inequalities in (1.4) are clear so we have only to check the last one. From (1.1) we obtain
$$\log(zB_1B_2\ldots B_r)/\log z < 1 + \sum_{n=1}^{\infty} e^{1-n/\beta} = 1 + \frac{e}{e^{1/\beta} - 1}.$$

To obtain (1.4) we require only that this does not exceed $s = \log D/\log z$. In fact
$$1 + \frac{e}{e^{1/\beta} - 1} < 1 + \frac{e}{\frac{1}{\beta} + \frac{1}{2\beta^2}} = 1 + \frac{\beta e}{1 + \frac{1}{2\beta}} < \beta e \leq s, \tag{1.5}$$
using (1.3), if β is large enough. Actually
$$1 + \frac{1}{2\beta} + \beta e < \beta e + \frac{e}{2} \quad \text{if} \quad 2\beta + 1 < \beta e,$$
so that (1.5) holds if $\beta > 1/(e-2)$, as is the case in (1.3). This establishes Lemma 1.

3.4.2 The Estimations

The next lemma follows in a more direct way than did the corresponding Lemma 3.3.5 in Sect. 3.3.

Lemma 2. *Suppose the function ρ has sifting density κ and that the parameters B_r are specified as in Lemma 1. Then the sum $T_n(P(z))$ appearing in Lemma 3.1.4 satisfies*

(i)
$$\frac{T_n(P(z))}{V(P(z))} \leq H a^n,$$

with
$$H = \frac{K^{1+\beta/\kappa}}{e^{\kappa+\beta+1}}, \quad a = \frac{e\kappa}{\beta} e^{\kappa/\beta},$$

(ii)
$$T_n(P(z)) = 0 \quad \text{if} \quad n \leq \beta,$$

(iii) *If a < 1 then*

$$\sum_{n=1}^{\infty} T_n(P(z)) \leq V(P(z)) \frac{Ha^\beta}{1-a}. \tag{2.1}$$

In Lemma 3.1.4, the conditions

$$\bar{\chi}^{(-)^{n+1}}(k) = 1, \quad \nu(k) = n$$

imply $p(k) > B_n$, because of the structure of $\bar{\chi}$ described in Lemma 3.1.2. Thus Lemma 3.3.1 applies with $z_n = B_n$ to give

$$\frac{T_n(P(z))}{V(P(z))} \leq \frac{1}{n!} \left\{ \log\left(K\left(\frac{\log z}{\log B_n}\right)^\kappa\right)\right\}^n K\left(\frac{\log z}{\log B_n}\right)^\kappa.$$

Since $\log z / \log B_n = e^{-1+n/\beta}$ in (1.1) and $1/n! < e^{-1}(e/n)^n$ from (3.2.2.4), we obtain

$$\frac{T_n(P(z))}{V(P(z))} \leq \frac{K}{e}\left\{\frac{e}{n}\left(\kappa\left(\frac{n}{\beta}-1\right)+\log K\right)\right\}^n \frac{e^{\kappa n/\beta}}{e^\kappa}$$

$$\leq \frac{Ke^{-\kappa}}{e}\left(\frac{e\kappa}{\beta}\right)^n \left(1+\frac{\beta\log(Ke^{-\kappa})}{\kappa n}\right)^n e^{\kappa n/\beta}$$

$$\leq \frac{Ke^{-\kappa}}{e}\left(\frac{e\kappa}{\beta}e^{\kappa/\beta}\right)^n \exp\left(\frac{\beta\log(Ke^{-\kappa})}{\kappa}\right)$$

$$= \frac{1}{e}(Ke^{-\kappa})^{1+\beta/\kappa}\left(\frac{e\kappa}{\beta}e^{\kappa/\beta}\right)^n,$$

as required in part (i) of the lemma.

Part (ii) is immediate because (1.1) gives $B_n > z$ if $n \leq \beta$, but in the definition (3.1.4.2) of T_n we have $B_n < p(k)$ because of the specification of $\bar{\chi}$ in Lemma 3.1.2, and $p(k) < z$ because $k|P$ in (3.1.4.2) and currently $P = P(z)$.

Part (iii) follows at once from parts (i) and (ii).

3.4.3 The Result

The principal result of Sect. 3.4 follows directly from Lemmas 1 and 2.

Theorem 1. *Let $c = 3.591...$ be the solution of the equation $e^{1/c} = c/e$. Suppose ρ has sifting density κ. Take $\beta = b\kappa$ and $a = b^{-1}e^{1+1/b}$, where $b > c$, so that $a < 1$. Suppose also that $s \geq e\beta$, and set $z = D^{1/s}$. Then*

$$S(\mathcal{A}, P(z)) = XV^\pm(D, P(z)) + \theta_2 \sum_{d \leq D} |r_\mathcal{A}(d)|,$$

3.4 Brun's Version of his Method

where
$$V^{\pm}(D, P(z)) = V(P(z))\left(1 + \theta_1 \frac{a^\beta}{1-a} \frac{K^{1+b}}{e^{\kappa+\beta+1}}\right), \tag{3.1}$$

with $0 \leq |\theta_i| \leq 1$.

This is the form taken by Lemma 3.1.3 when χ is as specified by (1.1), because Lemma 1 gives the condition $d \leq D$ and the estimate for $V^{\pm}$ is supplied via Lemma 3.1.4 by Lemma 2.

Theorem 1 gives the following corollary about the sifting limit $\beta(\kappa)$.

Corollary 1.1. *When $\kappa \geq 1$ the sifting limit satisfies*
$$\beta(\kappa) \leq e(c\kappa + 1). \tag{3.2}$$

In particular
$$\limsup_{\kappa \to \infty} \frac{\beta_\kappa}{\kappa} \geq ce. \tag{3.3}$$

For large κ this is weaker by a factor e than the result obtained in Corollary 3.3.1.3.

In Theorem 1, take $b = c + \delta$, where $0 < \delta < 1$. Then Lemma 3.3.7 gives
$$\frac{a^\beta}{1-a} < 1 \quad \text{provided} \quad \beta > \frac{8}{\delta} \log \frac{8}{\delta}.$$
Since $a < 1$ the conclusion (3.1) persists if $\beta > b\kappa = (c+\delta)\kappa$. If we specify
$$\beta = (c+\delta)\kappa + \frac{\log 8/\delta}{\delta/8},$$
then both provisos on β are satisfied and we obtain $V^{-}(D, P(z)) > 0$ if K is sufficiently close to 1 and if $D = z^{1/s}$ with $s \geq e\beta$, as required in Theorem 1. This is sufficient to derive the inference (3.3).

In Theorem 1, choose $\beta = c\kappa + 1$, so that $b = c + 1/\kappa$, and observe
$$a = \frac{e}{b} e^{1/b} < \frac{e}{b} e^{1/c} = \frac{c}{b},$$
so that $1/(1-a) < b/(b-c) = \kappa b = \beta$. Then for the θ_1 term in Theorem 1 we obtain, using the exact value for a^β,
$$\frac{K^{1+b}}{e^{\kappa+\beta+1}} \frac{a^\beta}{1-a} < \frac{K^{1+b}}{e^{\kappa+\beta+1}} b\kappa \left(\frac{e}{b} e^{1/b}\right)^{b\kappa} = \frac{K^{1+b}}{e^{\beta+1}} b\kappa \left(\frac{e}{b}\right)^{b\kappa}$$
$$= \frac{K^{1+b}}{e^{\beta+1}} (c\kappa + 1) \left(\frac{e\kappa}{c\kappa+1}\right)^{c\kappa+1} < \frac{K^{1+b}}{e^{\beta+1}} e\kappa \left(\frac{e}{c}\right)^{c\kappa} = \frac{K^{1+c+1/\kappa}}{e^{\beta+\kappa/\kappa}}$$
$$< 1$$

if K is sufficiently close to 1. In this case we obtain from Theorem 1 that $V^{-}(P(z)) > 0$ when $\beta = c\kappa + 1$, where we may choose $s = e\beta$, so that Corollary 1.1 follows.

3.5 Notes on Chapter 3

Sect. 3.1. The basic ideas of this section come, of course, from Brun (1920).

Sect. 3.2.2. The "Pure Sieve" appeared in [Brun 1915]. Subsequent accounts were given by [Landau (1927)], also by [Halberstam and Richert (1974)]. The version in Iwaniec's Rutgers lecture notes is explicit as regards the dependence on the constant c.

C. Hooley introduced an "almost pure" sieve in [Hooley (1994)]. This construction is similar to but more elaborate than the pure sieve of Brun, but is simpler than the full-blown version of Brun's method. It shares with the latter the feature that in, for example, an application to the upper bound considered in Theorem 2 the log log terms do not appear. It should be pointed out that Hooley's device was created for a specific application, in which a certain combination of power and simplicity was essential if the argument were to be useful for the matter in hand.

More recently, Hooley's construction has been revised somewhat in [Ford and Halberstam (2000)]. We will refer to this paper again in some remarks on the "vector sieve" of J. Brüdern and E. Fouvry in the notes on Chap. 7.

Sect. 3.2.4. Theorem 2 is, as stated in the text, Brun's. The idea that sifting processes could be applied to the problem of prime twins appeared already in [Merlin (1911, 1915)]. These papers were handsomely acknowledged in Brun's work, after Merlin's life had been cut tragically short when he was killed in World War I.

Sect. 3.3.1. Constructions like (1.1) played an important rôle in [Jurkat and Richert (1965)], in which Selberg's λ^2 method was also involved. The purely combinatorial construction described here appeared in print in [Iwaniec (1971)] and in [Selberg (1972)]. Selberg is one of the few people to have seen the manuscript of Rosser from around 1950 where the idea embodied in (3.1) appeared. There are some further remarks on Rosser's ms. in [Selberg (1991)].

Sect. 3.3.2. The estimations here and in Sect. 3.2.2 come directly from [Brun (1920)].

Sect. 3.3.3. The simplification used here appears in Iwaniec's 1995 Rutgers notes, which is a source drawn upon throughout this chapter.

Sect. 3.3.4. It follows from a discussion in [Selberg (1991)] that as $\kappa \to \infty$ the sifting limit $\beta(\kappa)$ for the sieve described here is asymptotically equal to the bound $c\kappa$ given by Corollary 1.1.

The reader can confirm that the choice of b which optimises this treatment of Corollary 1.2 is a certain function of u that is asymptotic as $u \to \infty$ to the value chosen here.

The Fundamental Lemma was obtained in a form not differing essentially from Corollary 1.2 by Halberstam and Richert (1974), using a construction related more closely to Brun's (for which see Sect. 3.4).

Sect. 3.3.5. Theorem 2 is slightly sharper than the corresponding result enunciated in the introduction that appeared in Brun (1920).

The total number of applications that have now been made of upper bound sieves by the methods of Brun (1920) or Selberg (1947) is enormous, and it would be an gigantic task to attempt to list them. Many of those appearing prior to 1974 appear in the bibliography in Halberstam and Richert (1974). Our bibliography includes a selection of papers appearing since 1974 that use sieve ideas, but no attempt has been made to include them all.

Sect. 3.4. The methods used here are Brun's, but the nature of the exposition is entirely different. The adventurous reader may consult [Brun (1920)] to examine the cumbersome notation and diagrams with which he worked.

Sect. 3.4.3. The asymptotic bound (as $\kappa \to \infty$) for the sifting limit of Brun's method is in an earlier 1989 version of Iwaniec's Rutgers notes.

4. Rosser's Sieve

This chapter supplies a version of the combinatorial sieve, based on the construction given in Sect. 3.3.1, that is more satisfactory than the analysis given in Sect. 3.3. It provides correct asymptotic estimates of the sums T_n that were dealt with in a more approximate fashion in Lemma 3.3.5. The terminology "Rosser's Sieve" follows established custom, although it could reasonably be applied to any discussion starting from Rosser's choice of the parameters B_r already used in Sect. 3.3.

We will in due course need to assume that the function ρ has sifting density κ in the stronger sense whereby the constant K in (1.3.5.2) satisfies an inequality $K \leq 1 + L/\log w$, as given in (1.3.5.4). Initially, however, we will proceed without this assumption, as in Chapter 3.

The symbol D is used as in the specification (3.3.1.1) of Rosser's construction, so that it denotes the level of the sifting functions $\lambda^{\pm}(d)$ employed, as in (1.3.1.3). The estimate in Theorem 3.3.1 will be replaced by improved inequalities of the shape

$$V^+(D, P(z)) \leq V(P(z))\Big(F(s) + O(\varepsilon)\Big),$$
$$V^-(D, P(z)) \geq V(P(z))\Big(f(s) + O(\varepsilon)\Big),$$

wherein $z = D^{1/s}$ as usual. Here $\varepsilon \to 0$ as $D \to \infty$, but in the interests of simplicity only a relatively weak quantitative expression of this fact, viz.

$$\varepsilon = \frac{(\log \log \log D)^3}{\log \log D},$$

will be derived in the first instance, in Sect. 4.4.1. A significantly stronger result will be obtained in Sect. 4.4.3, after a much more delicate treatment of the error terms involved. The precise form of the result will then depend on whether $\kappa > \frac{1}{2}$ or not.

The analysis in Sect. 4.2 of the leading terms $F(s)$ and $f(s)$ is of central importance. In particular, for the lower bound we wish to determine the "sifting limit", the least s_0 such that $f(s) > 0$ whenever $s > s_0$. The quantitative aspects of $F(s)$ and $f(s)$ are also of great significance.

The results assume a particularly simple and satisfactory form in the two cases $\kappa = 1$ and $\kappa = \frac{1}{2}$. We give a brief overview of the case $\kappa = 1$ first.

It will be seen that in the case $\kappa = 1$

$$F(s) = \frac{2e^\gamma}{s} \quad \text{if} \quad 1 \leq s \leq 3, \qquad f(s) = \frac{2e^\gamma}{s}\log(s-1) \quad \text{if} \quad 2 \leq s \leq 4,$$

while for larger s the function $F(s)$ decreases and $f(s)$ increases, to the common limit 1 as $s \to \infty$. The value 1 is of course in accordance with the Fundamental Lemma of Corollaries 3.3.1.1 and 3.3.1.2, and with Legendre's formula from Theorem 1.3.1.

The simple form assumed by the upper bound when $1 \leq s \leq 3$ then has a leading term

$$V\bigl(P(D^{1/s})\bigr) F(s) = \frac{2e^\gamma}{s} \prod_{p \mid P(D^{1/s})} \left(1 - \frac{\rho(p)}{p}\right)$$

$$\sim \frac{2}{\log D} \prod_{p \mid P(D^{1/s})} \left(\frac{1-\rho(p)/p}{1-1/p}\right) \quad \text{as} \quad D \to \infty,$$

because of Mertens' formula (1.1.4.2). At the point $s = 2$ (or when $1 \leq s \leq 2$) this agrees with the result provided by Selberg's λ^2 method. For larger values of s the result provided by Rosser's method turns out to be (in the present instance $\kappa = 1$) the better one.

As for the lower bound when $\kappa = 1$, it will appear that $f(s) > 0$ precisely for $s > 2$.

The λ^2-method of Selberg has however the convenient property that it leads directly to the upper bound expressed by a value of $F(2)$ without the intervention of considerations relating to $F(s)$ (or $f(s)$) for larger s. In contrast, to obtain even the result that $f(s) > 0$ when $s > 2$ it will be necessary to deal with $F(s)$ and $f(s)$ for all larger values of s. For these values of s the functions $F(s)$, $f(s)$ become more recondite, being given by the system of difference-differential equations enunciated in Sect. 4.2.1.

It would be very satisfactory if one could find a simple and more direct method of treating the lower bound involving $f(s)$ for small s without the necessity of referring to larger s. There appears, however, to be no immediate prospect of such a treatment being discovered.

There is an extremal example supplied by Selberg which shows that when $\kappa = 1$ the functions F and f are best possible, in that within the (wide) class of sequences to which the resulting bounds (1.3.1.7) for $S(A,P(z))$ are applicable there exist sequences for which the bounds represented by F and f are actually attained. We will discuss Selberg's example in Sect. 4.5. In the notes there are some remarks on the light that Selberg's example sheds on the details of Rosser's construction.

The other particularly simple and satisfactory instance of the results supplied by Rosser's sieve occurs when $\kappa = \frac{1}{2}$. In this case

$$F(s) = 2\sqrt{\frac{e^\gamma}{\pi s}} \quad \text{if} \quad 0 < s \leq 2, \qquad f(s) = \sqrt{\frac{e^\gamma}{\pi s}} \int_1^s \frac{dt}{\sqrt{t(t-1)}} \quad \text{if} \quad 1 \leq s \leq 3.$$

Again, an example provided by Selberg shows that in this case $\kappa = \frac{1}{2}$ the functions F and f are the best possible.

When $\kappa > 1$ the results found by Rosser's method are less satisfactory. It appears, for the larger values of κ and the smaller values of s, that the upper bound provided by Rosser's method is inferior to that given by Selberg's. As a corollary, Rosser's lower bound is also not the best for these κ; better results can be obtained using a combination of combinatorial ideas with Selberg's method. Some of the question that arise in this way have not been completely solved. We will discuss this area of the subject in Chapter 7.

In Sect. 4.1 we establish a sharp approximation by continuous functions f_n to the expressions T_n from (3.1.4.2) when χ is specified by Rosser's choice (3.3.1.1). Subject to questions of convergence, the resulting series (4.2.1.6) lead to the difference-differential relations for the functions F, f that are studied in Sect. 4.2. The associated convergence problem will be settled in Sect. 4.3, using the same circle of ideas.

This treatment may make the question of the existence of the functions f and F appear to be a somewhat delicate matter. It should be noted that the delicacy, if any, is associated not so much with this existence question as with the particular representation of the functions that we require.

The parameter β, which appeared first in Sect. 3.3.1, must be chosen in an appropriate way. Provided this is done, the results will be in accordance with the heuristic discussion of Sect. 3.3.1, in that the sifting limit (the greatest zero of f) will then be the chosen value of the parameter β. In particular, the actual choice of β will be $\beta = 1$ if $\kappa = \frac{1}{2}$ and $\beta = 2$ if $\kappa = 1$. It will also appear that $\beta \sim c\kappa$ as $\kappa \to \infty$, where $c = 3 \cdot 591 \ldots$ is the constant that has already appeared in Theorem 3.3.1, so that for large κ the result of Theorem 3.3.1 was already quite sharp.

Superior sifting limits when $\kappa > 1$ can, however, be obtained by the methods using Selberg's ideas that are discussed in Chap. 7.

Much of our work will, for technical reasons, proceed more easily if $\beta > 1$. This fact is inconvenient, as the value $\beta = 1$ is the one appropriate to the interesting case $\kappa = \frac{1}{2}$ (also if $0 < \kappa < \frac{1}{2}$). The reader may prefer, in the first place, to skip material relating to the case $\kappa \leq \frac{1}{2}$, or may be even more selective by paying attention initially only to the case $\kappa = 1$.

4.1 Approximations by Continuous Functions

We will provide a definitive analysis of the expressions $V^{\pm}(D, D^{1/s})$, when $\chi_D^{\pm} = \chi^{\pm}$ is defined as in Sect. 3.3.1. The parameters B_r of Lemma 3.3.1 are given by $p_r < B_r \iff p_1 p_2 \ldots p_{r-1} p_r^{\beta+1} < D$, where $\beta \geq 1$. As in (3.1.2.1) write $d = p_1 p_2 \ldots$, $p_1 > p_2 > \cdots$. The procedure rests on the fact that when $\chi^{\pm}$ is specified in this way the expressions $T_n(D, D^{1/s})$, which were estimated in a somewhat approximate way in Lemma 3.3.5, satisfy the recurrence relations in Lemma 1. We will use these recurrences to derive asymptotic inequalities for the quantities T_n in terms of certain continuous functions f_n, which will be defined inductively by the corresponding equations given in (3.2) below. These inequalities appear in Sect. 4.1.3.

4.1.1 The Recurrence Relations

First we derive (in Lemma 1) the recurrences for the quantities T_n on which the subsequent procedure depends.

Recall that in Lemma 3.1.3 the expression $S(\mathcal{A}, P)$ is estimated via Lemma 3.1.4 in terms of

$$V^-(P(z)) = V^-(D, P(z)) = V(P(z)) - T^-(P(z)) \qquad (1.1)$$
$$V^+(P(z)) = V^+(D, P(z)) = V(P(z)) + T^+(P(z)), \qquad (1.2)$$

where, once the specification of the parameters B_r has been made,

$$T^-(P(z)) = \sum_{j=1}^{\infty} T_{2j}(D, P(z)), \quad T^+(P(z)) = \sum_{j=0}^{\infty} T_{2j+1}(D, P(z)). \qquad (1.3)$$

Here (3.1.4.2) gives

$$T_n(D, P(z)) = \sum_{\nu(d)=n;\ d|P(z)} \frac{\bar{\chi}_D^{(-)^{n+1}}(d)\, \rho(d)}{d} V(P(q(d))), \qquad (1.4)$$

in which $q(d)$ denotes the least prime factor of d. Because of the specification (3.3.1.1) of B_r, Lemma 3.1.2 gives, for each $n \geq 1$ and with $p_1 > p_2 > \cdots$,

$$\bar{\chi}_D^-(p_1 \ldots p_{2n}) = 1 \quad \text{if} \quad \begin{cases} p_1 \ldots p_{2n-1} p_{2n}^{\beta+1} \geq D \\ p_1 \ldots p_{2j-1} p_{2j}^{\beta+1} < D \quad \text{when} \quad 1 \leq j < n, \end{cases}$$

$$\bar{\chi}_D^+(p_1 \ldots p_{2n-1}) = 1 \quad \text{if} \quad \begin{cases} p_1 \ldots p_{2n-2} p_{2n-1}^{\beta+1} \geq D \\ p_1 \ldots p_{2j-2} p_{2j-1}^{\beta+1} < D \quad \text{when} \quad 1 \leq j < n, \end{cases}$$

and in all other cases $\bar{\chi}_D^\pm(d) = 0$.

The recurrences in Lemma 1 below rest on the observations that for $n > 1$

$$\bar{\chi}_D^-(p_1 \ldots p_n) = \bar{\chi}_{D/p_1}^+(p_2 \ldots p_n)$$

$$\bar{\chi}_D^+(p_1 \ldots p_n) = \begin{cases} \bar{\chi}_{D/p_1}^-(p_2 \ldots p_n) & \text{if} \quad p_1^{\beta+1} < D \\ 0 & \text{if} \quad p_1^{\beta+1} \geq D \end{cases} \qquad (1.5)$$

$$\bar{\chi}_D^+(p_1) = \begin{cases} 1 & \text{if} \quad D \leq p_1^{\beta+1} \\ 0 & \text{if} \quad p_1^{\beta+1} \leq D. \end{cases}$$

4.1 Approximations by Continuous Functions

Note in passing the corresponding recurrences for χ : for $n \geq 1$

$$\chi_D^-(p_1 \ldots p_n) = \chi_{D/p_1}^+(p_2 \ldots p_n)$$

$$\chi_D^+(p_1 \ldots p_n) = \begin{cases} \chi_{D/p_1}^-(p_2 \ldots p_n) & \text{if } p_1^{\beta+1} < D \\ 0 & \text{if } p_1^{\beta+1} \geq D, \end{cases} \quad (1.6)$$

where when $n = 1$ the empty product $p_2 \ldots p_n$ is taken as 1.
For future convenience denote

$$\varepsilon_n = \begin{cases} 1 & \text{if } n \text{ is odd} \\ 0 & \text{if } n \text{ is even}. \end{cases} \quad (1.7)$$

Also define

$$y_n = D^{1/(\beta+n)}, \qquad z_n(s) = \min\{D^{1/s}, D^{1/(\beta+\varepsilon_n)}\}. \quad (1.8)$$

Lemma 1. *Write $z = D^{1/s}$ and use the notation (1.8). Then for $n \geq 2$ the quantities T_n given in (1.4) satisfy*

$$T_1(D, P(z)) = \sum_{y_1 \leq p < z} \frac{\rho(p)}{p} V(P(p)) \quad (1.9)$$

and when $n \geq 2$

$$T_n(D, P(z)) = \sum_{y_n \leq p < z_n(s)} \frac{\rho(p)}{p} T_{n-1}\left(\frac{D}{p}, P(p)\right) \quad \text{if } s \geq \beta - \varepsilon_n. \quad (1.10)$$

Conversely, the equations (1.9), (1.10) characterise $T_n(D, P(z))$ completely when $s \geq \beta - \varepsilon_n$.

The equation (1.9) is immediate from (1.4) and (1.5).
When $n > 1$ write (as in Sect. 3.1.2) $d = p_1 d_1$, where p_1 is the largest prime factor of d, so that $d_1 | P(p_1)$ (since d is squarefree). Then (1.4) and (1.5) give

$$T_n(D, P(z)) = \sum_{p_1 < z_n(s)} \frac{\rho(p_1)}{p_1} \sum_{d_1 | P(p_1)} \bar{\chi}_{D/p_1}^{(-)^{n+1}}(d_1) \frac{\rho(d_1)}{d_1} V(P(p(d_1))). \quad (1.11)$$

Here $p_1 < z_n(s)$ because (1.4) gives $p_1 < D^{1/s}$, while $p_1^{\beta+1} < D$ if n is odd, and if n is even then $p_1 < D^{1/s} \leq D^{1/\beta}$ because $\varepsilon_n = 0$ in (1.10).
In (1.11), the equations defining $\bar{\chi}_{D/p_1}(p_2 \ldots p_n)$ imply

$$D \leq p_1 \ldots p_{n-1} p_n^{\beta+1} \leq p_1^{n+\beta},$$

so that we may suppose $p_1 > y_n$. This proves (1.10).
The right side of (1.10) refers to $T_{n-1}(D_p, P((D_p)^{1/\sigma}))$, with $D_p = D/p$, only when $\sigma \geq \beta - \varepsilon_{n-1}$. This remark completes the proof of Lemma 1.

The restriction of Lemma 1 to $s \geq \beta - \varepsilon_n$ is not an inconvenience for the following reason. The motivation given in Sect. 3.3.1 for the choice (3.3.1.1) of B_r implied that the sifting limit ought to be the parameter β. Then $T_n(D, P(D^{1/s}))$ is of interest for even n for $s > \beta$, while for odd $n > 1$ it is constant when $s < \beta + 1$ because of the condition $p_1^{\beta+1} < D$ implicit in (1.4). Thus we need be concerned in the first place only with s for which $s \geq \beta - \varepsilon_n$.

4.1.2 Partial Summation

The application of partial summation in Lemma 2 is slightly more specialised than those recorded in Sect. 1.3. We use the standard assumptions that the function ρ has bounded sifting density κ to derive the upper estimates given in parts (i) and (ii) of Lemma 2. Initially it is only the weaker form (2.2) that will be used in the sequel. The corresponding estimate from below given in part (iii) is less important, from our point of view, but will be needed in due course.

Lemma 2. *Write $z = D^{1/s}$, $w = D^{1/\sigma}$, where $2 \leq w \leq z$. Suppose a function $H(t)$ is continuous and non-negative in the interval $s \leq t \leq \sigma$.*

(i) Suppose that ρ and K satisfy (1.3.5.4) and that $H(t)$ decreases in $s \leq t \leq \sigma$. Then

$$\sum_{w \leq p < z} \frac{\rho(p)}{p} \frac{V(P(p))}{V(P(z))} H\left(\frac{\log D}{\log p}\right) = \int_{s < t < \sigma} H(t) \frac{dt^\kappa}{s^\kappa} + E(w, z), \quad (2.1)$$

where

$$E(w, z) \leq (K - 1)\left(\frac{\sigma}{s}\right)^\kappa H(s). \quad (2.2)$$

(ii) Suppose ρ and L satisfy the stronger sifting density condition (1.3.5.3) whereby $K_t \leq 1 + L/\log t$ whenever $2 \leq t \leq z$. Suppose also that $t^\kappa H(t)$ decreases on $s \leq t \leq \sigma$. Then (2.1) holds with

$$E(w, z) \leq \frac{(\kappa + 1)L\,H(s)}{\log w}. \quad (2.3)$$

(iii) Suppose that instead of (1.3.5.2) the function ρ satisfies an analogous bound from below:

$$\left(1 - \frac{L'}{\log w}\right)\left(\frac{\log z}{\log w}\right)^\kappa \leq \prod_{w \leq p < z}\left(1 - \frac{\rho(p)}{p}\right)^{-1} \quad \text{when} \quad 2 \leq w < z,$$

and assume $t^\kappa H(t)$ decreases as in part (ii). Then the expression in (2.1) satisfies

$$E(w, z) \geq -\frac{(\kappa + 1)L'\,H(s)}{\log w}.$$

4.1 Approximations by Continuous Functions

The stronger hypothesis in part (ii) that $t^\kappa H(t)$ decreases is a natural one in this context, since (1.3.5.4) gives that the coefficient of $\rho(p)/p$ in (2.1) does not exceed

$$K\left(\frac{\log z}{\log p}\right)^\kappa H\left(\frac{\log D}{\log p}\right),$$

and the hypothesis in question is that this increases with p.

Write
$$t = \log D/\log x, \qquad H(t) = h(x), \tag{2.4}$$

so that $h(x)$ increases with x. Use partial summation, as given in Lemma 1.3.2. This description assumes that the derivative h' is e.g. continuous on (w, z), which will be the case in all our applications. Take

$$c_n = \frac{\rho(p)}{p}\frac{V(P(p))}{V(P(z))} \quad \text{if} \quad n = p$$

and $c_n = 0$ if n is not a prime p. Then $\sum c_n$ is expressed via an identity of Buchstab's type,

$$V(P(z)) = 1 - \sum_{p<z}\frac{\rho(p)}{p}V(P(p)), \tag{2.5}$$

which is the case $r = 1$ of Lemma 3.2.4. Thus, in the notation of Lemma 1.3.1,

$$C(x, z) = \sum_{x \leq p < z}\frac{\rho(p)}{p}\frac{V(P(p))}{V(P(z))} = \frac{V(P(x))}{V(P(z))} - 1. \tag{2.6}$$

The bound (1.3.5.4) gives $C(x, z) = \left(\log z/\log x\right)^\kappa - 1 + R_z(x)$, where

$$R_z(x) \leq (K-1)\left(\frac{\log z}{\log x}\right)^\kappa \tag{2.7}$$

is to be viewed as an error term. Observe that $R_z(z) = 0$.

The second equation in the "partial summation" Lemma 1.3.1(i) gives

$$\sum_{w \leq n < z} c_n h(n) = -\int_w^z h(x)\frac{d}{dx}\left(\left(\frac{\log z}{\log x}\right)^\kappa - 1\right)dx + E(w, z).$$

which in the notation (2.4) is exactly (2.1), where now

$$E(w, z) = -\int_w^z h(x)\,dR_z(x) = h(w)R_z(w) + \int_w^z R_z(x)h'(x)\,dx, \tag{2.8}$$

because $R_z(z) = 0$. When $R_z(x) \leq f(x)$ for $w \leq x \leq z$ this gives

$$E(w, z) \leq h(w)f(w) + \int_w^z f(x)h'(x)\,dx = h(z)f(z) - \int_w^z h(x)\,df(x), \tag{2.9}$$

because $h(x)$ is increasing and non-negative.

In part (i) of the lemma, (2.7) shows we can take the uniform bound $f(x) = (K-1)(\sigma/s)^\kappa$ in (2.9). This gives

$$E(w,z) \le h(z)f(z) = H(s)(K-1)(\sigma/s)^\kappa ,$$

which is (2.2).

Write $F(t) = f(x)$, where $t = \log D/\log x$ as in (2.4). In part (ii) the hypothesis (1.3.5.3) and (2.7) show we can take

$$f(x) = \frac{L}{\log x}\left(\frac{\log z}{\log x}\right)^\kappa \tag{2.10}$$

in (2.9), so that $F(t) = Lt^{\kappa+1}/s^\kappa \log D$. Then (2.9) gives

$$E(w,z) \le \frac{Ls\,H(s)}{\log D} + \frac{L}{s^\kappa \log D}\int_s^\sigma H(t)\,dt^{\kappa+1} .$$

Since $H(t)\,dt^{\kappa+1} = (\kappa+1)t^\kappa H(t)\,dt$ and $t^\kappa H(t)$ decreases on $s \le t \le \sigma$ this gives

$$E(w,z) \le \frac{Ls\,H(s)}{\log D} + \frac{(\kappa+1)(\sigma-s)L\,H(s)}{\log D} \le \frac{(\kappa+1)L\sigma\,H(s)}{\log D} ,$$

and part (ii) of the lemma follows.

For part (iii) of the lemma proceed as before, up to a point where we use $R_z(x) \ge -f(x)$ in (2.8), with

$$f(x) = \frac{L'}{\log x}\left(\frac{\log z}{\log x}\right)^\kappa$$

in place of (2.10). Instead of (2.9) we now have

$$-E(w,z) \le h(z)f(z) - \int_w^z h(x)\,df(x) ,$$

which is estimated as in part (ii).

4.1.3 The Leading Terms

From this point until the end of Sect. 4.4 we assume (1.3.5.2), of which we remind the reader from time to time. Thus Lemma 2(i) will apply.

Lemma 3 below will give an upper estimate of $T_n(D, D^{1/s})$ in the interval $s \ge \beta - \varepsilon_n$ defined via (1.7). For this we need the functions f_n defined by

$$s^\kappa f_1(s) = \int_{s<t<\beta+1} dt^\kappa , \tag{3.1}$$

$$s^\kappa f_n(s) = \int_{\max\{s,\beta+\varepsilon_n\}<t<\beta+n} f_{n-1}(t-1)\,dt^\kappa \quad \text{when } n > 1 . \tag{3.2}$$

These equations define $f_n(s)$ in the interval $s \geq \beta - \varepsilon_n$ because of reasoning exactly analogous to that applying to the equations for T_n in Lemma 1. The condition $t < \beta + n$ in (3.2) is actually redundant, because it would in any case be an easy induction that $f_n(s) = 0$ when $s > \beta + n$.

Lemma 3 exhibits a rather unpleasant n-dependence of an error expression Δ_n, which is a artefact of the inductive argument employed. In due course this feature will be improved by the more careful treatment supplied in Sect. 4.4.3.

A Special Case. The simplest instance of Lemma 3 will be when $K = 1$, so that the term $K - 1$ in Lemma 2(i) is replaced by 0. In this case it is easy to show

$$T_n(D, D^{1/s}) \leq V(P(D^{1/s})) f_n(s) \quad \text{if} \quad s \geq \beta - \varepsilon_n. \tag{3.3}$$

For (1.9) and Lemma 2(i) give (3.3) when $n = 1$, and if (3.3) holds when $n = N - 1$ then (1.10) gives

$$T_N(D, D^{1/s}) \leq \sum_{y_N \leq p < z_N(s)} \frac{\rho(p)}{p} V(P(p)) f_{N-1}\left(\frac{\log D}{\log p} - 1\right).$$

Then Lemma 2(i) gives first

$$T_N(D, D^{1/s}) \leq \left(\frac{s}{\max\{s, \beta + \varepsilon_N\}}\right)^\kappa V(P(z_N(s))) f_N(s),$$

from which (3.3) follows on using (1.3.5.2) when $z_N(s) < D^{1/s}$ in the case of odd N.

We will use the following bounds for $f_n(s)$, valid when $\beta > 1$, so that $\beta - \varepsilon_n > 0$ in (1.7). When $s \geq \beta - \varepsilon_n$ we have $f_n(s) \leq f_n(\beta - \varepsilon_n)$. Then the equations (3.1), (3.2) give

$$f_1(\beta - 1) \leq \left(\frac{\beta+1}{\beta-1}\right)^\kappa, \quad f_n(\beta - \varepsilon_n) \leq \left(\frac{\beta+n}{\beta-\varepsilon_n}\right)^\kappa f_{n-1}(\beta - \varepsilon_{n-1}).$$

Thus

$$f_n(\beta - \varepsilon_n) \leq \prod_{1 \leq m \leq n} \left(\frac{\beta+m}{\beta-\varepsilon_m}\right)^\kappa \leq \left(\frac{\beta+n}{\beta-1}\right)^{\kappa n} \quad \text{when} \quad \beta > 1. \tag{3.4}$$

The General Case. When $K \geq 1$ the estimate (3.3) does not apply, so must be replaced, for example by the following lemma.

Lemma 3. *Let T_n be as in (1.4) and K as in (1.3.5.4). Suppose $n \geq 1$ and $s > \beta - \varepsilon_n$, where ε_n is as in (1.7). Then*

$$T_n(D, D^{1/s}) \leq V(P(D^{1/s}))(f_n(s) + (K-1)K^n \Delta_n), \tag{3.5}$$

with

$$\Delta_n = n\left(\frac{\beta+n}{\beta-1}\right)^{\kappa n}. \tag{3.6}$$

From (1.9) and Lemma 2(i) we obtain, when $s > \beta - 1$,

$$T_1(D, D^{1/s}) = \sum_{D^{1/(\beta+1)} \leq p < D^{1/s}} \frac{\rho(p)}{p} V(P(p))$$

$$\leq V(P(z)) \left(\frac{1}{s^\kappa} \int_{s < t < \beta+1} dt^\kappa + (K-1)\left(\frac{\beta+1}{s}\right)^\kappa \right). \quad (3.7)$$

Since $s > \beta - 1$ this gives (3.5) in the case $n = 1$.

Proceed by induction on n. If (3.5) holds when $n = N - 1$ then (1.10) and Lemma 2(i) give

$$T_N(D, D^{1/s}) \leq \sum_{y_N \leq p < z_N(s)} \frac{\rho(p)}{p} V(P(p)) \left(f_{N-1}\left(\frac{\log D}{\log p} - 1\right) \right.$$

$$\left. + (K-1)K^{N-1}\Delta_{N-1} \right). \quad (3.8)$$

Since y_N, z_N are as in (1.8) a second use of Lemma 2(i) gives

$$\sum_{y_N \leq p < z_N(s)} \frac{\rho(p)}{p} V(P(p)) f_{N-1}\left(\frac{\log D}{\log p} - 1\right)$$

$$\leq \left(\frac{s}{\max\{s, \beta + \varepsilon_n\}}\right)^\kappa V\left(P(z_N(s))\right)$$

$$\times \left(f_N(s) + (K-1)f_{N-1}(\beta - \varepsilon_{N-1})\left(\frac{\beta+N}{\beta - \varepsilon_n}\right)^\kappa \right)$$

$$\leq V(P(D^{1/s})) \left(f_N(s) + (K-1)f_{N-1}(\beta - \varepsilon_{N-1})\left(\frac{\beta+N}{\beta - \varepsilon_N}\right)^\kappa \right), \quad (3.9)$$

wherein the treatment of the leading term $f_N(s)$ is as in the case $K = 1$ discussed in Sect. 4.1.3.

For the term involving Δ_{N-1} in (3.8) use the identity (2.5) to obtain

$$(K-1)K^{N-1}\Delta_{N-1} \sum_{y_N \leq p < z_N(s)} \frac{\rho(p)}{p} V(P(p))$$

$$\leq (K-1)K^{N-1}\Delta_{N-1} V\left(P(D^{1/(\beta+N)})\right)$$

$$\leq (K-1)K^N \Delta_{N-1} \left(\frac{\beta+N}{\beta - \varepsilon_n}\right)^\kappa V(P(D^{1/s})). \quad (3.10)$$

Assembling (3.8), (3.10) and (3.9) gives

$$T_n(D, D^{1/s}) \leq V(P(D^{1/s}))$$

$$\times \left(f_N(s) + (K-1)K^N \left(\frac{\beta+N}{\beta - \varepsilon_n}\right)^\kappa \left(f_{N-1}(\beta - \varepsilon_{N-1}) + \Delta_{N-1} \right) \right).$$

Because of (3.4) the induction will proceed provided

$$\Delta_N \geq \left(\frac{\beta+N}{\beta-\varepsilon_n}\right)^\kappa \left(\Delta_{N-1} + \prod_{1\leq m \leq N-1}\left(\frac{\beta+m}{\beta-\varepsilon_m}\right)^\kappa\right).$$

This is satisfied when

$$\Delta_n = n \prod_{1\leq m \leq n}\left(\frac{\beta+m}{\beta-\varepsilon_m}\right)^\kappa. \tag{3.11}$$

This is stronger than is needed to establish Lemma 3.

4.2 The Functions F and f

Lemma 4.1.3 about the sums T_n will lead to estimations of the quantities $V^\pm(P(D^{1/s}))$ in terms of corresponding functions $F(s)$, $f(s)$, which will be given by the series in (1.6) below. Subject to the validity of Proposition 1 about the convergence of these series, which will be established in Sect. 4.3, the recurrences defining the functions f_n given in Sect. 4.1.3 will induce a corresponding system of equations satisfied by F and f.

In this context the estimate to be obtained for the quantity V^-, as described by (4.1.1.1), will become trivial if $f(s) < 0$. Accordingly we will refer to the greatest zero of f as the "sifting limit". The consistency of this terminology with that of Sect. 1.3.6 rests on the results of Sect. 4.4.

In this section we study the properties of these functions F, f, taking their existence for granted meantime. The methods developed in Sects. 4.2.2 and 4.2.3 will be used again in Sects. 4.3 and 4.4.

In the case $\kappa = 1$, the existence and properties of F and f are implicit in results in the literature on functions introduced by Dickman and by Buchstab. The treatment given here applies for all κ, and is independent of these sources. The case of greatest interest, however, is that when $\kappa \leq 1$, in view of the fact that better results can be obtained when $\kappa > 1$ by using Selberg's ideas in the style introduced in Chap. 7.

4.2.1 The Difference-Differential Equations

The functions $F(s)$, continuous in $s > \beta - 1$, and $f(s)$, continuous in $s \geq \beta$, will satisfy the system

$$\frac{d}{ds}\left(s^\kappa F(s)\right) = \kappa s^{\kappa-1} f(s-1) \quad \text{if} \quad s > \beta+1, \tag{1.1}$$

$$\frac{d}{ds}\left(s^\kappa f(s)\right) = \kappa s^{\kappa-1} F(s-1) \quad \text{if} \quad s > \beta, \tag{1.2}$$

$$s^\kappa F(s) = C \quad \text{if} \quad 0 < s \leq \beta+1, \quad \beta^\kappa f(\beta) = B, \tag{1.3}$$

$$F(s) = 1 + o(s^{-2\kappa}), \quad f(s) = 1 + o(s^{-2\kappa}) \quad \text{as} \quad s \to \infty. \tag{1.4}$$

These equations imply

$$s^\kappa f(s) = B + C \int_\beta^s \frac{dx^\kappa}{(x-1)^\kappa} \quad \text{if} \quad \beta \le s \le \beta + 2. \tag{1.5}$$

It is required to determine the constants B, C, essentially by a transfer of information from (1.4), where $s \to \infty$, to the point $s = \beta + 1$. It will appear that B, C are determined by κ, β.

Our analysis of F will refer only to the domain $s \ge \beta - 1$. The extension into $s > 0$ given by (1.3) is a formal device that will be technically convenient later on, in particular in Chap. 5.

The existence of the functions F and f will be guaranteed by the following proposition.

Proposition 1. *Let f_n be defined as in Sect. 4.1.3. If $\kappa > 0$ then there exists $\beta \ge 1$, depending on κ and to be specified in (4.10), such that the series*

$$F(s) = 1 + \sum_{j=0}^{\infty} f_{2j+1}(s), \qquad f(s) = 1 - \sum_{j=1}^{\infty} f_{2j}(s) \tag{1.6}$$

converge boundedly in $s > \beta - 1$ and in $s \ge \beta$, respectively, to sums satisfying (1.1) – (1.4).

In particular (4.10) specifies $\beta = 1$ if $0 < \kappa \le \frac{1}{2}$ and $\beta = 2$ if $\kappa = 1$. For other values of κ the number β is somewhat more recondite, but it will appear that it is a certain algebraic number when 2κ is a positive integer.

For this β, Proposition 1 and the integral relations (4.1.3.1), (4.1.3.2) defining f_n imply that the functions F, f satisfy (1.1) and (1.2). Also, (1.3) will follow for some C, B, because (4.1.3.2) gives that for odd $n \ge 3$ and $s \le \beta + 1$ the expression $s^\kappa f_n(s)$ is constant, while from (4.1.3.1)

$$s^\kappa \bigl(1 + f_1(s)\bigr) = (\beta + 1)^\kappa \,.$$

The constants C and B are determined by Lemma 5 below. We will see that $B = 0$ when $\kappa \ge \frac{1}{2}$, and that for these κ the sifting limit is indeed β:

$$f(\beta) = 0, \; f(s) > 0 \text{ if } s > \beta \qquad \text{provided} \quad \kappa > \tfrac{1}{2}, \tag{1.7}$$

as was suggested by the heuristic discussion in Sect. 3.3.1.

It would be possible to establish the convergence in Proposition 1 for certain values of β other than that specified in (4.10). However, with such β the general air of correctness associated with the equation $B = 0$ would disappear. We will see in (4.9), at least in the case $\kappa = 1$, that the specified choice of β leads (the convergence question being taken for granted) to an optimal value of the sifting limit.

Reference to the Fundamental Lemma (Corollaries 3.3.1.1 and 3.3.1.2) leads one to suspect that the entries $o(s^{-2\kappa})$ in (1.4) could be replaced by (for example) $O(e^{-s})$. In fact a stronger estimate than this will be obtained via Lemma 6.

4.2 The Functions F and f

To make progress with the discussion of the system (1.1) – (1.4), pass from F, f to the combinations

$$P(s) = F(s) + f(s), \quad Q(s) = F(s) - f(s) \quad \text{when} \quad s \geq \beta. \tag{1.8}$$

Because of (1.1) – (1.5) these satisfy

$$\left.\begin{aligned}\frac{d}{ds}(s^\kappa P(s)) &= \kappa s^{\kappa-1} P(s-1) \\ \frac{d}{ds}(s^\kappa Q(s)) &= -\kappa s^{\kappa-1} Q(s-1)\end{aligned}\right\} \quad \text{if} \quad s \geq \beta+1, \tag{1.9}$$

$$P(s) = 2 + o(s^{-2\kappa}), \quad Q(s) = o(s^{-2\kappa}) \quad \text{as} \quad s \to \infty, \tag{1.10}$$

$$\left.\begin{aligned}s^\kappa P(s) &= C + B + C\int_\beta^s \frac{dx^\kappa}{(x-1)^\kappa} \\ s^\kappa Q(s) &= C - B - C\int_\beta^s \frac{dx^\kappa}{(x-1)^\kappa}\end{aligned}\right\} \quad \text{if} \quad \beta \leq s \leq \beta+1. \tag{1.11}$$

We wish to continue $P(s)$, $Q(s)$ into $s \geq \beta - 1$ by the equations (1.9). Then (1.11) leads to

$$s^\kappa P(s) = s^\kappa Q(s) = C \quad \text{if} \quad \beta - 1 < s < \beta. \tag{1.12}$$

There is no guarantee that the continued functions P, Q are continuous at β. Nevertheless, with the appropriate choice of β this will turn out to be the case, as a consequence of the equation $B = 0$ to be derived in Lemma 5.

The equations (1.9) can be rewritten

$$s P'(s) = -\kappa P(s) + \kappa P(s-1), \quad s Q'(s) = -\kappa Q(s) - \kappa Q(s-1). \tag{1.13}$$

It is in this form that they will be dealt with below.

4.2.2 The Adjoint Equation and the Inner Product

The equations (1.13) provide two instances $\langle a, b \rangle = \langle \kappa, \pm \kappa \rangle$ of a function R defined on $(\beta - 1, \infty)$, continuous in $[\beta, \infty)$ and satisfying

$$s R'(s) + a R(s) + b R(s-1) = 0 \quad \text{if} \quad s > \beta, \, s \neq \beta + 1. \tag{2.1}$$

The function R may have a discontinuity at β.

The equation *adjoint* to (2.1) is defined to be

$$\frac{d}{ds}(s r(s)) = a r(s) + b r(s+1). \tag{2.2}$$

4. Rosser's Sieve

In a notation consistent with (1.13) write

$$\frac{d}{ds}(s\,p(s)) = \kappa\, p(s) - \kappa\, p(s+1), \quad \frac{d}{ds}(s\,q(s)) = \kappa\, q(s) + \kappa\, q(s+1). \quad (2.3)$$

We will see in Sect. 4.2.3 that (2.2) has a relatively straightforward solution valid when $s > 0$. Its use will be to relate the values of $R(s)$ for large and small s in the following way.

Lemma 1. (i) *Suppose R satisfies (2.1), and suppose $r(s)$ satisfies the equation (2.2) when $s > \beta$. Then the inner product*

$$\langle R, r \rangle(s) = s\,r(s)R(s) - b\int_{s-1}^{s} r(x+1)R(x)\,dx \quad (2.4)$$

is constant in $s \geq \beta + 1$. Also this constant is

$$\langle R, r \rangle(\beta + 1) = \langle R, r \rangle(\beta), \quad (2.5)$$

where $R(\beta) = R(\beta+) = \lim_{s \to \beta+} R(s)$.
(ii) *If in addition*

$$s^a R(s) = C \quad \text{when} \quad \beta - 1 < s < \beta, \quad (2.6)$$

and $r(s)$ satisfies (2.2) in $s \geq \beta - 1$, then

$$\langle R, r \rangle(\beta) = \beta\, r(\beta)\left(R(\beta+) - \frac{C}{\beta^a}\right) + \lim_{s \to \beta+} \frac{C\,r(s-1)}{(s-1)^{a-1}}. \quad (2.7)$$

The value of this last limit is simply $C\,r(\beta-1)/(\beta-1)^{a-1}$, except in the more troublesome case when $\beta = 1$ and $a \geq 1$.

Observe that (2.1) and (2.2) give

$$\frac{d}{ds}(s\,r(s)R(s)) = b\,r(s+1)R(s) - b\,r(s)R(s-1), \quad (2.8)$$

that is to say

$$\frac{d}{ds}(\langle R, r \rangle(s)) = 0. \quad (2.9)$$

The constant nature of $\langle R, r \rangle(s)$ in $s \geq \beta+1$ follows. Also if (2.9) is integrated over $(\beta, \beta+1)$ then it gives (2.5) as asserted.

Next, use (2.2) in the form

$$\frac{d}{ds}(s^{1-a}r(s)) = \frac{b\,r(s+1)}{s^a}. \quad (2.10)$$

When (2.6) holds, (2.4) now gives

$$\langle R, r\rangle(\beta) = \beta\, r(\beta) R(\beta+) - C \int_{\beta-1}^{\beta} \frac{d}{dx}\left(x^{1-a} r(x)\right) dx\ ,$$

and (2.7) follows.

In this argument, an essential property of the adjoint equation is that it gives an equation (2.8) whose right side is of the form $\psi(s+1) - \psi(s)$.

It is of course possible to reformulate this argument so that it does not refer to $R(s)$ for $s < \beta$. To do this, use integration by parts. Write (2.1) in the form

$$\frac{d}{ds}\left(s^a R(s)\right) = -b s^{a-1} R(s-1)\ ,$$

and use (2.10). Then

$$b\int_{\beta}^{\beta+1} r(x+1) R(x)\, dx = \int_{\beta}^{\beta+1} x^a R(x)\, dx^{1-a} r(x)$$

$$= (\beta+1) R(\beta+1) r(\beta+1) - \beta R(\beta) r(\beta) - \int_{\beta}^{\beta+1} x^{1-a} r(x)\, dx^a R(x)\ ,$$

so that

$$\langle R, r\rangle(\beta+1) = \beta R(\beta) r(\beta) + \int_{\beta}^{\beta+1} x^{1-a} r(x)\, dx^a R(x)\ .$$

This is equivalent to (2.5).

4.2.3 Solutions of the Adjoint Equation

We will assemble some properties of solutions to the adjoint equation (2.2) that we will need in the sequel. There is a considerable amount that could be written about these solutions (and has been: see the references in the notes on this chapter). Observe that for our purposes it is not necessary to be concerned with uniqueness questions; the existence of one suitably behaved solution will suffice. The reader keen to progress to the discussion (in Sect. 4.2.4) of the sifting limit in the important case $\kappa = 1$ need at this stage note only the simplest of the polynomial solutions described below. The complete discussion of the case $\kappa = 1$ requires also the solution as a Laplace transform given in (3.6).

Polynomial Solutions. These occur when $a + b = n + 1$ is a positive integer, as occurs in (1.13) when 2κ is a positive integer, normally the case in applications to problems in Number Theory. In particular (2.2) has the solution $r(s) = 1$ if $a + b = 1$, and the solution $s - b$ if $a + b = 2$.

4. Rosser's Sieve

To see that such polynomials exist, write the proposed solution as

$$r(s) = r_{a,b}(s) = \sum_{0 \le l \le n} \binom{n}{l} c_l s^{n-l}, \tag{3.1}$$

and equate coefficients of powers of s between the two sides of (2.2), with $c_0 = 1$, say. The relation between the coefficients of s^n is satisfied automatically. The remaining relations then determine values of $c_0, \ldots, c_n$ that give a solution $r(s)$ in (2.2).

The coefficients c_l in (3.1) are then determined by the recurrence

$$c_0 = 1, \quad l c_l + b \sum_{0 \le j < l} \binom{l}{j} c_j = 0. \tag{3.2}$$

This fact will emerge more easily from the discussion leading to (3.12).

The recurrence (3.2) gives

$$c_0 = 1, \quad c_1 = -b, \quad c_2 = b^2 - \tfrac{1}{2}, \quad c_3 = -b^3 + \tfrac{3}{2}b^2 - \tfrac{1}{3},$$
$$c_4 = b^4 - 3b^3 + \tfrac{25}{12}b^2 - \tfrac{1}{4}b, \quad c_5 = -b^5 + 5b^4 - \tfrac{81}{12}b^3 + \tfrac{35}{12}b^2 - \tfrac{1}{5}b, \ldots$$

from which (3.1) gives the polynomials $r(s)$ when $a + b = n + 1$. The case $a = b = \kappa$ gives that (2.3) has solutions $q(s)$ given for $\kappa = \tfrac{1}{2}, 1, \tfrac{3}{2}, 2, \tfrac{5}{2}, 3$ respectively by

$$\begin{aligned}
q(s) &= 1 \\
q(s) &= s - 1 \\
q(s) &= s^2 - 3s + \tfrac{3}{2} \\
q(s) &= s^3 - 6s^2 + 9s - \tfrac{8}{3} \\
q(s) &= s^4 - 10s^3 + 30s^2 - \tfrac{85}{3}s + \tfrac{55}{12} \\
q(s) &= s^5 - 15s^4 + 75s^3 - 145s^2 + 90s - \tfrac{18}{5}.
\end{aligned} \tag{3.3}$$

Solutions as Laplace transforms. If $a + b < 1$ it is possible to find solutions of (2.2) in the form

$$r(s) = \int_0^\infty e^{-sx} \phi(x) \, dx \quad \text{if} \quad s > 0. \tag{3.4}$$

In particular this applies when $a = \kappa$, $b = -\kappa$, so will solve the equation for p in (2.3). For convergence reasons we require

$$\phi(x) = O(x^{-1+\varepsilon}) \quad \text{as} \quad x \to 0+, \qquad \phi(x) = O(x^A) \quad \text{as} \quad x \to \infty, \tag{3.5}$$

where ε, A are positive constants (small and large, respectively). The function ϕ will be differentiable on $(0, \infty)$.

4.2 The Functions F and f

The relation (3.4) gives

$$\frac{d}{ds}(s\,r(s) - \phi(0)) = \frac{d}{ds}\int_0^\infty e^{-sx}\phi'(x)\,dx = -\int_0^\infty e^{-sx}x\,\phi'(x)\,dx\;.$$

Thus a solution to (2.2) will be obtained provided

$$-x\,\phi'(x) = (a + be^{-x})\phi(x)\;.$$

This is satisfied if

$$\phi(x) = \frac{1}{\Gamma(1-a-b)}\exp\left(-(a+b)\int_1^x \frac{dt}{t} + b\int_0^x \frac{1-e^{-t}}{t}\,dt\right),$$

which does satisfy the conditions (3.5) when $a+b < 1$. Then

$$r(s) = \frac{1}{\Gamma(1-a-b)}\int_0^\infty e^{-sx}\Phi_b(-x)\frac{dx}{x^{a+b}}\;,\qquad (3.6)$$

where for future use we denote

$$\Phi_b(x) = \exp\left\{b\int_0^x \frac{1-e^t}{t}\,dt\right\}\;.\qquad (3.7)$$

In (3.6), the normalising constant $1/\Gamma(1-a-b)$ makes

$$r(s) \sim s^{a+b-1} \quad \text{as} \quad s \to \infty\;,\qquad (3.8)$$

because the substitution $sx = u$ gives

$$\frac{r(s)}{s^{a+b-1}} = \frac{1}{\Gamma(1-a-b)}\int_0^\infty e^{-u}\Phi_b(-u/s)\frac{du}{u^{a+b}}$$

$$\to \frac{1}{\Gamma(1-a-b)}\int_0^\infty e^{-u}\frac{du}{u^{a+b}} = 1 \quad \text{as} \quad s \to \infty\;.$$

Solutions as Contour Integrals. The Laplace transform solution fails if $a + b \geq 1$ because of the convergence question near 0. If $a+b$ is not a positive integer the polynomial solutions do not apply. The solutions described below apply when $a + b > 1$ or when $a + b$ is not an integer, and will also yield further information when it is, or when $a + b < 1$.

Lemma 2 may be passed over by the reader concerned primarily with the case $\kappa = 1$.

The point $x = 0$ can be avoided by replacing (3.6) by an expression

$$r(s) = r_{a,b}(s) = \frac{\Gamma(a+b)}{2\pi i}\int_P e^{sw}\Phi_b(w)\frac{dw}{w^{a+b}} \quad \text{if} \quad s > 0\;,\qquad (3.9)$$

where Φ_b is as in (3.7). Here P is a familiar keyhole-shaped contour which starts at $w = -\infty$, passes through the lower half-plane to encircle the origin (anti-clockwise, as usual), and returns to $-\infty$. Then r satisfies (2.2) because of the same integration by parts and differentiation that gave (3.6). For convenience the complex variable w in (3.9) corresponds to $-x$ in (3.6). The factor $\Gamma(a+b)$ will be seen to be appropriate (except when $a+b$ is a negative integer or zero, in which case refer to (3.6), or to Lemma 2 below).

The function in (3.7) can be expanded by Taylor's theorem:

$$\Phi_b(w) = \sum_{0 \le n < N} \frac{\Phi_b^{(n)}(0) w^n}{n!} + E_N(w), \qquad (3.10)$$

where $E_N(w) \ll w^N$.

Call (3.9) the *standard solution* to the adjoint difference-differential equation (2.2). It will be used to derive the following asymptotic expansion.

Lemma 2. *Let $r(s)$ be the standard solution (3.9) of the equation (2.2). Suppose $N > a + b - 1$. Then*

$$r(s) = \sum_{0 \le l < N} \binom{a+b-1}{l} \Phi_b^{(l)}(0) s^{a+b-1-l}$$

$$+ \frac{1}{\Gamma(1-a-b)} \int_0^\infty e^{-sz} E_N(-z) \frac{dz}{z^{a+b}}, \qquad (3.11)$$

where E_N is as in (3.10),

$$\binom{a+b-1}{l} = \frac{\Gamma(a+b)}{l! \, \Gamma(a+b-l)},$$

and $1/\Gamma(n)$ is to be replaced by 0 if n is 0 or a negative integer. The Taylor coefficients $\Phi_b^{(n)}(0)$ satisfy the recurrences

$$n \, \Phi_b^{(n)}(0) = -b \sum_{l=0}^{n-1} \binom{n}{l} \Phi_b^{(l)}(0). \qquad (3.12)$$

The solutions discussed earlier are special cases of this lemma. If $a + b$ is a positive integer then $r(s)$ reduces to a polynomial, and if $a + b < 1$ then we can take $N = 0$ and recover (3.6).

To obtain (3.11), substitute the Taylor expansion (3.10) in the contour integral (3.9). The coefficient of $\Phi_b^{(l)}(0)$ is

$$\frac{\Gamma(a+b)}{2\pi i l!} \int_P e^{sw} \frac{dw}{w^{a+b-l}} = \frac{\Gamma(a+b)}{l! \, \Gamma(a+b-l)},$$

by Hankel's integral for the Gamma-function. This gives the sum over l in (3.11). Since $N - a - b > -1$ the contour integral involving $E_N(w)$ can be

contracted to one over two copies of the negative real axis (with different orientations). Here
$$w^{a+b} = |w|^{a+b} e^{\pm i\pi(a+b)},$$
according as we approach the axis from above or below. The substitution $w = -z$ now gives
$$\frac{1}{2\pi i}\int_P e^{sw} E_N(w)\frac{dw}{w^{a+b}} = \frac{\sin(\pi(a+b))}{\pi}\int_0^\infty e^{-sz} E_N(-z)\frac{dz}{z^{a+b}}.$$

Use of the formula
$$\Gamma(a+b)\sin(\pi(a+b)) = \pi\,\Gamma(1-a-b)$$
leads to (3.11) as stated.

To obtain the recurrences (3.12), return to the differential equation
$$z\,\Phi_b'(z) = b(1-e^z)\Phi_b(z),$$
in the form
$$\sum_{n\geq 0}\frac{\Phi_b^{(n)}(0)z^n}{(n-1)!} = -b\sum_{l\geq 0}\frac{\Phi_b^{(l)}(0)z^l}{l!}\sum_{m\geq 1}\frac{z^m}{m!}.$$
Equating coefficients of $z^n/n!$ gives the identity (3.12) in Lemma 2.

Consequences of Lemma 2. Observe in particular that
$$r(s) \sim s^{a+b-1} \quad \text{as} \quad s \to \infty, \tag{3.13}$$
as already observed in (3.8) if $a+b < 1$. Also, if $a+b < 2$ the case $N = 1$ of Lemma 2 says
$$r(s) = s^{a+b-1} + \frac{1}{\Gamma(1-a-b)}\int_0^\infty (\Phi_b(-z)-1)e^{-sz}\frac{dz}{z^{a+b}}. \tag{3.14}$$

In (3.14), make the substitution $sz = u$, giving
$$\frac{r(s)}{s^{1-a-b}} = 1 + \frac{1}{\Gamma(1-a-b)}\int_0^\infty (\Phi_b(-u/s)-1)e^{-u}\frac{du}{u^{a+b}}. \tag{3.15}$$

Hence
$$r(s) < 0 \text{ as } s \to 0+ \quad \text{when} \quad 1 < a+b < 2, \tag{3.16}$$
since $\Gamma(1-a-b) < 0$ in this case and (3.7) shows that $\Phi_b(-u/s)$ becomes arbitrarily large as $s \to 0+$.

The Zeros of r. We assemble just about as much information about the real positive zeros of the function r as we will need. Lemma 3 guarantees that the choice of β specified in Lemma 4 below is in fact possible.

We use the full notation $r = r_{a,b}$, since we shall need to refer to more than one of these functions at a time. Thus the "standard" solutions of (2.3) are given by

$$p = r_{\kappa,-\kappa}, \quad q = r_{\kappa,\kappa}. \tag{3.17}$$

Lemma 3. *Let $r = r_{a,b}$ be as in Lemma 2. Then*
  (i) *the number of real positive zeros of $r_{a,b}$ is less than $a + b$*
  (ii) *such a zero exists if and only if $b > 0$ and $a + b > 1$.*

If $a + b$ is a positive integer then $r_{a,b}$ is a polynomial of degree $a+b-1$, so (i) follows in this case. If $a + b < 1$ then (3.6) gives $r_{a,b}(s) > 0$ when $s > 0$, and (i) again follows. Differentiation of (3.9) shows

$$c_{a,b}\, r_{a,b}^{(n)}(s) = r_{a-n,b}(s), \tag{3.18}$$

actually with $c_{a,b} = \Gamma(a+b-n)/\Gamma(a+b)$. When $a+b-1 < n < a+b$, (3.6) gives $r_{a-n,b}(s) > 0$ when $s > 0$, so part (i) of the lemma follows.

Proceed to part (ii). Observe that (3.13), (3.18) give

$$r_{a,b}(s) > 0, \quad r'_{a,b}(s) > 0 \quad \text{as} \quad s \to \infty. \tag{3.19}$$

First we show that $r_{a,b}(s) \neq 0$ when $s > 0$ in the case $b \leq 0$. For if $b = 0$ then (3.7) gives $\Phi_b(x) = 1$, so that Lemma 2 reduces to $r_{a,b}(s) = s^{a+b-1}$. If $b < 0$ and $r_{a,b}$ has a positive zero then we may, because of part (i), define α to be the largest one, so that $r'_{a,b}(\alpha) \geq 0$ and $r_{a,b}(\alpha+1) > 0$. Now (2.2) gives

$$\alpha\, r'_{a,b}(\alpha) = b\, r_{a,b}(\alpha+1),$$

a contradiction.

Secondly, $r_{a,b}(s) \neq 0$ if $a + b < 1$ because of (3.6), and $r_{a,b}(s) \neq 0$ if $a + b = 1$ because $r_{a,b}(s) = 1$ in this case.

Finally, we show by induction on $m \geq 1$ that $r_{a,b}$ has a positive zero when $b > 0$ and $m < a + b \leq m + 1$. When $m = 1$ the function $r_{a,b}$ has a positive zero because of (3.19) and (3.16). When $m > 1$ we will show that $r_{a,b}$ has a zero exceeding the greatest zero α_1 of $r'_{a,b}$, which is a zero of $r_{a-1,b}$ by (3.18), so that the induction proceeds. In fact $r_{a,b}(\alpha_1 + 1) \geq r_{a,b}(\alpha_1)$ because $r'_{a,b}(s) > 0$ when $s > \alpha_1$, so that (2.2) leads to

$$0 = (a-1)r(\alpha_1) + b\,r(\alpha_1+1) > (a+b-1)r(\alpha_1),$$

whence r has a zero exceeding α_1 because of (3.13).

4.2 The Functions F and f 123

Special Values of $r(s)$. The following special values and asymptotic properties of $r(s)$ will be needed in due course. Part (ii) of the next lemma is needed in the determination of $F(s)$ for small s in the important case $\kappa = 1$. Also, on taking $a = \frac{1}{2}$, $b = -\frac{1}{2}$ in part (i) we obtain in (2.3)

$$p(s) \sim \sqrt{\frac{\pi}{e^\gamma s}} \quad \text{as} \quad s \to 0+ \quad \text{if} \quad \kappa = \frac{1}{2}. \tag{3.20}$$

If $a = b = \kappa < 1$ then part (iii) shows that the (unique, by Lemma 3) zero of r satisfies

$$\alpha \sim \pi e^\gamma (2\kappa - 1)^2 \quad \text{as} \quad \kappa \to \frac{1}{2}+. \tag{3.21}$$

Here γ denotes Euler's constant.

Lemma 4. Let $r = r_{a,b}$ be as in Lemma 2.
(i) If $a + b < 1$ and $a < 1$ then

$$r(s) \sim \frac{e^{b\gamma} \Gamma(1-a)}{\Gamma(1-a-b)} s^{a-1} \quad \text{as} \quad s \to 0+.$$

(ii) If $a = 1$ and $b = -1$ then $p(1) = e^{-\gamma}$ in (2.3).
(iii) If $1 < a + b < 2$ and $b > 0$, let α denote the unique real positive zero of r provided by Lemma 3. Then

$$\alpha^b \sim \Gamma(1-a) e^{\gamma/2} (a+b-1) \quad \text{as} \quad a+b \to 1+.$$

For parts (i) and (ii) we need only the Laplace transform formula (3.6), which gives

$$r(s) = \frac{1}{\Gamma(1-a-b)} \int_0^\infty \exp\left(b \int_0^x \frac{1-e^{-t}}{t} dt - b \int_1^x \frac{dt}{t}\right) \frac{e^{-sx}}{x^a} dx.$$

The substitution $sx = u$ already used in (3.8) gives

$$r(s) = \frac{s^{a-1}}{\Gamma(1-a-b)} \int_0^\infty \exp\left(b \int_0^1 \frac{1-e^{-t}}{t} dt - b \int_1^{u/s} \frac{e^{-t}}{t} dt\right) \frac{e^{-u}}{u^a} du$$

$$\sim \frac{s^{a-1}}{\Gamma(1-a-b)} e^{b\gamma} \int_0^\infty u^{-a} e^{-u} du \quad \text{as} \quad s \to 0+, \tag{3.22}$$

as stated in part (i).
When $a = 1$, $b = -1$ the equation in (3.22) gives

$$r(1) = \int_0^\infty \exp\left(-\int_0^1 \frac{1-e^{-t}}{t} dt + \int_1^x \frac{e^{-t}}{t} dt\right) \frac{e^{-x}}{x} dx$$

$$= \exp\left(-\int_0^1 \frac{1-e^{-t}}{t} dt + \int_1^\infty \frac{e^{-t}}{t} dt\right) = e^{-\gamma},$$

which is part (ii).

Use (3.14) when $1 < a+b < 2$. Observe that (3.7) shows

$$\frac{\Phi_b(-z)}{z^b} = \exp\left(b\int_0^z \frac{1-e^{-u}}{u}\,du - b\int_1^z \frac{du}{u}\right)$$

$$= \exp\left(b\int_0^1 \frac{1-e^{-u}}{u}\,du - b\int_1^z \frac{e^{-u}}{u}\,du\right)$$

$$\to e^{b\gamma} \quad \text{as} \quad z \to \infty.$$

But $b > 0$ in part (iii), whence

$$\frac{\Phi_b(-z) - 1}{z^b} \to e^{b\gamma} \quad \text{as} \quad z \to \infty. \tag{3.23}$$

When $1 < a+b < 2$ use (3.14) with $s = \alpha$, so that $r(\alpha) = 0$, to obtain

$$\alpha^b = -\frac{1}{\Gamma(1-a-b)} \int_0^\infty \frac{\Phi_b(-u/\alpha) - 1}{(u/\alpha)^b} e^{-u} \frac{du}{u^a}. \tag{3.24}$$

In the first place this gives $\alpha \to 0$ as $a + b \to 1+$, because $1/\Gamma(x) \sim x$ as $x \to 1+$. Now using (3.23) in (3.24) gives

$$\alpha^b \sim -\frac{e^{b\gamma}\Gamma(1-a)}{\Gamma(1-a-b)} \quad \text{as} \quad a+b \to 1+,$$

and part (iii) of the lemma follows.

4.2.4 Particular Values of $F(s)$ and $f(s)$

Next, we determine the values of the constants C, B appearing in (1.3), once the parameter β has been specified in the way to be described in Lemma 5. We will also see that as $s \to \infty$ a much stronger statement than (1.4) holds. The procedure depends on a determination of the inner products appearing in Lemma 1, using the information collected in Sect. 4.2.3.

First choose $a = b = \kappa$ in Lemma 1, so that $R(s) = Q(s)$, as given in (1.8) and (1.12), and $q(s) = r_{\kappa,\kappa}(s)$ satisfies (2.3). Then (3.13) becomes

$$q(s) \sim s^{2\kappa - 1} \quad \text{as} \quad s \to \infty. \tag{4.1}$$

Because $Q(s) = o(s^{2\kappa})$ in (1.10) the inner product in (2.4) satisfies

$$\langle Q, q \rangle(s) \to 0 \quad \text{as} \quad s \to \infty.$$

So by Lemma 1(i)

$$\langle Q, q \rangle(s) = 0 \quad \text{if} \quad s \geq \beta + 1 \text{ or } s = \beta. \tag{4.2}$$

Second, choose $a = \kappa$, $b = -\kappa$ and proceed similarly. Then we can take $R(s) = P(s)$ in Lemma 1, and now $p(s) = r_{\kappa,-\kappa}(s)$ satisfies (2.3). Then

$$s\,p(s) \to 1 \quad \text{as} \quad s \to \infty \tag{4.3}$$

from (3.13), or equivalently using the inference (3.8) from the Laplace transform solution (3.6). Because $P(s) = 2 + o(s^{-2\kappa})$ in (1.10), Lemma 1 gives

$$\langle P, p \rangle(s) = 2 \quad \text{if} \quad s \geq \beta + 1 \text{ or } s = \beta. \tag{4.4}$$

$F(s)$ and $f(s)$ for Small s. These are determined by obtaining the values of the constants C and B appearing in (1.3). For this purpose note the values of $P(s)$, $Q(s)$ relevant in (4.4) and (4.2), as described by (1.11) and (1.12):

$$s^\kappa Q(s) = C \quad \text{if} \quad \beta - 1 < s < \beta, \qquad \beta^\kappa Q(\beta+) = C - B, \tag{4.5}$$
$$s^\kappa P(s) = C \quad \text{if} \quad \beta - 1 < s < \beta, \qquad \beta^\kappa P(\beta+) = C + B. \tag{4.6}$$

In this situation, (4.2), (4.4) and the equation (2.7) in Lemma 1 give

$$B \frac{q(\beta)}{\beta^{\kappa-1}} = \lim_{s \to \beta+} \frac{C q(s-1)}{(s-1)^{\kappa-1}}, \quad B \frac{p(\beta)}{\beta^{\kappa-1}} + \lim_{s \to \beta+} \frac{C p(s-1)}{(s-1)^{\kappa-1}} = 2. \tag{4.7}$$

The sifting limit when $\kappa = 1$. We look at this case briefly first. As noted in (3.3), $q(s) = s - 1$ is now our standard solution to (2.3). Consequently (4.7) gives

$$(\beta - 1)B = C(\beta - 2). \tag{4.8}$$

The choice $\beta = 2$ is actually the best as well as the simplest, and gives $B = 0$. Then (1.5) gives

$$s f(s) = C \int_2^s \frac{dx}{x-1} \quad \text{if} \quad 2 < x < 4.$$

It follows from (1.1)-(1.3) that provided $C > 0$ (which we establish below) $s f(s)$ increases with s, so that the sifting limit described in Sect. 4.2.1 (the largest zero of f) is actually $\beta = 2$.

For the other values of $\kappa > \frac{1}{2}$ the structure of this part of the argument is similar. The details follow in Lemma 5, where we also determine the value of the constant C.

If we chose $\beta > 1$ possibly different from 2 then we would obtain from (1.5) and (4.8)

$$s f(s) = C \left(\frac{\beta - 2}{\beta - 1} + \log \frac{s-1}{\beta - 1} \right) \quad \text{when} \quad \beta < s < \beta + 2,$$

with its zero at

$$s_0 = 1 + (\beta - 1)e^{-1+1/(\beta-1)}, \tag{4.9}$$

which is indeed minimised at $\beta = 2$.

The General Case. The specification of β is as follows. Let α be the greatest zero of the function $q = r_{\kappa,\kappa}$ given in (3.17), so that α is as in Lemma 4. Let

$$\beta = \begin{cases} \alpha + 1 & \text{if } \kappa > \frac{1}{2} \\ 1 & \text{if } 0 < \kappa \leq \frac{1}{2}. \end{cases} \tag{4.10}$$

In particular $\beta > 1$ when $\kappa > \frac{1}{2}$, and $\beta = 1$ when $\kappa = \frac{1}{2}$. When $\kappa = 1$, (4.10) says $\beta = 2$, because of (3.3).

126   4. Rosser's Sieve

Lemma 5. *Let β be as in (4.10). Then the constants in (1.3) determining $F(s)$, $f(s)$ for small s are as follows.*

(i) *When $\kappa > \frac{1}{2}$*

$$C = \frac{2(\beta - 1)^{\kappa - 1}}{p(\beta - 1)}, \quad B = 0, \tag{4.11}$$

where $p = r_{\kappa, -\kappa}$ as in (3.17). In particular, $C = 2e^\gamma$ when $\kappa = 1$, where γ is Euler's Constant. Also

$$\beta - 1 \sim \pi e^\gamma (2\kappa - 1)^2, \quad C \to 2\sqrt{\frac{e^\gamma}{\pi}} \quad \text{as} \quad \kappa \to \tfrac{1}{2}+. \tag{4.12}$$

(ii) *When $\kappa = \frac{1}{2}$*

$$C = 2\sqrt{\frac{e^\gamma}{\pi}}, \quad B = 0. \tag{4.13}$$

(iii) *When $0 < \kappa < \frac{1}{2}$*

$$C = \frac{e^{\gamma\kappa}}{\Gamma(1-\kappa)} \frac{2 I_\kappa(\kappa)}{I_\kappa(\kappa) + I_\kappa(-\kappa)}, \quad B = \frac{2 e^{\gamma\kappa}}{I_\kappa(\kappa) + I_\kappa(-\kappa)}, \tag{4.14}$$

where

$$I_\kappa(l) = \int_0^\infty \frac{e^{-x}}{x^\kappa} \exp\left(k \int_x^\infty \frac{e^{-u}}{u} du\right) dx. \tag{4.15}$$

When $\kappa = 1$ this lemma and (1.3), (1.5) give

$$F(s) = \frac{2e^\gamma}{s} \quad \text{if } 0 < s \leq 3, \quad f(s) = \frac{2e^\gamma}{s} \log(s - 1) \quad \text{if } 2 \leq s \leq 4. \tag{4.16}$$

When $\kappa = \frac{1}{2}$ (so $\beta = 1$) it gives

$$f(s) = \sqrt{\frac{e^\gamma}{\pi}} \int_1^s \frac{dt}{\sqrt{t(t-1)}} \quad \text{if } 1 < s < 3. \tag{4.17}$$

Thus the sifting limit (where $f(s) = 0$) is 1 ($= \beta$). If $\kappa < \frac{1}{2}$ there is no sifting limit in the usual sense, because $B > 0$, so that $f(s) > 0$ whenever $s > \beta$.

To establish part (i), proceed as in the derivation of (4.8), using the facts that $q(\beta - 1) = q(\alpha) = 0$ and $\beta > 1$. Then (4.7) gives $B = 0$, and as a consequence $2 = C p(\beta - 1)/(\beta - 1)^{\kappa - 1}$, as stated in Lemma 5.

In the case $\kappa = 1$ we have $q(s) = s - 1$ and $\beta = 2$, so that (4.11) expresses C in terms of $p(1)$. Thus $C = 2e^\gamma$ because of part (ii) of Lemma 4.

The relation (3.21) says exactly that β satisfies (4.12). Then (4.11) and (3.20) give $C \sim 2(\beta - 1)^{\kappa - 1}\sqrt{(e^\gamma(\beta - 1)/\pi)}$ as $\kappa \to \frac{1}{2}+$, and (4.12) follows. This proves Lemma 5(i).

When $\kappa \leq \frac{1}{2}$ we specified $\beta = 1$. In this case the limits as $s \to 1+$ appearing in (4.7) were evaluated in Lemma 4(i). Thus

$$C e^{-\gamma\kappa}\Gamma(1-\kappa) + B\,p(1) = 2, \qquad (4.18)$$
$$C e^{\gamma\kappa}\Gamma(1-\kappa) - B\,q(1)\Gamma(1-2\kappa) = 0. \qquad (4.19)$$

If $\kappa = \frac{1}{2}$ then (4.19) should be read as saying $B = 0$, as follows more directly from (4.7) since the standard (and obvious) solution in (2.3) is $q(s-1) = 1$. Then (4.18) gives $C = 2e^{\gamma/2}/\Gamma(\frac{1}{2})$, and part (ii) of the lemma follows.

When $\kappa < \frac{1}{2}$ the Laplace transform representation (3.6) gives

$$p(1) = \int_0^\infty \exp\left(-x - \kappa \int_0^x \frac{1-e^{-t}}{t}\,dt\right) dx,$$
$$q(1)\Gamma(1-2\kappa) = \int_0^\infty \exp\left(-x + \kappa \int_0^x \frac{1-e^{-t}}{t}\,dt\right) \frac{dx}{x^{2\kappa}}.$$

Since

$$e^\gamma = \exp\left(\int_0^1 \frac{1-e^{-t}}{t}\,dt - \int_1^\infty \frac{e^{-t}}{t}\right),$$

it follows that

$$p(1)e^{\gamma\kappa} = \int_0^\infty \exp\left(-x - \kappa\log x - \kappa\int_x^\infty \frac{e^{-t}}{t}\,dt\right) dx,$$
$$q(1)\Gamma(1-2\kappa)e^{-\gamma\kappa} = \int_0^\infty \exp\left(-x + \kappa\log x + \kappa\int_x^\infty \frac{e^{-t}}{t}\,dt\right) \frac{dx}{x^{2\kappa}}.$$

Thus, in the notation (4.15),

$$p(1)e^{\gamma\kappa} + q(1)\Gamma(1-2\kappa)e^{-\gamma\kappa} = I_\kappa(\kappa) + I_\kappa(-\kappa),$$

and the equations (4.18), (4.19) have the solution (3.4). This completes the proof of Lemma 5.

$F(s)$ and $f(s)$ for Large s. The condition (1.4) was chosen as the weakest convenient one sufficient for the arguments presented in Sect. 4.2.4. In fact substantially more than (1.4) follows.

Lemma 6. *If $F(s)$, $f(s)$ satisfy (1.1)–(1.3) then*

$$F(s) = 1 + O(e^{-s\log s}), \quad F(s) = 1 + O(e^{-s\log s}) \quad \text{as} \quad s \to \infty, \qquad (4.20)$$

where the implied constant may depend on κ.

From (2.4), the inner product equations (4.2), (4.4) say

$$s\,q(s)Q(s) = \kappa\int_{s-1}^s q(x+1)Q(x)\,dx, \quad s\,p(s)P(s) + \kappa\int_{s-1}^s p(x+1)P(x)\,dx = 2,$$

128   4. Rosser's Sieve

since p, q are as in (3.17) and the relevant inner products are as in (4.2) and (4.4). The second of these holds also when $P(s) = 2$ since this is another solution of the conditions for P in (1.10) and (1.13). Hence

$$s\, p(s)\bigl(P(s) - 2\bigr) + \kappa \int_{s-1}^{s} p(x+1)\bigl(P(x) - 2\bigr)\, dx = 0\,.$$

Since (4.1), (4.3) say $q(s) \sim s^{2\kappa-1}$, $p(s) \sim 1/s$ as $s \to \infty$ this gives

$$s\, Q(s) \ll \int_{s-1}^{s} |Q(x)|\, dx\,, \qquad s\bigl(P(s) - 2\bigr) \ll \int_{s-1}^{s} |P(x) - 2|\, dx\,,$$

if s is large enough.

The conclusion of Lemma 6 then follows from (1.8) and Lemma 7, when used with $U(s) = Q(s)$ and $U(s) = P(s) - 2$. Lemma 7, which is used again in Sect. 3.3, is deduced from Lemmas 8 and 9, both of which are arranged so as to be useful again in Chap. 7.

In the application in this chapter the function U will be continuous. In the situation arising in Chap. 7 the expression $U(x)$ arising in Lemma 8 will be non-negative, but may have jump discontinuities at which it decreases. Accordingly the function u, when considered in Lemma 9, is allowed to have discontinuties of this type.

Lemma 7. *Suppose a function U, continuous in $s > s_0 - 1$, satisfies*

$$s\, |U(s)| \le M \int_{s-1}^{s} |U(x)|\, dx \quad \text{if} \quad s > s_0\,,$$

for some constant M. Then

$$U(s) \ll e^{-s \log s}\,, \tag{4.21}$$

where the implied constant may depend on M.

Lemma 8. *Suppose $U(s)$ is bounded and integrable on $s \ge s_0$, and define $u(x) = |U(x)|e^{\phi(x)}$, where $\phi(x) = x \log x - Nx$ for some constant N. Then*

$$\frac{\kappa}{s} \int_{s-1}^{s} |U(x)|\, dx < \tfrac{1}{2} e^{-\phi(s)} \sup_{s-1 \le x \le s} u(x)\,,$$

if s/e^N exceeds a suitable constant depending on κ.

Lemma 9. *Suppose $u(x)$ is continuous on $x \ge s_0$ apart from jump discontinuities at which it may decrease. If $u(x)$ is bounded when $s_0 - 1 \le x \le s_0$ and*

$$u(s) < \tfrac{1}{2} + \tfrac{1}{2} \sup_{s-1 \le x \le s} u(x) \quad \text{when} \quad s > s_0\,, \tag{4.22}$$

then $u(s)$ is bounded in $s > s_0$.

Proof of Lemma 7. To deduce Lemma 7 from Lemmas 8 and 9, take $N = 0$ in Lemma 8. The hypothesis in Lemma 7 then shows that $u(x)$, defined as in Lemma 8, satisfies the hypothesis of Lemma 9, which then gives (4.21) as stated.

Proof of Lemma 8. This rests on the fact that $\phi'(x) = \log x + 1 - N$ increases with x. Consequently

$$\phi(s) - \phi(x) \leq (s-x)\phi'(s) \quad \text{if} \quad s-1 \leq x \leq s\,.$$

Hence the expression to be estimated in Lemma 8 does not exceed

$$e^{-\phi(s)}\frac{\kappa}{s}\int_{s-1}^{s} e^{\phi(s)-\phi(x)} u(x)\,dx$$

$$\leq e^{-\phi(s)}\frac{\kappa}{s}\int_{s-1}^{s} e^{(s-x)\phi'(s)}\,dx \sup_{s-1\leq x\leq s} u(x)$$

$$\leq e^{-\phi(s)}\frac{\kappa\, e^{\phi'(s)}}{s\,\phi'(s)} \sup_{s-1\leq x\leq s} u(x)\,.$$

Here

$$\frac{\kappa\, e^{\phi'(s)}}{s\,\phi'(s)} \leq \frac{\kappa\, e^{1-N}}{\log s + 1 - N} < \tfrac{1}{2}$$

if $\log s - N$ is large enough, and Lemma 8 follows.

Proof of Lemma 9. We show that $u(s) \leq A$ when $s \leq s_0 + n$ by induction on n, for those $A > 1$ for which it holds at $n = 0$. Then, if $u(s) > A$ for some $s \leq n+1$, let s_1 be the infimum of such s. Then s_1 is not a point of discontinuity of u, because u decreases at such discontinuities. Thus $u(s_1) = A$, while $u(x) \leq A$ for $s_1 - 1 \leq x \leq s_1$. In (4.22), take $s = s_1$ to obtain $A < \tfrac{1}{2} + \tfrac{1}{2}A$, contrary to the hypothesis $A > 1$. So no such s exists, and Lemma 9 follows.

4.2.5 Asymptotic Analysis as $\kappa \to \infty$

When κ becomes large we can say a little more about the sifting limit $\beta(\kappa)$ in Rosser's sieve. For present purposes, this sifting limit is simply the number β specified in (4.10). Once Proposition 1 (about the existence of the function f) has been established in Sect. 4.3, it will follow from (1.5) and Lemma 5 that $f(\beta) = 0$ and $f(s) > 0$ when $s > \beta$.

The sifting limit $\beta(\kappa)$ has a richer interpretation in the light of results from Sects. 3.3 and 4.4. The results in Sect. 4.4 (Corollary 4.4.1.1 is sufficient) show that $\liminf_{D\to\infty} V(D, P(z))/V(P(D^{1/s})) > 0$ when $s > \beta$. In Sect. 3.3 we examined the results that followed from Rosser's construction using a somewhat approximate analysis derived from that used by Brun. Thus the lower bound for $V(D, P(z))$ established in Theorem 3.3.1 will imply an upper bound for the number β appearing in the current chapter. In particular Corollary 3.3.1.3 gives that the sifting limit $\beta(\kappa)$ satisfies $\limsup_{\kappa\to\infty} \beta(\kappa)/\kappa \le c$, where $c = 3 \cdot 591\ldots$ is given by the equation $e^{1/c} = c/e$.

Corollary 1.1, which follows directly from Theorem 1, gives a sharper version of this proposition, derived from information about the expression $q(s)$ defined by (3.9) and (3.17). Actually Theorem 1 can be developed so that Corollary 1.1 holds with equality, for which see the reference given in the notes on this chapter.

Throughout this section it is of course crucial that all constants, such as those implied by the O notation, are absolute, in particular independent of κ.

Theorem 1. *Let $q(s) = r_{\kappa,\kappa}(s)$ be as given in (3.9), and let $c = 3 \cdot 591\ldots$ be the solution of $c \log c - c = 1$. Then there exists a constant c_1 such that $q(s) > 0$ whenever $s > c\kappa + c_1 \kappa^{1/3}$.*

Corollary 1.1. *The greatest zero α of the function $q(s)$, and the sifting limit $\beta = \alpha + 1$ appearing in (4.10), satisfy*

$$\beta = \alpha + 1 \le c\kappa + O\!\left(\kappa^{1/3}\right).$$

The standard representation (3.9) of the function q can be written as

$$q(s) = \frac{\Gamma(2\kappa)}{2\pi i} \int_P e^{l(z)} \frac{dz}{z^{2\kappa}}, \qquad (5.1)$$

where

$$l(z) = sz + \kappa \int_0^z \frac{1 - e^u}{u}\, du. \qquad (5.2)$$

We may take P to consist of the lower and upper sides of a cut from $-\infty$ to $-r$, together with a circle of radius r, to be specified. Then we can write

$$q(s) = \frac{\Gamma(2\kappa)}{\pi} I_1 + \frac{\Gamma(2\kappa)}{2\pi} r^{1-2\kappa} I_2, \qquad (5.3)$$

where I_1 is a contribution from the real axis and I_2 is that from the circle, on which we use the usual substitution $z = re^{i\theta}$:

$$I_2 = \int_{-\pi}^{\pi} \exp\!\left(l(re^{i\theta}) - (2\kappa - 1)i\theta\right) d\theta. \qquad (5.4)$$

On the negative real axis set $z = -t$, so that $t > 0$. Note that $(1 - e^{-t})/t$ decreases as t increases, its derivative having the sign of $(t + 1)e^{-t} - 1 < 0$ because $e^t > t + 1$. Thus $(1 - e^{-t})/t < 1$, so $l'(-t) > s - \kappa$. Consequently

$$|I_1| \le e^{l(-r)} \int_r^\infty \frac{dt}{t^{2\kappa}} \le e^{l(-r)} \frac{r^{1-2\kappa}}{2\kappa - 1} \quad \text{if} \quad s > \kappa > \tfrac{1}{2}. \qquad (5.5)$$

4.2 The Functions F and f

In (5.2) and (5.4), write $L(\theta) = l(z) = l(re^{i\theta})$. Then the first few derivatives of L satisfy

$$L'(\theta) = iz\frac{d}{dz}L(\theta) = i(sz + \kappa(1 - e^z)), \quad L'(0) = i(sr + \kappa(1 - e^r)),$$
$$L''(\theta) = iz\frac{d}{dz}L'(\theta) = -z(s - \kappa e^z), \quad L''(0) = -r(s - \kappa e^r), \quad (5.6)$$
$$L'''(\theta) = iz\frac{d}{dz}L''(\theta) = -i(sz - \kappa(z + z^2)e^z),$$

with a similar but more complicated expression for $L^{(4)}(\theta)$.

The choice of r used in the proof of Theorem 1 will satisfy

$$r < \log c, \quad (5.7)$$

so that in particular r is bounded above. Then the expression $L(\theta) = l(re^{i\theta})$ occurring in (5.4) can be expanded in the forms

$$L(\theta) = L(0) + \theta L'(0) + \tfrac{1}{2}\theta^2 L''(0) + O((s + \kappa)\theta^3), \quad (5.8)$$
$$L(\theta) = L(0) + \theta L'(0) + \tfrac{1}{2}\theta^2 L''(0) + \tfrac{1}{6}L'''(0) + O((s + \kappa)\theta^4).$$

These expansions will be useful when θ is relatively small.

In estimating the size of the integrand in (5.4) we will also be concerned with the real part of $L(\theta)$ for all θ.

Lemma 10. *The expression $L(\theta) = l(re^{i\theta})$ given by (5.2) satisfies*

$$\operatorname{Re} L(\theta) \leq L(0) + 2L''(0)\sin^2 \tfrac{1}{2}\theta.$$

In particular

$$l(-r) \leq L(0) + 2L''(0). \quad (5.9)$$

From (5.2) we obtain

$$l(z) = sz - \kappa \int_0^z \sum_{n=1}^{\infty} \frac{u^{n-1}}{n!} du = sz - \kappa \sum_{n=1}^{\infty} \frac{z^n}{n!n}.$$

Hence

$$\operatorname{Re} L(\theta) = sr\cos\theta - \kappa \sum_{n=1}^{\infty} \frac{r^n \cos n\theta}{n!n} = l(r) - 2sr\sin^2 \tfrac{1}{2}\theta + 2\kappa \sum_{n=1}^{\infty} \frac{r^n \sin^2 \tfrac{1}{2}n\theta}{n!n}.$$

An easy induction shows $|\sin n\phi| \leq n|\sin \phi|$ for integers $n \geq 1$, whence

$$\operatorname{Re} L(\theta) \leq L(0) - 2\sin^2 \tfrac{1}{2}\theta \left(sr - \kappa \sum_{n=1}^{\infty} \frac{r^n}{(n-1)!}\right) = L(0) + 2L''(0)\sin^2 \tfrac{1}{2}\theta.$$

The inequality (5.9) follows on taking $\theta = \pi$.

Proof of Theorem 1. These preliminaries completed, we invoke the saddle-point method, the principle of which is to approximate to $q(s)$ in the neighbourhood of its stationary points. We aim to choose r so that the integrand in (5.4) has a maximum at $\theta = 0$ when s is close to or larger than $c\kappa$. Observe that (5.6) gives

$$L'(0) - i(2\kappa - 1) = i\kappa \left(\frac{s}{\kappa} r - e^r - 1 + \frac{1}{\kappa}\right), \qquad L''(0) = -\kappa r \left(\frac{s}{\kappa} - e^r\right). \quad (5.10)$$

The limiting case of this process would be when $L'(0) = i(2\kappa - 1)$ and $L''(0) = 0$. Write $s/\kappa = b$. Note that if the term $1/\kappa$ in (5.10) is neglected then this asks $b = e^r$, $br - e^r = 1$, so that $b \log b - b = 1$, and b would be the constant c appearing in Theorem 1.

From (5.10), the equation $L'(0) = i(2\kappa - 1)$ actually reads

$$sr - \kappa e^r = \kappa - 1, \quad (5.11)$$

having two roots, separated by the zero of its derivative $s - \kappa e^r = -L''(0)/r$. Choose r to be the smaller root, so that $e^r < s/\kappa$, and $r > 0$ because $\kappa > 1$. Then

$$L'(0) = i(2\kappa - 1), \qquad L''(0) < 0. \quad (5.12)$$

We need to check that r is bounded as indicated in (5.7). Observe

$$\frac{s}{\kappa} > e^r, \qquad \frac{s}{\kappa} r - e^r - 1 < 0,$$

so $0 > re^r - e^r - 1$. This expression increases with r, since $r > 0$, and vanishes when $r = \log c$. Hence our choice of r satisfies (5.7).

In the opposite direction it appears from (5.11) that when κ and s/κ are large the number r will be rather small, close to $2\kappa/s$.

Dissect the integral in (5.4) as $I_2 = I_3 + I_4$, where

$$I_3 = \int_{-\delta}^{\delta} \exp\left(L(\theta) - (2\kappa - 1)i\theta\right) d\theta,$$

with $\delta < \pi$, and I_4 denotes the integral over the remainder of $(-\pi, \pi)$. We will take $\delta = (s + \kappa)^{-1/3}$. In I_3, use the expansion (5.8), to obtain

$$I_3 = \int_{-\delta}^{\delta} \exp\left(L(0) + \theta(L'(0) - i(2\kappa - 1)) + \tfrac{1}{2}\theta^2 L''(0)\right)\left(1 + O((s+\kappa)\theta^3)\right) d\theta,$$

since $(s + \kappa)\delta^3$ is bounded. In I_4, use Lemma 10 to obtain

$$|I_4| \le \int_{\delta < |\theta| < \pi} \exp\left(L(0) + 2L''(0) \sin^2 \tfrac{1}{2}\theta\right) \frac{\theta^3}{\delta^3} d\theta.$$

4.2 The Functions F and f

Here $\frac{1}{2}\theta^2 \geq 2\sin^2\frac{1}{2}\theta$, so use of (5.12) gives

$$I_2 = \int_{-\delta}^{\delta} \exp\left(L(0) - \tfrac{1}{2}\theta^2|L''(0)|\right) d\theta$$
$$+ O\left(\int_{-\pi}^{\pi}(s+\kappa)\exp\left(L(0) - 2|L''(0)|\sin^2\tfrac{1}{2}\theta\right)\theta^3\, d\theta\right).$$

Then use of $\sin^2\frac{1}{2}\theta \geq \theta^2/\pi^2$ when $-\pi < \theta < \pi$ leads to

$$I_2 = \int_{-\infty}^{\infty} \exp\left(L(0) - \tfrac{1}{2}\theta^2|L''(0)|\right) d\theta$$
$$+ O\left(\int_{-\infty}^{\infty}(s+\kappa)\exp\left(L(0) - 2|L''(0)|\theta^2/\pi^2\right)\theta^3\, d\theta\right),$$

the effect of extending the first integration to $(-\infty, \infty)$ being absorbed by the stated O-term. Thus

$$I_2 = \frac{\sqrt{2\pi}e^{L(0)}}{\sqrt{|L''(0)|}} + O\left(\frac{(s+\kappa)e^{L(0)}}{|L''(0)|^2}\right).$$

The estimate for $q(s)$ follows on combining this estimate with that in (5.5), in which we use the estimate $l(-r) \leq L(0) - 2|L''(0)|$ following from (5.9) and (5.12). Then (5.3) gives

$$q(s) = \frac{\Gamma(2\kappa)}{\sqrt{2\pi}}r^{1-2\kappa}e^{L(0)}\left(\frac{1}{\sqrt{|L''(0)|}} + O\left(\frac{s+\kappa}{|L''(0)|^2} + \frac{e^{-2|L''(0)|}}{2\kappa - 1}\right)\right). \quad (5.13)$$

First, consider the extreme case where $s = c\kappa + c_1\kappa^{1/3}$ and c_1 is a suitable constant, independent of κ. In the first place (5.11) gives

$$cr - e^r + \frac{c_1 r}{\kappa^{2/3}} = 1 - \frac{1}{\kappa}.$$

Write $r = \log c + x$. Then $e^r = ce^x$, and now

$$c\log c - c + cx - c(e^x - 1) + \frac{c_1\log c}{\kappa^{2/3}} + \frac{c_1 x}{\kappa^{2/3}} = 1 - \frac{1}{\kappa},$$

of which the negative root x is to be taken since r is the smaller root of (5.11). Since $c\log c - c = 1$, this root satisfies

$$\tfrac{1}{2}cx^2 = \frac{c_1\log c}{\kappa^{2/3}} + \frac{c_1 x}{\kappa^{2/3}} + \frac{1}{\kappa} + O(x^3) \sim \frac{c_1\log c}{\kappa^{2/3}} \quad \text{as} \quad \kappa \to \infty.$$

The number $L''(0)$ is now found from (5.6), to satisfy

$$-L''(0) = \kappa r(c - e^r) + c_1 r\kappa^{2/3} = \kappa(\log c + x)\left(c(1 - e^x) + \frac{c_1}{\kappa^{1/3}}\right)$$
$$= \kappa\left(-cx\log c + \frac{c_1\log c}{\kappa^{1/3}} + O(x^2)\right) \sim A\sqrt{c_1}\kappa^{2/3} \quad \text{as} \quad \kappa \to \infty,$$

where $A = \sqrt{2c}\log^{3/2} c + c_1 \log c$. In (5.13) we now have

$$\frac{1}{\sqrt{|L''(0)|}} \sim \frac{1}{\sqrt{A}c_1^{1/4}\kappa^{1/3}}, \quad \frac{s+\kappa}{|L''(0)|^2} \sim \frac{c+1}{A^2 c_1 \kappa^{1/3}} \quad \text{as} \quad \kappa \to \infty,$$

so that $q(s) > 0$ when $s = c\kappa + c_1 \kappa^{1/3}$ if c_1 is chosen sufficiently large.

We need to check that this conclusion persists when s takes values larger than $c\kappa + c_1 \kappa^{1/3}$. Actually (5.11) shows that $dr/ds = -r/(s - \kappa e^r) < 0$, because $e^r < s/\kappa$, so that r decreases with increasing s. Then (5.10) and (5.11) give

$$|L''(0)| = rs - \kappa r e^r = \kappa e^r + \kappa - 1 - \kappa r e^r,$$

so that $|L''(0)|$ increases with increasing s, and the conclusion $q(s) > 0$ again follows. This establishes Theorem 1, and Corollary 1.1 follows.

4.3 The Convergence Problem

In this section we establish Proposition 4.2.1, from which it follows that the sums $F(s)$, $f(s)$ do in fact possess the properties given in Sect. 4.2. Essentially the procedure is to establish inequalities $f_n(s) \le \psi_n(s)$ by an induction, where the majorant $\psi_n(s)$ is to be chosen in a suitable way. These inequalities will then establish the required convergence.

In the cases $\kappa = 1$ and $\kappa = \frac{1}{2}$ it is possible to make progress with ψ_n chosen to be of some explicit form resembling $\psi_n(s) = A\alpha^n e^{-s}$, using the fact that the appropriate choice of the parameter β is also easy to describe explicitly ($\beta = 2$ when $\kappa = 1$, and $\beta = 1$ when $\kappa = \frac{1}{2}$).

When 2κ is not an integer the choice of β made in Lemma 4.2.5 does not admit such a simple description, but involves non-trivial functions of the type described in Lemma 4.2.2. We use majorants $h(s)$ and $H(s)$ defined in a similar way. Even when 2κ is an integer this procedure seems to be essential for the stronger version of the result of Sect. 4.4 recorded as Theorem 4.4.2.

Other approaches to the convergence problem when $\kappa = 1$ are referred to in the notes.

The Proof of Proposition 4.2.1. This statement about the convergence of the series for $F(s)$ and $f(s)$ will be obtained as follows. Choose β as in Lemma 4.2.5. When $\kappa > \frac{1}{2}$ the partial sums of these series are the quantities g_n satisfying the estimates given in Lemmas 1 and 2 below. The equations (4.2.1.1)–(4.2.1.4) then follow from (4.1.3.1) and (4.1.3.2), using term by term integration. In particular the asymptotic relations (4.2.1.4) follow in the stronger form noted in Lemma 4.2.6.

When $0 < \kappa \le \frac{1}{2}$ it is necessary to use Lemma 6 in place of Lemma 2.

4.3.1 The Auxiliary Functions

Our concern is not so much with the individual functions f_n as with the partial sums

$$g_n(s) = \sum_{\substack{1 \le m \le n \\ m \equiv n, \bmod 2}} f_m(s) \qquad (1.1)$$

of the series appearing in (4.2.1.6). We will estimate these in terms of functions H, h given by the equations

$$\frac{d}{ds}\left(s^{\kappa+1} H(s)\right) = -\kappa s^\kappa h(s-1) \quad \text{if} \quad s > \beta + 1, \qquad (1.2)$$

$$\frac{d}{ds}\left(s^{\kappa+1} h(s)\right) = -\kappa s^\kappa H(s-1) \quad \text{if} \quad s > \beta, \qquad (1.3)$$

$$s^{\kappa+1} H(s) = D \quad \text{if} \quad 0 < s \le \beta + 1, \qquad (1.4)$$

$$s^{\kappa+1} h(s) = E \quad \text{if} \quad 0 < s \le \beta, \qquad (1.5)$$

where we will specify

$$D = (\beta - 1)^\kappa, \quad E = \beta^\kappa. \qquad (1.6)$$

These equations imply

$$s^{\kappa+1} h(s) - E = -D \int_\beta^s \frac{\kappa t^\kappa}{(t-1)^{\kappa+1}}\, dt \quad \text{if} \quad \beta \le s \le \beta + 1. \qquad (1.7)$$

Observe that there is no difficulty surrounding the existence of $H(s)$ and $h(s)$; they are given by successive integrations of the equations (1.2) and (1.3). On the other hand Lemma 2 below is not obvious.

The reason for the choice of this particular system of equations can be best appreciated after a reading of the proofs of the lemmas that follow, in particular of Lemma 1. The choice (1.6) of the constants D and E is crucial, for example to the validity of Lemma 2.

To make the notation more compact, define $h^+ = h$, $h^- = H$, so that

$$h^{(-)^n}(s) = \begin{cases} H(s) & \text{if } 2 \nmid n \\ h(s) & \text{if } 2 \mid n, \end{cases} \qquad (1.8)$$

where H, h are given by (1.2) – (1.6). As in (4.1.3.6) denote

$$\varepsilon_n = 1 \quad \text{if} \quad n \text{ is odd}, \qquad \varepsilon_n = 0 \quad \text{if} \quad n \text{ is even}. \qquad (1.9)$$

Lemma 1. *When $\kappa > \frac{1}{2}$, and $\beta > 1$ is chosen as in Lemma 4.2.5, there is a constant $\mu > 0$ so that the sum (1.1) satisfies*

$$g_n(s) \le \mu h^{(-)^n}(s) \quad \text{if} \quad s \ge \beta - \varepsilon_n, \qquad (1.10)$$

where $h^\pm$ are given by (1.8).

4. Rosser's Sieve

The proof of Lemma 1 will depend on the following properties of the functions $h^\pm$.

Lemma 2. *Under the hypotheses of Lemma 1 the functions $h^\pm$ given in (1.8) satisfy the inequalities*

$$h^\pm(s) = O(e^{-s \log s}) \quad \text{as} \quad s \to \infty, \tag{1.11}$$

$$h^\pm(s) > 0 \quad \text{if} \quad s > 0. \tag{1.12}$$

Also, they satisfy the recursion

$$\frac{1}{s} \int_s^\infty h^{(-)^{n-1}}(t-1) \kappa t^\kappa \, dt = s^\kappa h^{(-)^n}(s) \quad \text{if} \quad s \geq \beta + \varepsilon_n. \tag{1.13}$$

Lemma 2 will be obtained using the techniques developed in Sect. 4.2 in connection with the functions F, f. In this instance we infer the behaviour of $h^\pm(s)$ for large s from that for small s, rather than the other way round.

When the bounds (1.11) are known, the recursion (1.13) follows immediately on integrating (1.2) and (1.3).

Proof of Lemma 1. This follows quickly once Lemma 2 is established. Proceed by induction on n. The equations (1.1) and (4.1.3.1) show that $g_1(s)$ is bounded and supported in $s \leq \beta + 1$. Then we can choose μ sufficiently large so that (1.10) holds when $n = 1$, since $H(s) > 0$ in (1.4).

Suppose now that (1.10) holds when $n = N - 1$. When $n \geq 2$ the definitions (1.1) and (4.1.3.2) induce

$$g_n(s) = \frac{1}{s^\kappa} \int_s^\infty g_{n-1}(t-1) \, dt^\kappa \quad \text{if} \quad s \geq \beta + \varepsilon_n$$
$$g_n(s) = g_n(\beta + 1) + f_1(s) \quad \text{if} \quad 2 \nmid n, \ \beta - 1 < s \leq \beta + 1, \tag{1.14}$$

wherein $g_n(s) \geq 0$. So the induction will proceed if we secure

$$\int_s^\infty h^{(-)^{n-1}}(t-1) \, dt^\kappa \leq s^\kappa h^{(-)^n}(s) \quad \text{if} \quad s \geq \beta + \varepsilon_n, \tag{1.15}$$

$$\mu(\beta + 1)^\kappa H(\beta + 1) + s^\kappa f_1(s) \leq \mu s^\kappa H(s) \quad \text{if} \quad \beta - 1 \leq s \leq \beta + 1. \tag{1.16}$$

Because of (1.4) and (4.1.3.1) the inequality (1.16) follows provided

$$\frac{\mu D}{\beta + 1} + (\beta + 1)^\kappa - s^\kappa \leq \frac{\mu D}{s} \quad \text{if} \quad \beta - 1 \leq s < \beta + 1.$$

This says

$$\frac{(\beta + 1)^\kappa - s^\kappa}{\beta + 1 - s} \leq \frac{\mu D}{s(\beta + 1)} \quad \text{if} \quad \beta - 1 \leq s < \beta + 1.$$

Thus (1.16) follows provided $\kappa(\beta+1)^{\kappa-1} \leq \mu D/(\beta^2-1)$, which holds if μ is large enough.

It remains to check that (1.15) follows from (1.2) and (1.3). If $s \geq \beta + \varepsilon_n$ we obtain, after using (1.11),

$$\int_s^\infty h^{(-)^{n-1}}(t-1)\,dt^\kappa = \int_s^\infty h^{(-)^{n-1}}(t-1)\kappa t^{\kappa-1}\,dt$$

$$\leq \frac{1}{s}\int_s^\infty h^{(-)^{n-1}}(t-1)\kappa t^\kappa\,dt = s^\kappa h^{(-)^n}(s)\,, \quad (1.17)$$

which is (1.15). Thus Lemma 1 will follow from Lemma 2.

4.3.2 Adjoints and Inner Products

As in Sect. 4.2.1 we pass from $H(s)$, $h(s)$ to the functions

$$U(s) = H(s) + h(s)\,, \quad V(s) = H(s) - h(s) \quad \text{if} \quad s \geq \beta\,. \quad (2.1)$$

Because of (1.2) – (1.7) these satisfy

$$\left.\begin{aligned}\frac{d}{ds}\left(s^{\kappa+1}U(s)\right) &= -\kappa s^\kappa U(s-1) \\ \frac{d}{ds}\left(s^{\kappa+1}V(s)\right) &= \kappa s^\kappa V(s-1)\end{aligned}\right\} \quad \text{if} \quad s > \beta+1\,, \quad (2.2)$$

$$\left.\begin{aligned}s^{\kappa+1}U(s) &= D + E - D\int_\beta^s \frac{\kappa t^\kappa}{(t-1)^{\kappa+1}}\,dt \\ s^{\kappa+1}V(s) &= D - E + D\int_\beta^s \frac{\kappa t^\kappa}{(t-1)^{\kappa+1}}\,dt\end{aligned}\right\} \quad \text{if} \quad \beta \leq s \leq \beta+1\,. \quad (2.3)$$

Continue $U(s)$, $V(s)$ into $\beta - 1 < s < \beta$ by requiring that (2.2) holds if $\beta \leq s \leq \beta+1$. Then from (2.3) we have

$$U(s) = V(s) = \frac{D}{s^{\kappa+1}} \quad \text{if} \quad \beta-1 < s < \beta\,, \quad (2.4)$$

and U, V are not necessarily continuous at β.

Lemma 2 will be deduced, using (2.1), from parts (i) and (ii) of Lemma 3.

Lemma 3. *Under the conditions of Lemma 1 we obtain*

(i) $\quad U(s) = O(e^{-s\log s})\,, \quad V(s) = O(e^{-s\log s}) \quad \text{as} \quad s \to \infty\,,$

not uniformly in β

(ii) *there exists a constant η with $0 < \eta < 1$ such that*

$$|V(s)| < \eta U(s) \quad \text{if} \quad s \geq \beta$$

(iii) $\quad U(s-1) \ll s^2 U(s) \quad \text{when} \quad s > \beta\,.$

Before we prove part (ii) of Lemma 3 there is a very special case that we will need to consider first. As is evident from (1.6) and (1.7), it is actually the case $h(\beta+1) > 0$ of Lemma 2.

Lemma 4. When $\kappa > \frac{1}{2}$, let $\beta-1$ be the largest zero of q, as in Lemma 4.2.5. Then
$$\int_\beta^{\beta+1} \frac{\kappa t^\kappa}{(t-1)^{\kappa+1}} \, dt < \left(\frac{\beta}{\beta-1}\right)^\kappa. \tag{2.5}$$

Proof of Lemma 2. To deduce Lemma 2 from Lemma 3, note that parts (i) and (ii) give
$$0 < U(s) \pm V(s) < 2U(s) \ll e^{-s \log s} \quad \text{when} \quad s \geq \beta + \varepsilon_n,$$

Now (1.11) and (1.12) follow from (2.1), because when $s \leq \beta$ the inequality (1.12) is a consequence of (1.4)–(1.6). The treatment of Lemma 2 will then be complete, the recursion (1.13) having already been discussed at (1.17).

Proof of Lemma 3(i). The equations (2.2) and (2.4) give
$$\left.\begin{array}{l} sU'(s) = -(\kappa+1)U(s) - \kappa U(s-1) \\ sV'(s) = -(\kappa+1)V(s) + \kappa V(s-1) \end{array}\right\} \quad \text{if} \quad s > \beta,\ s \neq \beta+1,$$

which are the cases $(a,b) = (\kappa+1, \kappa)$ and $(\kappa+1, -\kappa)$ of (4.2.2.1). Thus the adjoint equations (4.2.2.2) are, respectively,
$$\frac{d}{ds}(s u(s)) = (\kappa+1) u(s) + \kappa u(s+1), \tag{2.6}$$
$$\frac{d}{ds}(s v(s)) = (\kappa+1) v(s) - \kappa v(s+1). \tag{2.7}$$

The standard solutions of (2.6) and (2.7) are $u(s) = r_{\kappa+1,\kappa}(s)$, $v(s) = 1$, in the notation of Sect. 4.2.3. Thus (4.2.3.18) gives $u'(s) = 2\kappa r_{\kappa,\kappa}(s) = 2\kappa q(s)$, where $q(s)$ is as in (4.2.3.17). Note that (4.2.3.13) gives
$$u(s) \sim s^{2\kappa-1}, \quad u'(s) \sim 2\kappa s^{2\kappa-1} \quad \text{as} \quad s \to \infty. \tag{2.8}$$

Since $\beta-1$ is the greatest zero of $q(s) = r_{\kappa,\kappa}(s) = u'(s)/2\kappa$, the function u satisfies
$$u'(\beta-1) = 0,\ u'(s) > 0 \quad \text{if} \quad s > \beta-1. \tag{2.9}$$

Then (2.6) gives
$$u(\beta) + u(\beta-1) = 0, \tag{2.10}$$

and if $s > \beta-1$ we find $0 < su'(s)/\kappa = u(s) + u(s+1) < 2u(s+1)$. Thus
$$u(s) > 0 \quad \text{if} \quad s > \beta. \tag{2.11}$$

Because of (2.3), Lemma 4.2.1(ii) applies with $C = D$, $a = \kappa + 1$, $R = U$, $r = u$, and gives

$$\langle U, u \rangle(s) = \langle U, u \rangle(\beta) = \frac{E\,u(\beta)}{\beta^\kappa} + \frac{D\,u(\beta-1)}{(\beta-1)^\kappa} = 0 \quad \text{if} \quad s \geq \beta + 1, \quad (2.12)$$

after use of (2.10) and the choice (2.7) of D and E. In a similar way

$$\langle V, v \rangle(\beta) = -\frac{E}{\beta^\kappa} + \frac{D}{(\beta-1)^\kappa} = 0,$$

so that

$$\left.\begin{array}{l} s\,u(s)U(s+) = \kappa \displaystyle\int_{s-1}^{s} u(x+1)U(x)\,dx \\[1em] s\,V(s) = -\kappa \displaystyle\int_{s-1}^{s} V(x)\,dx \end{array}\right\} \quad \text{if} \quad s \geq \beta+1 \text{ or } s = \beta. \quad (2.13)$$

Because of (2.8) and (2.13) there exists M such that

$$s\,U(s) < M \int_{s-1}^{s} |U(x)|\,dx, \quad s\,V(s) < M \int_{s-1}^{s} |V(x)|\,dx\,,$$

and part (i) of Lemma 3 follows by appealing to Lemma 4.2.7, in the fashion previously used to derive Lemma 4.2.6.

Proof of Lemma 4. Because of (2.4) the equations (2.13) give

$$\beta\,u(\beta)U(\beta+) = D\kappa \int_{\beta}^{\beta+1} \frac{u(t)}{(t-1)^{\kappa+1}}\,dt\,,$$

$$\beta\,V(\beta+) = -D\kappa \int_{\beta}^{\beta+1} \frac{dt}{(t-1)^{\kappa+1}}\,.$$

Use this information in the form

$$\kappa \int_{\beta}^{\beta+1} \frac{u(t) + u(\beta)}{(t-1)^{\kappa+1}}\,dt = \frac{\beta\,u(\beta)}{D}(U(\beta+) - V(\beta+))$$

$$= \frac{2E\beta\,u(\beta)}{D\beta^{\kappa+1}} = \frac{2\,u(\beta)}{D}, \quad (2.14)$$

obtained using (2.3) and (1.6).

The connection with (2.5) is that because of (2.6) the expression

$$\delta(t) = \frac{\beta^\kappa}{2\,u(\beta)} \frac{u(t) + u(\beta)}{t^\kappa}$$

has

$$\delta'(t) = \frac{\beta^\kappa}{2\,u(\beta)} \frac{t\,u'(t) - \kappa(u(t) + u(\beta))}{t^{\kappa+1}} = \frac{\kappa\beta^\kappa}{2\,u(\beta)}(u(t+1) - u(\beta))\,,$$

so that $\delta(t) > \delta(\beta) = 1$ when $t > \beta$ because of (2.9). Thus (2.14) gives

$$\int_\beta^{\beta+1} \frac{\kappa t^\kappa}{(t-1)^\kappa} \, dt < \int_\beta^{\beta+1} \delta(t) \frac{\kappa t^\kappa}{(t-1)^\kappa} \, dt$$

$$= \frac{\beta^\kappa}{2\,u(\beta)} \cdot \frac{2\,u(\beta)}{D} = \left(\frac{\beta}{\beta-1}\right)^\kappa,$$

as required in Lemma 4, because D is as in (1.6).

Proof of Lemma 3(ii). Set

$$\gamma = \left(\frac{\beta}{\beta-1}\right)^\kappa - \int_\beta^{\beta+1} \frac{\kappa t^\kappa}{(t-1)^{\kappa+1}} \, dt \, ,$$

so that $0 < \gamma < 1$, the inequality $\gamma < 1$ following by replacing t^κ by β^κ in the integral, while $\gamma > 0$ because of Lemma 4. Since $E/D = \left(\beta/(\beta-1)\right)^\kappa$ because of (1.6), the equations (2.3) give

$$\frac{(\beta-1)^\kappa - \beta^\kappa}{(\beta-1)^\kappa + \beta^\kappa} = \frac{D+E}{D-E} \leq \frac{V(s)}{U(s)} \leq \frac{1-\gamma}{1+\gamma} \quad \text{if} \quad \beta \leq s \leq \beta+1 \, ,$$

so that for a certain η with $0 < \eta < 1$ we obtain

$$|V(s)| < \eta U(s) \quad \text{if} \quad \beta \leq s \leq \beta+1 \, . \tag{2.15}$$

We show that (2.15) holds for all $s \geq \beta$ with the same constant η. For if not then there would exist $s > \beta$ so that $|V(s)| = \eta U(s)$ but $|V(t)| < \eta U(t)$ whenever $\beta < t < s$. Then using (2.9) and (2.13) we find

$$s|V(s)| \leq \kappa \int_{s-1}^s |V(t)| \, dt < \eta\kappa \int_{s-1}^s U(t) \, dt$$

$$< \frac{\eta\kappa}{u(s)} \int_{s-1}^s U(t) u(t+1) \, dt = \eta s\, U(s) \, ,$$

a contradiction. This completes the proof of Lemma 3(ii), so that the proofs of Lemmas 1 and 2, and hence of Proposition 4.2.1 in the case $\kappa > \frac{1}{2}$, are complete.

Proof of Lemma 3(iii). This is not required for Lemmas 1 or 2, but will be needed for Lemma 4.4.3. From part (ii) we infer $U(s) > 0$ if $s > \beta$, so (2.2) and (2.4) show that $s^{\kappa+1}U(s)$, and hence $U(s)$, are decreasing in $s > \beta - 1$. Also $u(s)$ increases in $s > \beta - 1$, from (2.9). We observed $\langle U, u \rangle(s) = 0$ if $s \geq \beta + 1$ in (2.12), and $u(s) > 0$ if $s > \beta$ in (2.11). Hence if $s \geq \beta + 1$

$$s\,u(s)U(s) = \kappa \int_{s-1}^s u(x+1)U(x) \, dx$$

$$> \kappa\,u(s) \int_{s-\frac{1}{2}}^s U(x) \, dx \geq \tfrac{1}{2}\kappa\,u(s)U\left(s - \tfrac{1}{2}\right) \, ,$$

the last step following because $U(s)$ decreases. This shows $U(s-\tfrac{1}{2}) \ll s\,U(s)$, whence $U(s-1) \ll s^2 U(s)$ as required. This completes the proof of Lemma 3.

4.3.3 The Case $\kappa \leq \frac{1}{2}$

The construction of the auxiliary functions h and H in the preceding sections is valid only when $\kappa > \frac{1}{2}$ because (in common with much of the previous material in this chapter) it depended on the associated property $\beta > 1$. Our treatment of Proposition 4.2.1 when $\kappa \leq \frac{1}{2}$ depends on Lemma 6 below, which associates the functions g_n, constructed with this κ, with the auxiliary functions h and H constructed using a certain $\kappa_0 > \frac{1}{2}$. The parameter β obtained from κ_0 via Lemma 4.2.5 will be denoted β_0, so that $\beta_0 > 1$.

Lemma 5 provides an extension of the inequality (1.15) occurring in the proof of Lemma 1. Note from (1.4) that $H(s) = D/s^{\kappa_0+1}$ near $s = 0$. Thus in Lemma 5 one effect of the factor $(1 - 1/t)^{\kappa_0+\delta}$ that has been introduced is to secure convergence of the integral (3.1) near $t = 1$.

Lemma 6 will extend Lemma 1 to the context $0 < \kappa \leq \frac{1}{2}$, thereby completing the treatment of Proposition 4.2.1. The details are more involved than in Sect. 4.3.2, but the general structure of the argument is similar.

Lemma 5. *Fix $\delta > 0$, and let $h^+ = h$, $h^- = H$ be as in (1.8), but constructed with κ set equal to $\kappa_0 > \frac{1}{2}$. Then*

$$s^{\kappa_0+1} h^{(-)^n}(s) > \frac{1}{2} \int_{1+\varepsilon_n}^{\infty} \left(\frac{t-1}{t}\right)^{\kappa_0+\delta} t^{\kappa_0} h^{(-)^{n-1}}(t-1)\, dt, \quad (3.1)$$

provided κ_0 is chosen sufficiently close to $\frac{1}{2}$.

Recall that the parameter $\beta_0 > 1$ was specified as in Lemma 4.2.5, which gave

$$\beta_0 \to 1 \quad \text{as} \quad \kappa_0 \to \tfrac{1}{2}+, \quad (3.2)$$

actually $\beta_0 \sim 1 + \pi e^{\gamma}(2\kappa_0 - 1)^2$. In particular we will suppose $\frac{1}{2} < \kappa_0 < 1$ and $1 < \beta_0 < 2$ throughout the ensuing argument.

With (1.4) and (1.5) the recurrences in Lemma 2 give

$$s^{\kappa_0+1} h^{(-)^n}(s) = \kappa_0 \int_{t>s;\, t>\beta_0+\varepsilon_n} t^{\kappa_0} h^{(-)^{n-1}}(t-1)\, dt, \quad (3.3)$$

from which (3.1) follows if $s \geq \beta_0 + \varepsilon_n$. Since the expressions on the left sides are the constants D and E appearing in (1.3) and (1.4) it is now necessary only to obtain (3.1) when $s = 1 + \varepsilon_n$. Split the integral in (3.1) at the point $\beta_0 + \varepsilon$ and use (3.3). In this way Lemma 5 will follow provided $I_1(h) \leq J_1(h)$, $I_0(H) \leq J_0(H)$, where for $k = h = h^+$ or $k = H = h^-$ we denote

$$I_{\varepsilon_n}(k) = \int_{1+\varepsilon_n}^{\beta_0+\varepsilon_n} \left(\frac{t-1}{t}\right)^{\kappa_0+\delta} t^{\kappa_0} k(t-1)\, dt, \quad (3.4)$$

$$J_{\varepsilon_n}(k) = \int_{\beta_0+\varepsilon_n}^{\infty} \left(1 - \left(\frac{t-1}{t}\right)^{\kappa_0+\delta}\right) t^{\kappa_0} k(t-1)\, dt. \quad (3.5)$$

To establish $I_0(H) \leq J_0(H)$, note that (1.4) gives $H(s) = D/s^{\kappa_0+1}$ when $s \leq \beta_0 + 1$, actually with $D = (\beta_0 - 1)^{\kappa_0}$ as in (1.6). Then

$$I_0(H) = \int_1^{\beta_0} \frac{D}{t^\delta (t-1)^{1-\delta}} \, dt \leq \frac{D(\beta_0 - 1)^\delta}{\delta}.$$

Meantime

$$J_0(H) > \int_{\beta_0}^{\beta_0+1} \left(1 - \left(\frac{\beta_0}{\beta_0+1}\right)^{\kappa_0+\delta}\right) \beta_0^{\kappa_0} \frac{D}{(t-1)^{\kappa_0+1}} \, dt$$

$$= \left(1 - \left(\frac{\beta_0}{\beta_0+1}\right)^{\kappa_0+\delta}\right) \frac{D\beta_0^{\kappa_0}}{\kappa_0} \left(\frac{1}{(\beta_0-1)^{\kappa_0}} - \frac{1}{\beta_0^{\kappa_0}}\right).$$

Hence $I_0(H) < J_0(H)$ if κ_0 is sufficiently close to $\frac{1}{2}$ because $\beta_0 \to 1$ as in (3.2).

This argument depended on the values of $h^-(s) = H(s)$ for $s \leq \beta_0 + 1$. For $h^+(s) = h(s)$ the related observation from (1.2), (1.4) is

$$\int_{\beta_0+1}^\infty \kappa_0 t^{\kappa_0} h(t-1) \, dt = (\beta_0+1)^{\kappa_0+1} H(\beta_0+1) = D = (\beta_0-1)^{\kappa_0},$$

where the choice (1.6) of D will now be relevant to the argument. We will show slightly more, that

$$\int_{\beta_0+1}^{\beta_0+2} t^{\kappa_0} h(t-1) \, dt > c_1 (\beta_0 - 1)^{\kappa_0}, \tag{3.6}$$

for some absolute constant $c_1 > 0$ independent of β_0. Lemma 2 and (1.3) show that $h(s)$ decreases, so for (3.6) to follow it will suffice to show

$$h(\beta_0 + \tfrac{1}{2}) > c_2 (\beta_0 - 1)^{\kappa_0},$$

for some absolute constant $c_2 > 0$.

Consider the equation (1.7) with $s = \beta_0 + \frac{1}{2}$ and $s = \beta_0 + 1$. Take the difference of the two results. Since $h(\beta_0 + 1) > 0$ (again from Lemma 2) this gives

$$(\beta_0 + \tfrac{1}{2})^{\kappa_0+1} h(\beta_0 + \tfrac{1}{2}) > D \int_{\beta_0+\frac{1}{2}}^{\beta_0+1} \frac{\kappa_0 t^{\kappa_0}}{(t-1)^{\kappa_0+1}} \, dt > \frac{D\kappa_0 (\beta_0 + \tfrac{1}{2})^{\kappa_0}}{2\beta_0^{\kappa_0+1}}.$$

This gives (3.6), because $D = (\beta_0 - 1)^{\kappa_0}$ from (1.6).

It is now straightforward to verify that $I_1(h) < J_1(h)$ in (3.4) and (3.5). Using (1.5) and (1.6) it follows that

$$I_1(h) < \beta_0^{\kappa_0} \int_2^{\beta_0+1} \frac{dt}{t^\delta (t-1)^{1-\delta}} < c_3 (\beta_0 - 1).$$

4.3 The Convergence Problem

When $t < \beta_0 + 1$ and $\beta_0 < 2$ we have $1 - (1 - 1/t)^{\kappa_0 + \delta} > c_4$. Then (3.5) and the inequality (3.6) give

$$J_1(h) > c_2 c_4 (\beta_0 - 1)^{\kappa_0} > c_2 c_4 (\beta_0 - 1)^{1/2}.$$

Because of (3.2) the required inequality $I_1(h) < J_1(h)$ follows if κ_0 is sufficiently close to $\frac{1}{2}$. This completes the proof of Lemma 5.

Proceed now to the extension of the "convergence" Lemma 1 to the situation $0 < \kappa \le \frac{1}{2}$. An additional factor s appears in the bounds enunciated in the next lemma. In Lemma 6, g_n is as in (1.1), where the functions f_n are given by (4.1.3.1) and (4.1.3.2) with $\beta = 1$.

Lemma 6. *Let $h^{\pm}$ be as in Lemma 5, using the parameter $\kappa_0 > \frac{1}{2}$. Suppose $0 < \kappa \le \frac{1}{2}$, and take $\beta = 1$. Then there is a constant $\mu > 0$ so that*

$$g_n(s) \le \mu s\, h^{(-)^n}(s) \quad \text{if} \quad s > \beta - \varepsilon_n.$$

As with Lemma 1, proceed by induction on n. If μ is chosen sufficiently large then the result holds when $n = 1$. Suppose now that the result holds when $n = N - 1$. If $s \ge 1 + \varepsilon_n$ then the recurrences (1.14) for g_n give

$$s^{\kappa} g_N(s) \le \tfrac{1}{2} \mu \int_s^{\infty} (t-1) h^{(-)^{N-1}}(t-1) t^{\kappa - 1}\, dt,$$

in which $\kappa = \frac{1}{2}$. After multiplication by $(t/(t-1))^{1-\kappa_0 - \delta}$ and by $(t/s)^{\kappa_0 - \kappa}$ this gives

$$s^{\kappa} g_N(s) \le \tfrac{1}{2} \mu \int_s^{\infty} \left(\frac{t-1}{t}\right)^{\kappa_0 + \delta} h^{(-)^{n-1}}(t-1) \frac{t^{\kappa_0}}{s^{\kappa_0 - \kappa}}\, dt.$$

Now Lemma 1 gives $g_N(s) < s\, h^{(-)^{n-1}}(s)$, so that the induction proceeds.

If N is odd and $0 < s < 2$ then the definitions (1.1), (4.1.3.1), (4.1.3.2) now lead to

$$s^{\kappa} g_N(s) = 2^{\kappa} g_N(2) + 2^{\kappa} - s^{\kappa}$$
$$\le 2^{\kappa + 1} \mu H(2) + 2^{\kappa} - s^{\kappa}.$$

This is required not to exceed $s^{\kappa + 1} \mu H(s)$ if $1 \le s \le 2$. So the induction will proceed provided

$$2^{\kappa} - s^{\kappa} \le \mu \left(\frac{s^{\kappa_0 + 1} H(s)}{s^{\kappa_0 - \kappa}} - \frac{2^{\kappa_0 + 1} H(2)}{2^{\kappa_0 - \kappa}} \right) \quad \text{if} \quad 1 \le s \le 2,$$

wherein $\kappa_0 > \kappa = \frac{1}{2}$. Since $s^{\kappa_0 + 1} H(s)$ is constant in $1 \le s \le 2$ this requirement says

$$2^{\kappa_0 + 1} \mu H(2) \ge (2s)^{\kappa_0 - \kappa} \left(\frac{2^{\kappa} - s^{\kappa}}{2^{\kappa_0 - \kappa} - s^{\kappa_0 - \kappa}} \right) \quad \text{if} \quad 1 \le s \le 2. \tag{3.7}$$

By Cauchy's Mean Value Theorem the expression on the right does not exceed

$$4^{\kappa_0-\kappa} \max_{1\leq s\leq 2} \frac{\kappa s^\kappa}{(\kappa_0-\kappa)s^{\kappa_0-\kappa}} \leq 4^{\kappa_0-\kappa} \frac{\kappa}{\kappa_0-\kappa} \max_{1\leq s\leq 2}\left\{1, s^{2\kappa-\kappa_0}\right\},$$

so (3.7) holds provided μ is large enough. This completes the proof of Lemma 6, so that the proof of Proposition 4.2.1 is now complete.

4.4 A Sieve Theorem Following Rosser

In this section we improve the results of Sect. 3.3 to theorems of the type that is reported to have first appeared (for the case $\kappa = 1$) in an unpublished manuscript of J. B. Rosser. Our theorems are of the type described, generically, by Proposition 1 below. We will need to assume the sifting density hypothesis in the stronger form (1.3.5.3), which is more precise than the weaker version that sufficed for the arguments used in Chap. 3.

Proposition 1. *Suppose that ρ has sifting density $\kappa > 0$. Write $z = D^{1/s}$, as usual, where $s > 0$. Then, in the language of Theorem 1.3.1, there exists a pair of sifting functions $\lambda_D^\pm$ such that $\left|\lambda^\pm(d)\right| \leq 1$ and*

$$V^+(D, P(z)) \leq V(P(z))\{F(s) + O(\eta(D,s))\}$$
$$V^-(D, P(z)) \geq V(P(z))\{f(s) + O(\eta(D,s))\},$$

where the functions F, f are as in Sect. 4.2. The term $\eta(D,s)$ is to be described, but will satisfy $\eta(D,s) \to 0$ as $s \to \infty$.

In the earlier Theorem 3.3.1 of this type the number 1 appeared in place of $F(s)$ and $f(s)$, and the expression for $\eta(D,s)$ was, for the smaller values of s, a correspondingly weak one. Theorems 1 and 2 provide estimates for which $\eta(D,s) \to 0$ as $D \to \infty$, for each s. In these theorems, the construction of the functions $\lambda^\pm$ is of the combinatorial type described in Lemma 3.1.3, but will not be exactly as given in (4.1.1.1) and (4.1.1.2).

The enunciation in the range $0 < s < 1$ is no more than a convenient formal device. The lower bound is trivial in this range, since $f(s) = 0$ when $s < \beta$. The upper bound follows from that when $s = 1$ because the sifting density property (1.3.5.2) gives

$$V(P(D))F(1) \leq V(P(D^{1/s}))\left(\frac{\log D^{1/s}}{\log D}\right)^\kappa F(1) = V(P(D^{1/s}))F(s),$$

since $F(s)$ is as in (4.2.1.3).

We will establish various versions of Proposition 1 in the theorems that follow. Of these, the simplest is Theorem 1, which does not derive a particularly sharp estimate for the error term η. In the first instance our treatments of Theorems 1 and 2 assume $\kappa > \frac{1}{2}$.

4.4 A Sieve Theorem Following Rosser

The case $\kappa = \frac{1}{2}$ is however of considerable interest, arising for example in problems involving sums of two squares in the manner indicated in Sect. 1.2.5. Some details of applications of this nature are given in the notes. It is therefore of some importance that the theorems should be extended to cover the case $\kappa = \frac{1}{2}$.

One way of dealing with the case $\kappa = \frac{1}{2}$ is to use the result with a certain $\kappa > \frac{1}{2}$, in the style suggested (for rather different reasons) in Sect. 1.3.5. If this is done, then we can allow κ to tend to $\frac{1}{2}+$ afterwards. Then Lemma 4.2.5 shows that the sifting limit β (such that $f(s) > 0$ when $s > \beta$) then tends to 1 and that the constant A appearing in the expression (4.2.1.3) of $F(s)$ tends to $2\sqrt{\pi/e^\gamma}$. It is more satisfactory, however, to extend the validity of Theorem 1 directly to the case $\kappa = \frac{1}{2}$ and, indeed, to the case $0 < \kappa \leq \frac{1}{2}$. This is done essentially by allowing κ to tend to $\frac{1}{2}+$ (so that $\beta \to 1$) in the estimates obtained for the auxiliary functions H, h of Sect. 4.3.

The definitive result in this section is Theorem 2, which gives a much better estimate for the error term η, and does so for all $\kappa > 0$. The reader who prefers to proceed directly to Theorem 2 need not take much notice of the easier but weaker Theorem 1.

4.4.1 The Case $\kappa > \frac{1}{2}$: a First Result

The case of Proposition 1 that follows fairly directly using the discussions in Sects. 4.1 and 4.3 is Theorem 1 below. Initially we assume $\kappa > \frac{1}{2}$, so that $\beta > 1$ from Lemma 4.2.5. This inequality is required so that Lemma 4.1.3 can be used. This procedure leads to an estimate for the error term $\eta(D, s)$ that is not particularly sharp.

Theorem 1. *Let $\varepsilon = \varepsilon_D > 0$ be sufficiently small. Suppose*

$$\log D > L, \qquad \log D > e^{4\kappa\varepsilon^{-1}\log^2\varepsilon}, \qquad (1.1)$$

where $L \geq 1$ is the constant in the sifting density hypothesis (1.3.5.3). Then Proposition 1 holds with

$$\eta(D, s) = O(\varepsilon L^{11}), \qquad (1.2)$$

where the implied constant may depend upon κ.

The relation between D and ε_D in Theorem 1 has the following consequence.

Corollary 1.1. *In Theorem 1, we may take*

$$\varepsilon = \varepsilon_D = \frac{(\log\log\log D)^3}{\log\log D}$$

if $D > e^L$ and $D > D_0(\kappa)$, where $D_0(\kappa)$ is sufficiently large.

With ε, D as stated we obtain $4\kappa\varepsilon^{-1}\log^2\varepsilon^{-1} < \log\log D$ if D is large enough, so that Corollary 1.1 follows from Theorem 1.

The dependence of ε_D on D in Corollary 1.1 is not particularly strong. On the other hand in some situations, and in particular the in application to Theorem 6.1.1, Corollary 1.1 already gives as much as might be desired.

Here and elsewhere in this chapter we will use a composition of two sieves as described in Sect. 1.3.7. Set

$$D_1 = D^\varepsilon, \quad D_2 = D^{1-\varepsilon}, \quad w = D^{1/v}, \quad P_1 = P(w), \quad P_2 = P(z)/P_1, \quad (1.3)$$

where $0 < \varepsilon < \frac{1}{2}$ and $v > 1$, to be specified. The first preliminary sieving will rest on the "Fundamental Lemma" described in Sect. 3.3, using the product P_1 of the smallest primes from P and the comparatively small level of distribution D_1. The second sieve will be constructed using the principles of this chapter, using the larger level of distribution $D_2 = D^{1-\varepsilon}$ and the product P_2 of the remaining primes from $P(z)$.

The following statement on such a composition of sieves follows easily from the Lemma of Sect. 1.3.7.

Lemma 1. *Suppose we have constructed sieves so that*

$$V(P_1)(1+\zeta_1) \geq V^+(D_1,P_1) \geq V^-(D_1,P_1) \geq V(P_1)(1-\zeta_1) > 0$$
$$V(P_1)(F+\zeta_2) \geq V^+(D_2,P_2) \geq V^-(D_2,P_2) \geq V(P_2)(f-\zeta_2) > 0.$$

Then the composite sieves described in Lemma 1.3.7 satisfy

$$V^+(D,P) \leq V(P)(F+\zeta_2)(1+\zeta_1),$$
$$V^-(D,P) \geq V(P)\big((f-\zeta_2)(1-\zeta_1) - 2\zeta_1(F-f+2\zeta_2)\big). \quad (1.4)$$

For V^+ the result is immediate, because Lemma 1.3.7 gave

$$V^+(D,P) = V^+(D_1,P_1)V^+(D_2,P_2).$$

For V^- arrange the expression given in Lemma 1.3.5 as

$$V^-(D,P) = V^-(D_2,P_2)V^-(D_1,P_1)$$
$$- \big(V^+(D_2,P_2) - V^-(D_2,P_2)\big)\big(V^+(D_1,P_1) - V^-(D_1,P_1)\big).$$

Then (1.4) follows.

In Lemma 1, the requirements $1 - \zeta_1 > 0$, $f - \zeta_2 > 0$ ask that the lemma is used only with non-trivial lower estimates. The requirements $V^+ \geq V^-$ are satisfied by the combinatorial sieves used in this chapter, because the quantities $T^\pm$ first appearing in Lemma 3.1.4 are non-negative.

4.4 A Sieve Theorem Following Rosser

Notice also that a sharp estimate for $V^+(D_2, P_2)$ is not crucial for the procedure described here, because it appears only when multiplied by a quantity bounded by a small amount $2\zeta_1 V(P)$.

Theorem 1 when $\kappa > \frac{1}{2}$. The proof is along the indicated lines, arranging the procedure so that Lemma 4.1.3 is not used with a value of n so large that the estimate (4.1.3.6) for Δ_n becomes unacceptable.

For the preliminary sieve, with level D_1, Corollary 3.3.1.1 gives

$$V^\pm(D_1, P_1) = V(P_1)\bigl(1 + O(e^{-s_1}L^{10})\bigr), \quad (1.5)$$

where $s_1 = \log D_1 / \log w = v\varepsilon$, from (1.3). Here we used the inequality $K \ll L$; actually (1.3.5.3) gives $K < 3L$ since now $L \geq 1$. Make $e^{-s_1} = O(\varepsilon)$, by choosing

$$v = \varepsilon^{-1} \log \varepsilon^{-1} . \quad (1.6)$$

In the main sieve of level D_2 the procedure was described in Sect. 4.1. The sum $T_n(D_2, P_2)$ described by (4.1.1.4) will vanish unless

$$p_1 p_2 \cdots p_{n-3} p_{n-2}^{\beta+1} \leq D_2 .$$

Here $p_i | P_2$, so $p_i > w = D^{1/v}$ as in (1.3). This gives $D > D_2 \geq w^{n-2+\beta}$ and

$$n \leq v + 2 - \beta \leq v + 1 . \quad (1.7)$$

Lemma 4.1.3 now gives that the sums $V^\pm(D_2, P_2)$ given in (4.1.1.1) and (4.1.1.2) satisfy

$$V^+(D_2, P_2) \leq V(P_2)\bigl(F(s_2) + \delta\bigr), \quad V^-(D_2, P_2) \geq V(P_2)\bigl(f(s_2) - \delta\bigr), \quad (1.8)$$

where

$$\delta = \sum_{n \leq v+2-\beta} (K-1) K^n \Delta_n . \quad (1.9)$$

Here $s_2 = \log D_2 / \log z$, and the sums $\sum f_n(s_2)$ which arise have been estimated from above by the series appearing in the expressions (4.2.1.6) for $F(s_2)$ and $f(s_2)$.

The distinction between s_2 and s will be unimportant because

$$F(s_2) = F(s) + O(\varepsilon), \quad f(s_2) = f(s) + O(\varepsilon) . \quad (1.10)$$

In fact $s_2 = (1 - \varepsilon)s$, so

$$F(s) - F(s_2) = O\bigl(|s - s_2||F'(s_2)|\bigr) = O(\varepsilon) , \quad (1.11)$$

because (4.2.1.1)–(4.2.1.4) show that $sF'(s) = O(1)$. The corresponding inequalities for f are similar.

For the sum in (1.9), the estimate (4.1.3.6) in Lemma 4.1.3 gives

$$\delta < (K-1)K^{v+1}v^2 \left(\frac{v+2}{\beta-1}\right)^{\kappa(v+1)} . \quad (1.12)$$

Hence use of (1.3.5.3), (1.1) and (1.3) give

$$K \le 1 + \frac{L}{\log w} = 1 + \frac{vL}{\log D} \le 1 + v. \qquad (1.13)$$

Since (1.6) implies $\varepsilon v > 1$ when $\varepsilon < 1/e$ this gives

$$\delta \le \frac{L\varepsilon}{\log D} \frac{v^2(v+2)^{(\kappa+1)(v+1)}}{(\beta-1)^{\kappa(v+1)}}.$$

If $\varepsilon > 0$ is sufficiently small (so that v is large) then use of (1.6) gives

$$\log \frac{\delta \log D}{L\varepsilon} < (\kappa+1)(v+1)\log(v+2) + 2\log v - \kappa(v+1)\log(\beta-1)$$
$$\le 4\kappa\varepsilon^{-1} \log^2 \varepsilon^{-1}, \qquad (1.14)$$

where we used $\kappa + 1 \le 3\kappa$. Thus $\delta \le \varepsilon L$ when (1.1) holds.

We may use Lemma 1 with $\zeta_1 = O(\varepsilon L^{10})$ following (1.5) and (1.6), and with $\zeta_2 = \varepsilon L$ because of (1.8). This gives Proposition 1 when $\kappa > \frac{1}{2}$, with $\eta(D, s)$ as stated in Theorem 1.

4.4.2 Theorem 1 when $\kappa \le \frac{1}{2}$

The treatment follows the same general lines as in Sect. 4.4.1. The only complication arises from the fact that the parameter β involved in the construction of the functions F, f now takes the value $\beta = 1$, so that the estimate (1.9) does not apply. We replace Lemma 4.1.3 by a suitable modification valid also in the case $\beta = 1$. Lemma 2 is exceedingly weak for small w, but it is planned to use it only for larger $w = D^{1/v}$ with v chosen rather as in Sect. 4.4.1.

The guiding principle behind Lemma 2 is that in Lemma 4.1.3 the entries $1/(\beta - 1)$, which become illegal when $\beta = 1$, would in that case arise only from the contribution from very small primes. However, we plan to use this portion of the argument in a context where all the primes concerned exceed a moderately large bound w, so that this illegality should be avoidable.

Lemma 2. *Suppose $\beta \ge 1$, $n \ge 1$ and $s > \beta - \varepsilon_n$, where ε_n is as in (4.1.1.7). Suppose also that all the primes in the product P satisfy $w \le p < z$, for a certain w with $2 \le w < z$. Let*

$$\beta^* = 1 + \frac{\log w}{\log z},$$

so that $\beta^* > 1$. Then

$$T_n(D, D^{1/s}) \le V(P(D^{1/s}))\left(f_n(s) + (K-1)K^n \Delta_n\right),$$

as in Lemma 4.1.3, provided Δ_n is replaced by

$$\Delta_n = n\left(\frac{1}{\beta^* - 1}\right)^{\kappa n}.$$

Once Lemma 2 is established, Theorem 1 now follows as in Sect. 4.4.1. The only modification is that the error term δ in (1.9) is now estimated via Lemma 2 as

$$\delta < (K-1)K^{v+1}v^2\left(\frac{1}{\beta^*-1}\right)^{\kappa(v+1)}$$

in place of (1.12). Since $\beta^* = 1 + \log w / \log z = 1 + 1/v$, this leads to the estimate (1.14), where the term involving $\log(\beta-1)$ is now to be omitted. Theorem 1 now follows by the same argument as before.

Proof of Lemma 2. Actually we prove

$$T_n(D, D^{1/s}) \leq V(P(D^{1/s})) \Big(h_n(s) + (K-1)K^n \Delta_n \Big), \qquad (2.1)$$

where

$$s^\kappa h_1(s) = \int_{\substack{s < t < \beta+1 \\ t < s/(\beta^*-1)}} dt^\kappa ,$$

$$s^\kappa h_n(s) = \int_{\substack{\max\{s, \beta+\varepsilon_n\} < t < \beta+n \\ t < s/(\beta^*-1)}} h_{n-1}(t-1) \, dt^\kappa \quad \text{when} \quad n > 1 .$$

Thus, since f_n is as in (4.1.3.2),

$$0 \leq h_n(s) \leq f_n(s) \quad \text{when} \quad s \geq \beta - \varepsilon_n . \qquad (2.2)$$

Also, the bound (4.1.3.4) can be replaced by

$$h_n(s) \leq \left(\frac{1}{\beta^*-1}\right)^{\kappa n}, \qquad (2.3)$$

as follows by an easy induction. For after ignoring the conditions $t < \beta + n$ we obtain

$$s^\kappa g_1(s) \leq \left(\frac{s}{\beta^*-1}\right)^\kappa ,$$

and, once (2.3) is known for $n \leq N-1$,

$$s^\kappa h_n(s) \leq \left(\frac{1}{\beta^*-1}\right)^{\kappa(N-1)} \left(\frac{s}{\beta^*-1}\right)^\kappa ,$$

which gives (2.3) with $n = N$.

Write $z = D^{1/s}$ as usual, so that $w = D^{(\beta^*-1)/s}$, and proceed as in Lemma 4.1.3. First

$$T_1(D, D^{1/s}) = \sum_{\substack{D^{1/(\beta+1)} \leq p < D^{1/s} \\ D^{(\beta^*-1)/s} \leq p}} \frac{\rho(p)}{p} V(P(p))$$

$$\leq V(P(D^{1/s})) \left(h_1(s) + (K-1)\left(\frac{1}{\beta^*-1}\right)^\kappa \right) ,$$

and (2.1) follows when $n = 1$.

If (2.1) holds with $n = N - 1$ then

$$T_n(D, D^{1/s}) \leq \sum_{\substack{y_N \leq p < z_N(s) \\ D^{(\beta^* - 1)/s} \leq p}} \frac{\rho(p)}{p} V(P(p)) \left(h_{N-1}\left(\frac{\log D}{\log p} - 1\right) \right.$$
$$\left. + (K - 1) K^{N-1} \Delta_{N-1} \right).$$

Using Lemma 4.1.2 a second time gives

$$\sum_{\substack{y_N \leq p < z_N(s) \\ D^{(\beta^* - 1)/s} \leq p}} \frac{\rho(p)}{p} V(P(p)) h_{N-1}\left(\frac{\log D}{\log p} - 1\right)$$
$$\leq V\left(P(D^{1/s})\right) \left(h_N(s) + (K - 1)\left(\frac{1}{\beta^* - 1}\right)^{\kappa(N-1)} \left(\frac{1}{\beta^* - 1}\right)^{\kappa} \right)$$

since $\log z_N(s)/\log w \leq \log D^{1/s}/\log D^{(\beta^* - 1)/s}$. Also

$$(K - 1) K^{N-1} \Delta_{N-1} \sum_{\substack{y_N \leq p < z_N(s) \\ D^{(\beta^* - 1)/s} \leq p}} \frac{\rho(p)}{p} V(P(p))$$
$$\leq (K - 1) K^{N-1} \Delta_{N-1} V\left(P(D^{(\beta^* - 1)/s})\right)$$
$$\leq (K - 1) K^N \Delta_{N-1} \left(\frac{1}{\beta^* - 1}\right)^{\kappa} V\left(P(D^{1/s})\right),$$

by (1.3.5.2), and the induction proceeds. This establishes Lemma 2.

4.4.3 An Improved Version of Proposition 1

In Sect. 4.4.1 the stronger form (1.3.5.3) of the sifting density hypothesis was used at (1.13) in a minimal way, only for the value of w described in (1.3). By using this hypothesis systematically a substantially better expression for the error term $\eta(D, s)$ can be derived than was stated in Theorem 1. However, we do not try to derive absolutely the best possible result in this direction.

Theorem 2. *Fix $0 < \Delta < 1$, and make $0 < \gamma < \Delta/20$. If $\kappa \leq \frac{1}{2}$, then suppose $\Delta > \frac{1}{2}$. Then in Proposition 1 we may take*

$$\eta(D, s) = \frac{e^{\sqrt{L}} \psi(s)}{(\log D)^{1-\Delta}},$$

where

$$\psi(s) = \exp\bigl(-s \log s + O(\log s)\bigr) \quad \text{if} \quad 1 \leq s \leq (\log D)^{\gamma/2} \tag{3.1}$$
$$\psi(s) = \exp\bigl(-s \log s + s \log \log Ls + O(s)\bigr) \quad \text{if} \quad s \geq (\log D)^{\gamma/2}. \tag{3.2}$$

The O-constants may depend on κ and Δ.

4.4 A Sieve Theorem Following Rosser

Recall that in the definitions of sifting density given in Sect. 1.3 we obtain in particular from (1.3.5.4)

$$K \leq 1 + \frac{L}{\log 2} \leq 3L,$$

because $L \geq 1$.

When $s > (\log D)^{\gamma/2}$ the result (3.2) follows from the Fundamental Lemma given in Corollary 3.3.1.2, because in this case

$$(\log D)^{1-\Delta} \leq \exp(\log \log D) \leq \exp(O(s)).$$

In the remainder of the treatment we may therefore suppose $s \leq (\log D)^{\gamma/2}$.

Use a composition of two sieves as in Lemma 1. In (1.3) take

$$v = 1/\varepsilon^2, \qquad \varepsilon = \frac{1}{\log^\gamma D}, \qquad (3.3)$$

so that $P_1 = P(w)$ with $w = D^{\varepsilon^2}$, and $D_1 = D^\varepsilon$.

Here it is sufficient and technically more convenient to use the weaker Corollary 3.3.1.1, as in Sect. 4.1. Thus we may use (1.5), but where now

$$s_1 = \log D_1/\log w = \varepsilon^{-1} = \log^\gamma D \geq s^2.$$

Note $e^{s \log s}(\log D)^{1-\Delta} \leq \exp(\log^\gamma D + O(1)) = O(e^{s_1})$. In this way we obtain

$$V^\pm(D_1, P_1) = V(P_1)\left(1 + O\left(\frac{L^{10} e^{-s \log s}}{(\log D)^{1-\Delta}}\right)\right). \qquad (3.4)$$

With D_2 and P_2 as in (1.3) it remains to deal with $V^\pm(D_2, P_2)$, as described in terms of T_n in (4.1.1.1)–(4.1.1.3). The difficulties associated with the n-dependence of the estimate for T_n in Lemma 4.1.3 are avoided by dealing with the partial sums

$$\Sigma_n(D_2, P_2(D^{1/s})) = \sum_{\substack{1 \leq m \leq n \\ m \equiv n, \bmod 2}} T_m(D_2, P_2(D^{1/s})), \qquad (3.5)$$

but at the cost of introducing a more delicate induction argument. We will use the next lemma in the case $y = D_2$, but it is convenient to state it for other values. The entry $\sqrt{L}$ in Lemma 2 is a convenient choice which is not optimal, and which arises from considerations arising when y is relatively small in terms of L.

Lemmas 2–4 use a notation from Sect. 4.1.1:

$$\varepsilon_n = 1 \text{ if } n \text{ is odd}, \quad \varepsilon_n = 0 \text{ if } n \text{ is even}. \qquad (3.6)$$

4. Rosser's Sieve

Lemma 3. *Suppose that the hypotheses of Theorem 2 hold, and that P_2 is as in (1.3), with v and ε as in (3.3). Then for $2 \leq y \leq D$ the expression defined in (3.5) satisfies*

$$\Sigma_n\bigl(y, P_2(y^{1/s})\bigr) \leq V\bigl(P_2(y^{1/s})\bigr)\left(g_n(s) + \frac{c_0 s\, e^{-s\log s + \sqrt{L}}}{(\log y)^{1-\Delta}}\right) \quad \text{if} \quad s \geq \beta - \varepsilon_n ,$$

where c_0 is a constant that depends at most on Δ and κ.

In Lemma 3, the continuous functions corresponding to $\Sigma_n\bigl(y, P_2(y^{1/s})\bigr)$ are as defined in (4.3.1.1):

$$g_n(s) = \sum_{\substack{1 \leq m \leq n \\ m \equiv n,\, \text{mod } 2}} f_m(s) . \tag{3.7}$$

Thus (4.2.1.6) says

$$F(s) = 1 + \lim_{j \to \infty} g_{2j+1}(s), \qquad f(s) = 1 - \lim_{j \to \infty} g_{2j}(s),$$

the question of convergence when $\kappa > \tfrac{1}{2}$ having been dealt with in Sect. 4.3.

Proof of Theorem 2. This follows at once after Lemma 3 is established. In the notations of (4.1.1.1)–(4.1.1.3) and (3.5), Lemma 3 gives

$$V^+(D_2, P_2) \leq V(P_2)\bigl(F(s_2) + \delta\bigr), \qquad V^-(D_2, P_2) \geq V(P_2)\bigl(f(s_2) - \delta\bigr),$$

as in (1.8), but now with $\delta \ll e^{-s\log s + \sqrt{L}}/(\log D_2)^{1-\Delta}$. Theorem 2 now follows using (3.4) and Lemma 1.

Proof of Lemma 3. The lemma is trivial if $y^{1/s} < w$ because then there are no primes in P_2, so that the sum in (3.5) is empty. Since $\gamma < \tfrac{1}{2}$ in Theorem 2, (3.3) shows

$$w = D^{\varepsilon^2} = \exp\bigl((\log D)^{1-2\gamma}\bigr) \geq \exp\bigl((\log y)^{1-2\gamma}\bigr) = y^{1/(\log y)^{2\gamma}} .$$

Thus
$$\Sigma_n\bigl(y, P_2(y^{1/s_0})\bigr) = 0 \quad \text{where} \quad s_0 = (\log y)^{2\gamma} . \tag{3.8}$$

Furthermore, in establishing the lemma we may suppose $s \leq s_0$, since it is trivial otherwise.

For small values of y we can again appeal to the Fundamental Lemma (Corollary 3.3.1.2). Since $K \leq 3L$ it gives

$$\Sigma_n\bigl(y, P_2(y^{1/s})\bigr) = V\bigl(P_2(y^{1/s})\bigr)\bigl(1 + O\bigl(L^5 e^{-s\log s + s\log\log 3Ls + O(s)}\bigr)\bigr) .$$

4.4 A Sieve Theorem Following Rosser

So provided $s \log \log 3Ls + 5 \log L + O(s) + 10 \log L < \sqrt{L} + O(1)$ the result of Lemma 3 will follow. In fact

$$s \log \log 3Ls + O(s) \leq (\log y)^{2\gamma} \left(\log(\log 3L + 2\gamma \log \log y) + O(1) \right),$$

because $s \leq (\log y)^{2\gamma}$. If $\log^\gamma y \ll L^{1/5}$ this gives

$$s \log \log 3Ls + O(s) + 5 \log L \leq \sqrt{L} + O(1),$$

as required. Thus in proving Lemma 3 we may now assume (3.8) and

$$s \leq s_0 = (\log y)^{2\gamma} < c_1 L^{2/5}, \qquad (3.9)$$

where the constant c_1 may be chosen arbitrarily large.

As earlier, define h^+, h^- by

$$h^{(-)^n}(s) = \begin{cases} H(s) & \text{if } 2 \nmid n \\ h(s) & \text{if } 2 \mid n, \end{cases}$$

where H, h are the functions introduced in Sect. 4.3.1.

Because of (3.9) and the estimates in Lemma 4.3.2, to establish Lemma 3 it will suffice to obtain the following estimate for Σ_n, will be done using an induction.

Lemma 4. *Suppose that (3.8), (3.9) and the hypotheses of Proposition 1 hold, and that P_2 and y are as in Lemma 3. If $\kappa > \frac{1}{2}$ let $h^\pm$ be as in Lemma 4.3.2, but if $\kappa \leq \frac{1}{2}$ let $h^\pm$ be constructed with κ replaced by a suitable $\kappa_0 > \frac{1}{2}$ as in Lemma 4.3.5. Suppose $s > \beta - \varepsilon_n$, where ε_n is as in (3.6). Then the expression Σ_n defined in (3.5) satisfies*

$$\Sigma_n\bigl(y, P_2(y^{1/s})\bigr) \leq V\bigl(P_2(y^{1/s})\bigr) \left(g_n(s) + \frac{c_0 Ls\, U(s)}{(\log y)^{1-\Delta}} \right), \qquad (3.10)$$

where c_0 is a constant that depends at most on κ and Δ, and $U(s) \ll e^{-s \log s}$ is as in Lemma 4.3.3.

When $n = 1$ (so that (3.5) gives $\Sigma_1 = T_1$) use Lemma 4.1.2(i) as in Sect. 4.1.4. We obtain

$$\Sigma_1\bigl(y, P_2(y^{1/s})\bigr) \leq V\bigl(P_2(y^{1/s})\bigr) \left(f_1(s) + \frac{c_2 L}{s^\kappa \log D} \right), \qquad (3.11)$$

with f_1 as in (4.1.3.1), because we can now take $K - 1 = L/\log(D^{1/(\beta+1)})$ in (4.1.4.12). We may suppose $s < \beta + 1$ (otherwise T_1 vanishes, e.g. from Lemma 4.1.1), so this gives (3.10) when $n = 1$, because $s\, h^-(s) = s\, H(s)$ and $1/s^\kappa = s\, H(s)/(\beta - 1)^\kappa$ in (4.3.1.4).

The recurrences in Lemma 4.1.1 lead through (3.5) to

$$\Sigma_n(y, P_2(y^{1/s})) = \sum_{p<y^{1/s}} \frac{\rho(p)}{p} \Sigma_{n-1}\left(\frac{y}{p}, P_2(p)\right) \quad \text{if} \quad s \geq \beta + \varepsilon_n \quad (3.12)$$

$$\Sigma_n(y, P_2(y^{1/s})) = \Sigma_n(y, P_2(y^{1/(\beta+1)})) + \Sigma_1(y, P_2(y^{1/s}))$$
$$\text{if} \quad 2 \nmid n, \ \beta - 1 < s \leq \beta + 1. \quad (3.13)$$

Analogously, (4.1.3.1), (4.1.3.2) give from (3.7)

$$g_n(s) = \frac{1}{s^\kappa} \int_s^\infty g_n(t-1) \, dt^\kappa \quad \text{if} \quad s \geq \beta + \varepsilon_n \quad (3.14)$$

$$g_n(s) = g_n(\beta + 1) + f_1(s) \quad \text{if} \quad 2 \nmid n, \ \beta - 1 < s \leq \beta + 1,$$

in which $g_n(s) \geq 0$.

We may suppose (3.10) already proved when $n = N - 1$. In the first place we will obtain (3.10) under the extra hypothesis

$$s \geq \beta + \varepsilon_n, \quad (3.15)$$

which will be easily removed at the end. For these s, (3.12) gives

$$\Sigma_n(y, P_2(y^{1/s})) = \Sigma_n(y, P_2(y^{1/s_0}))$$
$$+ \sum_{y^{1/s_0} \leq p < y^{1/s}} \frac{\rho(p)}{p} V(P_2(p)) \Sigma_{n-1}\left(\frac{y}{p}, P_2(p)\right),$$

where actually $\Sigma_n(y, P_2(y^{1/s_0})) = 0$, as noted at (3.8). The inductive hypothesis now gives

$$\Sigma_N(y, P_2(y^{1/s})) \leq \Sigma_N^{(1)}(y, P_2(y^{1/s})) + \Sigma_N^{(2)}(y, P_2(y^{1/s})), \quad (3.16)$$

where

$$\Sigma_N^{(1)}(y, P_2(y^{1/s})) = \sum_{y^{1/s_0} \leq p < y^{1/s}} \frac{\rho(p)}{p} V(P_2(p)) g_{N-1}\left(\frac{\log y}{\log p} - 1\right), \quad (3.17)$$

$$\Sigma_N^{(2)}(y, P_2(y^{1/s}))$$

$$= \sum_{y^{1/s_0} \leq p < y^{1/s}} \frac{\rho(p)}{p} V(P_2(p)) \frac{\left(\frac{\log y}{\log p} - 1\right) h^{(-)^{N-1}}\left(\frac{\log y}{\log p} - 1\right)}{\left(\log \frac{y}{p}\right)^{1-\Delta}}. \quad (3.18)$$

Since (3.14) shows that $s^\kappa g_n(s)$ decreases when $s \geq \beta + \varepsilon_n$, we may use Lemma 4.1.2(ii) to obtain

$$\frac{\Sigma_N^{(1)}(y, P_2(y^{1/s}))}{V(P_2(y^{1/s}))} \leq \frac{1}{s^\kappa} \int_s^{s_0} g_n(t-1) \, dt^\kappa + \frac{(\kappa+1)(s-1) L \, h^{(-)^{N-1}}(s-1)}{\log(y^{1/s_0})},$$

where we used the fact that g_n is bounded as in Lemma 4.3.1 (or, if $\kappa < \frac{1}{2}$, as in Lemma 4.3.6). Since $s \le s_0 = (\log y)^{2\gamma}$, and $h(s-1) \ll s^2 U(s)$ from Lemma 4.3.3(iii), this gives

$$\frac{\Sigma_N^{(1)}(y, P_2(y^{1/s}))}{V(P_2(y^{1/s}))} \le g_n(s) + O\left(\frac{Ls\,U(s)}{(\log y)^{1-6\gamma}}\right), \qquad (3.19)$$

where we used (3.14).

Initially we suppose $\kappa > \frac{1}{2}$, because the treatment of $\Sigma^{(2)}$ becomes more involved when $\kappa \le \frac{1}{2}$. Before proceeding note

$$\frac{(t-1)h^\pm(t-1)}{(1-1/t)^{1-\Delta}} = \left(1 - \frac{1}{t}\right)^\Delta t\,h^\pm(t-1). \qquad (3.20)$$

Since Lemma 4.3.2 gave $h^\pm(s) \ge 0$ if $s \ge \beta$ the defining equations (4.3.1.2)–(4.3.1.5) show that $s^{\kappa+1}h^\pm(s)$, and therefore $s\,h^\pm(s)$, is decreasing in $s > 0$. Hence the expression on the left of (3.20) decreases in $t > 1$. Thus we may apply Lemma 4.1.2(ii) to (3.18) and use (3.20) to obtain

$$\Sigma_N^{(2)}(y, P_2(y^{1/s})) \le \frac{c_0 V(P_2(y^{1/s}))}{(\log y)^{1-\Delta}} (M(s) + E(s)),$$

where

$$M(s) = \frac{1}{s^\kappa} \int_s^{s_0} \left(1 - \frac{1}{t}\right)^\Delta t\,h^{(-)^{N-1}}(t-1)\,dt^\kappa, \qquad (3.21)$$

$$E(s) = \frac{(\kappa+1)L}{\log(y^{1/s_0})} \left(1 - \frac{1}{s}\right)^\Delta s\,h^{(-)^{N-1}}(s-1). \qquad (3.22)$$

Using Lemma 4.3.2 for $M(s)$ and Lemma 4.3.3(iii) for $E(s)$ gives

$$M(s) \le \left(1 - \frac{1}{s_0}\right)^\Delta \int_s^{s_0} \kappa t^\kappa h^{(-)^{N-1}}(t-1)\,dt \le \left(1 - \frac{1}{s_0}\right)^\Delta h^{(-)^N}(s)$$

$$E(s) \ll \frac{Ls^3 U(s)}{(\log y)^{1-2\gamma}} \le \frac{Ls\,U(s)}{(\log y)^{1-6\gamma}},$$

since (3.9) applies in Lemma 4.

With (3.16) and (3.19) this gives

$$\Sigma_N(y, P_2(y^{1/s})) \le V(P_2(y^{1/s}))(g_n(s) + c_0 U(s) R_N(y,s)),$$

where

$$R_N(y,s) = \frac{\left(1 - \frac{1}{s_0}\right)^\delta}{(\log y)^{1-\Delta}} + O\left(\frac{L}{(\log y)^{1-10\gamma}}\right), \qquad (3.23)$$

actually with $\delta = \Delta$. In (3.9) we ensured $\log y > c_1 L^{1/(5\gamma)}$, where c_1 may be arbitrarily large. Specify γ suitably small ($0 < \gamma < \Delta/20$ is sufficient) and choose c_1 large enough. This gives $R_N(y,s) < 1/(\log y)^{1-\Delta}$, because $L \geq 2$. This completes the inductive proof of (3.10) when (3.15) holds.

It remains to deduce (3.10) when (3.15) is false, so that (3.13) applies. We can apply (3.10) to the first term on the right of (3.13), which gives the required estimate with the term $f_1(s)$ replaced by $f_1(\beta+1) = 0$. The other term in (3.13) was estimated in (3.11), which supplies the term $f_1(s)$ and an extra contribution which can be assessed as

$$\frac{c_2 L}{s^\kappa \log D} \ll \frac{c_2 L s\, H(s)}{\log D}$$

using (4.3.1.4), which is acceptable provided we choose c_0 appropriately. This establishes Lemma 4 when $\kappa > \frac{1}{2}$.

When $\kappa \leq \frac{1}{2}$ the treatment of $\Sigma^{(2)}$ is somewhat different, and succeeds only when $\Delta > \frac{1}{2}$. The functions $h^\pm$ are now constructed using the parameter $\kappa_0 > \frac{1}{2}$ as described in Lemma 4. Proceeding rather as in Lemma 4.3.6 we now have from (3.21)

$$M(s) \leq \kappa \int_s^{s_0} \left(1 - \frac{1}{t}\right)^\Delta \left(\frac{t}{s}\right)^\kappa h^{(-)N-1}(t-1)\, dt$$
$$\leq \tfrac{1}{2} \int_s^{s_0} \left(1 - \frac{1}{t}\right)^{\kappa_0 + 2\delta} \left(\frac{t}{s}\right)^{\kappa_0} h^{(-)N-1}(t-1)\, dt\,,$$

if we choose $\delta > 0$ and $\kappa_0 - \frac{1}{2} > 0$ small enough. Then Lemma 4.3.5 gives

$$M(s) < \left(1 - \frac{1}{s_0}\right)^\delta s\, h^{(-)N}(s)\,,$$

from which we obtain (3.23) and proceed as before. This proves Lemma 4, so that Lemma 3 and Theorem 2 follow as described earlier.

4.4.4 A Two-Sided Estimate

In this volume we have taken some trouble to ensure that, whenever possible, our results require only our standard one-sided sifting density hypothesis from Sect. 1.3.5. However, there will remain some occasions when we will be forced to capitulate and assume the two-sided variant given in Definition 1.3.8. An instance will arise in the discussion of the sieve with weights dealt with in Chapter 5. Thus we assume $2 \leq L \leq L'$ and

$$\left(1 - \frac{L'}{\log w}\right)\left(\frac{\log z}{\log w}\right)^\kappa < \prod_{w \leq p < z} \left(1 - \frac{\rho(p)}{p}\right)^{-1} < \left(1 + \frac{L}{\log w}\right)\left(\frac{\log z}{\log w}\right)^\kappa \quad (4.1)$$

when $2 \leq w < z$.

4.4 A Sieve Theorem Following Rosser

Theorem 3 provides estimates for V^+ from below and for V^- from above. These supplement the estimates provided by Theorem 2, so that the expressions on the right of (4.2) actually yield asymptotic estimates (from above and below) for the quantities $V^\pm$.

Theorem 3. *Assume that the function ρ has a two-sided sifting density κ. Suppose $0 < \Delta < 1$ and $0 < \gamma < \Delta/20$, with $\Delta > \frac{1}{2}$ if $\kappa \leq \frac{1}{2}$, as in Theorem 2. Then*

$$V^+(D, P_2(z)) \geq V(P_2(z))\Big(F(s) + O(\eta(D,s))\Big),$$
$$V^-(D, P_2(z)) \leq V(P_2(z))\Big(f(s) + O(\eta(D,s))\Big), \quad (4.2)$$

where

$$\eta(D,s) = \frac{L' e^{\sqrt{L}} \psi(s)}{(\log D)^{1-\Delta}}$$

with $\psi(s)$ as in Theorem 2.

Observe that (in contrast to Theorem 2) this result depends upon both sides of the estimate (4.1), and that the nature of the dependences on L' and L are quite different. The need for this feature, even in the case where s, L' and L are bounded uniformly in y, will become clear on studying the structure of the proof of Lemma 5 below.

The following treatment is along lines similar to those used for Theorem 2, but there are some additional matters of detail to be attended to.

Begin by using Lemma 1 as before, but with the rôles of V^+ and V^- interchanged. Invoke the Fundamental Lemma from Sect. 3.3.1 as in Sect. 4.4.3. Further, use it to deal with the cases where $s \leq s_0$ or $\log^\gamma y \ll L^{1/5}$ as in Sect. 4.4.3. Then Theorem 3 will follow from the following analogue of Lemma 4.

Lemma 5. *Assume (3.9) and the other conditions of Lemma 4, so that in particular $\beta - \varepsilon_n < s \leq s_0 = (\log y)^{2\gamma}$, where ε_n is as in (3.6). Then the quantity Σ_n defined in (3.5) satisfies*

$$\Sigma_n(y, P_2(y^{1/s})) \geq V(P_2(y^{1/s}))\left(g_n(s) - \frac{c_3 L' e^{\sqrt{L}} s U(s)}{(\log y)^{1-\Delta}}\right), \quad (4.3)$$

where c_3 may depend on κ and Δ.

The general plan of the proof of Lemma 5 closely resembles that for Lemma 4, so we merely draw attention to those features that differ. As before, proceed by induction on n. The result holds when $n = 1$, this time because of Lemma 4.1.2(iii). If it holds when $n = N$ and $s \geq \beta + n$ then

$$\Sigma_N(y, P_2(y^{1/s})) \geq \Sigma_N^{(1)}(y, P_2(y^{1/s})) - \Sigma_N^{(2)}(y, P_2(y^{1/s})),$$

where $\Sigma_n^{(1)}$ and $\Sigma_n^{(2)}$ are as in (3.17) and (3.18). Thus we need an upper bound for $\Sigma^{(2)}$, as provided in the proof of Lemma 4. This is one reason why Lemma 5 necessarily uses both sides of the estimate (4.1), so that the parameters L and L' both appear in the enunciation. In fact, via (3.21), (3.22) and (3.23) we derived

$$\Sigma_N^{(2)}\bigl(y, P_2(y^{1/s})\bigr) < c_0 \frac{V\bigl(P(y^{1/s})\bigr) h^{(-)^N}(s)}{(\log y)^{1-\Delta}}.$$

For $\Sigma_N^{(1)}$ use of Lemma 4.1.2(iii) gives

$$\frac{\Sigma_N^{(1)}\bigl(y, P_2(y^{1/s})\bigr)}{V\bigl(P_2(y^{1/s})\bigr)} \geq g_n(s) - g_n(s_0) + O\!\left(\frac{L's\, h^{(-)^N}(s)}{(\log y)^{1-6\gamma}}\right), \qquad (4.4)$$

analogously to (3.19). The term $g_n(s_0)$ cannot simply be dropped on this occasion as it has the wrong sign. However, Lemmas 4.3.1 and 4.3.2 show

$$g_n(s_0) \leq g_n(s) \ll h^{(-)^n}(s),$$

because $s \leq s_0$. Furthermore $L \gg s_0^{5/2} = (\log y)^{5\gamma}$ at this point in the argument, so that $e^{\sqrt{L}} \gg \log y$, the implied constant depending on γ. This shows

$$g_n(s_0) \ll \frac{e^{\sqrt{L}} h^{(-)^n}(s)}{\log y},$$

which is majorised by the error term stated in Lemma 5.

The proof of Lemma 5 is now completed along the lines used for Lemma 4.

4.5 Extremal Examples

The question arises whether the functions $f = f_\kappa$ and $F = F_\kappa$ appearing in Proposition 4.4.1 can be improved upon. When $\kappa > 1$ this can be done, using Selberg's ideas from Chapter 2; for the lower bound sifting problem we may use one of the ways introduced in Chap. 7. On the other hand, in the two special cases $\kappa = 1$ and $\kappa = \frac{1}{2}$ there exist sequences $\mathcal{A}$ which show that the expressions $f(s)$ and $F(s)$ are best possible, in that the inequalities in Proposition 4.4.1 are, for these sequences, actually asymptotic equalities. We will discuss the cases $\kappa = 1$ and $\kappa = \frac{1}{2}$ in turn.

4.5.1 The Linear Case

For the case $\kappa = 1$ suitable extremal examples $\mathcal{A}^+$ and $\mathcal{A}^-$ are given by

$$\mathcal{A}^{(-)^r} = \mathcal{A}^{(-)^r}(X) = \bigl\{a : 2X \leq a < 4X,\ \Omega(n) \equiv r+1,\ \operatorname{mod} 2\bigr\}, \qquad (1.1)$$

4.5 Extremal Examples

where $\Omega(n)$ denotes the total number of prime factors of n, counted according to multiplicity:

$$\Omega(n) = \sum_{\substack{p,\alpha: \\ p^\alpha \| n}} 1 \, .$$

We need the following result, closely related to the Prime Number Theorem: the Liouville function $\lambda(d) = (-1)^{\Omega(d)}$ satisfies the estimate

$$\sum_{d \leq X} \lambda(d) = O\left(X e^{-c\sqrt{\log X}}\right), \qquad (1.2)$$

for some positive constant c. The corresponding relation for the Möbius function μ is perhaps more well known. Then (1.2) follows by an argument based on the relationship

$$\sum_{n<X} \lambda(n) = \sum_{m^2 n < X} \mu(n) = \sum_{m<\sqrt{X}} \sum_{n<X/m^2} \mu(n) \, ,$$

in which the inner sum is estimated trivially by X/m^2 when $m \geq e^{c'\sqrt{\log X}}$.

There is a good estimate of how many $a \in \mathcal{A}^{(-)^r}(X)$ are divisible by a sifting modulus d. The values of $\Omega(n)$, mod 2 are picked out by the relation

$$\frac{1 + (-1)^r \lambda(n)}{2} = \begin{cases} 1 & \text{if } \Omega(n) \equiv r, \text{ mod } 2 \\ 0 & \text{otherwise.} \end{cases}$$

The function Ω is totally multiplicative, so the numbers divisible by d in the sequence (1.1) are precisely those for which

$$a = dn \, , \quad \Omega(n) \equiv r + 1 + \Omega(d), \text{ mod } 2 \, , \quad 2X \leq a < 4X \, .$$

Thus, in the standard notation (1.2.1.3),

$$|A_d^\pm| = \sum_{2X/d \leq n < 4X/d} \tfrac{1}{2}(1 \pm \lambda(n)) = \frac{X}{d} + r_A(d) \, ,$$

where

$$r_A(d) = O\left(\frac{X}{d} e^{-c\sqrt{\log X/d}}\right) . \qquad (1.3)$$

Thus (1.3.1.1) is satisfied with $\rho(d) = 1$.

This allows a level of distribution $D = X^{1-\varepsilon}$, provided $\varepsilon = \varepsilon(X)$ tends to 0 sufficiently slowly as $X \to \infty$. For (1.3) gives

$$\sum_{d < X^{1-\varepsilon}} |r_A(d)| \ll X \log X \, e^{-c\sqrt{\varepsilon \log X}} . \qquad (1.4)$$

We may take $\varepsilon = A(\log \log X)^6 / \log X$, with a suitable constant A, such that the estimate (1.4) becomes $X/\log^2 X$.

160   4. Rosser's Sieve

A Special Situation. Our initial use of the example (1.1) is to show that in the Linear Sieve Theorem (the case $\kappa = 1$ of Proposition 4.4.1) the sifting limit 2 is best possible. Take $r = 1$ in (1.1).

Suppose Proposition 4.4.1 applied with $s < 2$ and $f(s) > 0$. Let P be the product of all primes not exceeding D. Then the estimate (1.4) leads, with $D = X^{1-\varepsilon}$ and $\eta(D,s) \to 0$ as $D \to \infty$, to

$$S(\mathcal{A}^-(X), P(D^{1/s}))$$
$$\geq X V(P(D^{1/s}))(f(s) + O(\eta(D,s))) - O(X/\log^2 X), \quad (1.5)$$

the r-term being treated as in Corollary 1.3.1.1. Now $a < 4X$ for all a in (1.1), and $D^{1/s} > \sqrt{4X}$ as soon as X is large enough. But the members of $\mathcal{A}^-(X)$ have an even total number $\Omega(a)$ of prime factors. Thus the left side of (1.5) is zero, contradicting the proposition that $f(s) > 0$ with $s < 2$ on the right.

The following similar argument using $\mathcal{A}^+(X)$ shows that the value of $F(s)$ when $1 \leq s < 3$ is also optimal, so that when $\kappa = 1$ the upper bound in Proposition 4.4.1 can also not be improved for these s. Because of (1.4) the proposition gives

$$S(\mathcal{A}^+, P(D^{1/s})) \leq X V(P(D^{1/s}))(F(s) + O(\eta(D,s))) - O(X/\log^2 X), \quad (1.6)$$

where $\eta(D,s) \to 0$ as $D \to \infty$. When $a \in \mathcal{A}^+$ the number $\Omega(a)$ is odd, so when $1 < s < 3$ only the primes between $2X$ and $4X$ are counted in (1.6). Thus

$$S(\mathcal{A}^+, P(D^{1/s})) = \frac{2X}{\log X} + O\left(\frac{1}{\log^2 X}\right) \quad \text{if} \quad 1 < s < 3.$$

But Mertens' formula shows

$$V(P(z)) = \frac{e^{-\gamma}}{\log z}\left(1 + O\left(\frac{1}{\log z}\right)\right).$$

For these purposes the distinction between $D = X^{1-\varepsilon}$ and X is unimportant. In fact $X^{1/s_1} = D^{1/s_2}$ where $s_1 \to s_2$ (since $\varepsilon \to 0$) as $X \to \infty$. Consequently

$$V(P(D^{1/s_2}))F(s_2) \sim V(P(X^{1/s_1}))F(s_1) \quad \text{as} \quad X \to \infty,$$

so that in (1.6) the entry $F(s) = 2e^{\gamma}/s$ when $1 < s < 3$ cannot be improved.

The General Result. A similar argument shows that Theorem 1, as enunciated below, implies that when $\kappa = 1$ the functions F and f in Proposition 4.4.1 are optimal throughout their range. In Theorem 1 and the more general Lemma 1 denote

$$\phi^+ = F, \quad \phi^- = f, \quad \phi_r = \phi^{(-)^r}, \quad (1.7)$$

where $F = F_\kappa$ and $f = f_\kappa$ are the functions of Sect. 4.2. Thus $F = \phi_0 = \phi_2$, for example. Take $P = P(X)$ to be the product of all primes not exceeding X.

4.5 Extremal Examples

Theorem 1. *Let $\mathcal{A}^{\pm}$ be as in (1.1), and suppose that $1 \leq s \leq \log X$. Then*

$$S(\mathcal{A}^{(-)^r}(X), P(X^{1/s})) = XV(P(X^{1/s}))\left(\phi_r(s) + O\left(\frac{s^2 \log s}{\log X}\right)\right), \quad (1.8)$$

where ϕ_r is as in (1.7).

The s-dependence of the error term is rather poor, an artefact of the method of proof employed. The restriction to $s \leq \log X$ is then not a serious one, since the stated O-term is already unsatisfactorily large if $s \geq \sqrt{\log X}$.

Theorem 1 will be established by an inductive process depending on Buchstab's identity (3.1.3.3):

$$S(\mathcal{A}, P(X^{1/\sigma})) = S(\mathcal{A}, P(X^{1/s})) + \sum_{X^{1/\sigma} \leq p < X^{1/s}} S(\mathcal{A}_p, P(p)) \quad \text{if} \quad s \leq \sigma.$$

This will be used when $\mathcal{A} = \mathcal{A}^{(-)^r}(X)$, so that $\mathcal{A}_p = \mathcal{A}^{(-)^{r+1}}(X/p)$. Adopt the notation

$$\psi_r(X, s) = \psi^{(-)^r}(X, s) = S(\mathcal{A}^{(-)^r}(X), P(X^{1/s})) . \quad (1.9)$$

Thus, with currently $\rho(p) = 1$ for all p, the expression ψ_r in (1.9) satisfies

$$\psi_r(X, \sigma) = \psi_r(X, s) + \sum_{X^{1/\sigma} \leq p < X^{1/s}} \rho(p)\psi_{r+1}\left(\frac{X}{p}, \frac{\log X}{\log p} - 1\right) \quad (1.10)$$

when $\sigma \geq s$. This expresses $\psi^{\pm}(X, \sigma)$ when $\sigma < s+1$ in terms of values of $\psi^{\pm}(y, t)$ with $t < s$.

Theorem 1 has already been established when $1 \leq s < 2$, so will then follow from the following generalisation, which will also be used in the context $\kappa = \frac{1}{2}$. The restriction of Lemma 1 to the case $\kappa \leq 1$ is a technical convenience which simplifies the discussion slightly.

Lemma 1. *Let $0 < \delta \leq 1$. Suppose that $0 < \kappa \leq 1$ and the function ρ satisfies a formula of Mertens' type,*

$$V(P(z)) = \frac{C_\kappa}{\log^\kappa z}\left(1 + O\left(\frac{1}{\log z}\right)\right), \quad (1.11)$$

where C_κ is a constant depending only on κ, so that in particular ρ has sifting density κ. Suppose that ψ satisfies the Buchstab identity (1.10). Assume that, for each r and when $1 \leq s < 2$,

$$\psi_r(X, s) = XV(P(X^{1/s}))\left(\phi_r(s) + O\left(\frac{s^2 \log s}{\log^\delta X}\right)\right). \quad (1.12)$$

Then ψ_r satisfies (1.12) when $1 \leq s \leq \log^\delta X$.

4. Rosser's Sieve

Define $E_r(X,s)$ by

$$\psi_r(X,s) = XV(P(X^{1/s}))\left(\phi_r(s) + \frac{E_r(X,s)}{\log^\delta X}\right). \tag{1.13}$$

so that we require to show $E_r(X,s) \ll s^2 \log s$. In Lemma 1, this result is given when $1 \leq s < 2$. We will establish the result when $s < n$, by induction on $n \geq 2$. For this purpose denote

$$\Delta_N(X) = \sup_{y \leq X} \sup_{t \leq N} \max_r |E_r(y,t)|, \tag{1.14}$$

so that $\Delta_2(X) = O(1)$. We will prove

$$\Delta_N(X) \ll N^2 \log N \quad \text{when} \quad N \ll \log^\delta X, \tag{1.15}$$

so that Lemma 1 will follow on making $s \leq N < s+1$.

Once (1.15) has been established for $n = N$, proceed as follows. In (1.10), the inductive hypothesis gives that when $\sigma < N+1$

$$\sum_{X^{1/\sigma} \leq p < X^{1/N}} \rho(p)\psi_{r+1}\left(\frac{X}{p}, \frac{\log X}{\log p} - 1\right)$$

$$= \sum_{X^{1/\sigma} \leq p < X^{1/N}} \frac{X\rho(p)}{p} V(P(p))\left(\phi_{r+1}\left(\frac{\log X}{\log p} - 1\right) + E_{r+1}\left(\frac{X}{p}, t\right)\right), \tag{1.16}$$

in which $t = \log x/\log p - 1 < N$.

Use parts (ii) and (iii) of Lemma 4.1.2, in which the numbers L and L_- are now certain absolute constants. The leading term in (1.16) then contributes

$$XV(P(X^{1/N}))\left(\int_{N<t<\sigma} \phi_{r+1}(t-1)\frac{dt^\kappa}{N^\kappa} + O\left(\frac{1}{\log^\delta X^{1/\sigma}}\right)\right)$$

$$= XV(P(X^{1/N}))\left(\frac{1}{N^\kappa}(\sigma^\kappa \phi_r(\sigma) - N^\kappa \phi_r(N)) + O\left(\frac{N+1}{\log^\delta X}\right)\right), \tag{1.17}$$

where we used the equations (4.2.1.1)–(4.2.1.3) for the expressions ϕ_r given in (1.7), and the implied constant is absolute. Meantime, the first term on the right of (1.10) is estimated by (1.13), as

$$\psi_r(X,N) = XV(P(X^{1/N}))\left(\phi_r(N) + \frac{E_r(X,N)}{\log^\delta X}\right). \tag{1.18}$$

Thus the leading term in the resulting estimate for (1.10) is

$$XV(P(X^{1/N}))\frac{\sigma^\kappa}{N^\kappa}\phi_r(\sigma) = XV(P(X^{1/\sigma}))\phi_r(\sigma)\left(1 + O\left(\frac{1}{\log^\delta X}\right)\right), \tag{1.19}$$

because of (1.11). This is suitable for the current purpose.

4.5 Extremal Examples

When $N \geq 2$ the contribution from the term $E_r(X/p, t)$ in (1.16) has absolute value

$$\ll \frac{X \Delta_N(X)}{\log^\delta X} \sum_{X^{1/\sigma} \leq p < X^{1/N}} \frac{\rho(p)}{p} V(P(p))$$

$$= \frac{X \Delta_N(X)}{\log^\delta X} \left(V(P(X^{1/\sigma})) - V(P(X^{1/N})) \right)$$

$$\ll \frac{X \Delta_N(X)}{N \log^\delta X} V(P(X^{1/N})) \left(1 + O\left(\frac{1}{\log X}\right)\right), \quad (1.20)$$

where we used (4.1.2.5), (1.11), and the inequality $\sigma^\kappa/N^\kappa - 1 < 1/N$ when $\sigma < N+1$ and $0 \leq \kappa \leq 1$.

The error term in the estimate for (1.10), as accumulated from (1.18), (1.20) and (1.17), now has absolute value bounded by

$$\frac{X V(P(X^{1/N}))}{\log^\delta X} \left(\Delta_N(X) \left(1 + \frac{1}{N} + O\left(\frac{1}{N \log^\delta X}\right)\right) + O(N) \right). \quad (1.21)$$

But $V(P(X^{1/N})) \leq V(P(X^{1/\sigma})(1 + N^{-1} + O(1/\log^\delta X))$ when $\sigma < N+1$. Hence (1.19) and (1.21) give an estimate for (1.10) when $\sigma < N+1$ implying that Δ, as defined by (1.14) and (1.13), satisfies

$$\Delta_{N+1}(X) \leq \Delta_N(X)\left(1 + \frac{2}{N} + \frac{1}{N^2} + \frac{C}{\log^\delta X}\right) + CN,$$

with a suitable constant C. We remarked earlier that $\Delta_2(X) = O(1)$.

Suppose, then, that numbers δ_n satisfy

$$\delta_{n+1} \leq \delta_n \left(A + \frac{2}{n} + \frac{1}{n^2}\right) + Bn,$$

where $A > 1$ and we make $B \geq \delta_2 > 0$. In our application A will be close to 1. Then little is lost by saying

$$\frac{\delta_{n+1}}{A^{n+1}} \leq \frac{\delta_n}{A^n}\left(1 + \frac{1}{n}\right)^2 + Bn,$$

so that

$$\frac{\delta_{n+1}}{(n+1)^2 A^{n+1}} \leq \frac{\delta_n}{n^2 A^n} + \frac{B}{n}.$$

Successive applications of this inequality show

$$\delta_n \leq n^2 A^n (\delta_2 + B \log n) \ll B A^n n^2 \log n.$$

When applied to $\Delta_N(X)$ this shows that for all N

$$\Delta_N(X) \ll C N^2 \log N \left(1 + \frac{C}{\log^\delta X}\right)^N \leq C N^2 \log N \exp\left(\frac{CN}{\log^\delta X}\right).$$

This implies (1.15), so that the proof of Lemma 1 is complete.

4.5.2 The Case $\kappa = \frac{1}{2}$

For this value of κ suitable extremal examples are provided by

$$\mathcal{A}^{(-)^r} = \mathcal{A}^{(-)^r}(X) = \left\{n : 1 \leq n < 2X, n \equiv (-1)^r, \bmod 4\right\}. \quad (2.1)$$

In this case it is entirely elementary, as in Sect. 1.2.2, that

$$|\mathcal{A}_d^{\pm}| = \frac{X}{d} + r_{\mathcal{A}^{\pm}}(d) \quad \text{with} \quad |r_{\mathcal{A}^{\pm}}(d)| \leq 1, \quad (2.2)$$

so that

$$\sum_{d \leq D} |r_{\mathcal{A}^{\pm}}(d)| = O\left(\frac{X}{\log X}\right) \quad \text{when} \quad D = \frac{X}{\log X}.$$

In this section we take

$$P(z) = \prod_{\substack{p < z \\ p \equiv 3, \bmod 4}} p. \quad (2.3)$$

We will discuss the estimate for $S(\mathcal{A}^+(X), P(z))$ provided by Proposition 4.4.1. In view of (2.2) we therefore take

$$\rho(p) = \begin{cases} 1 & \text{if } p \equiv 3, \bmod 4 \\ 0 & \text{otherwise.} \end{cases} \quad (2.4)$$

We will establish the following theorem, to the general effect that when $\kappa = \frac{1}{2}$ it is again the case that the functions f, F appearing in Proposition 4.4.1 cannot be improved.

Theorem 2. *Suppose $\mathcal{A}^{\pm}(X)$ is as in (2.1) and that $1 \leq s \leq \sqrt{\log X}$. Let $\phi^+ = F$ and $\phi^- = f$ be as in (1.7), now with $\kappa = \frac{1}{2}$. Then, with $P(z)$ as in (2.3),*

$$S(\mathcal{A}^{(-)^r}(X), P(X^{1/s})) = X\, V(P(X^{1/s})) \left(\phi_r(s) + O\left(\frac{s^2 \log s}{\sqrt{\log X}}\right)\right),$$

analogously to the result of Theorem 1.

Theorem 2 will be deduced from Lemma 1, in which we take $\delta = \frac{1}{2}$, ρ as in (2.4) and ψ_r as in (1.9), but now with $\mathcal{A}^{\pm}$ and $P(z)$ as stated. Thus

$$\psi_r(X, s) = S(\mathcal{A}^{(-)^r}(X), P(z)). \quad (2.5)$$

There are some preliminary remarks to be made.

Define

$$c = \frac{1}{2\sqrt{2}} \prod_{p \equiv 3, \bmod 4} \left(1 - \frac{1}{p^2}\right)^{1/2}.$$

4.5 Extremal Examples

We will require the formula

$$\sum_{\substack{1 \leq n < x \\ p|n \Rightarrow p \equiv 1 \bmod 4}} 1 = \frac{cx}{\sqrt{\log x}} + O\left(\frac{x}{\log x}\right), \tag{2.6}$$

in which the precise power of $\log x$ appearing in the O-term is not crucial. This is an analogue of the Prime Number Theorem for the numbers counted on the left. These numbers are just those expressible as $n = u^2 + v^2$ with n odd and $(u, v) = 1$.

If $n \equiv 1$, mod 4 then the total number of prime factors p of n with $p \equiv 3$, mod 4 is even, when counted according to multiplicity. If $z > \sqrt{2X}$ then the numbers n counted by $S(\mathcal{A}^+(X), P(z))$ have $n < z^2$, $n \equiv 1$, mod 4 and $p \geq z^2$ for each such p, so this even number is necessarily zero. Thus these n are precisely those counted in (2.6), with $x = 2X$, so that

$$\psi_0(X, s) = c\frac{2X}{\sqrt{\log X}} + O\left(\frac{X}{\log X}\right) \quad \text{if } z > \sqrt{2X}. \tag{2.7}$$

We need a formula of Mertens' type for the product (2.3). Since

$$\prod_{\substack{p<z \\ p\equiv 1, \bmod 4}} \left(1 - \frac{1}{p}\right) \prod_{\substack{p<z \\ p\equiv 3, \bmod 4}} \left(1 + \frac{1}{p}\right) = \prod_{p<z}\left(1 - \frac{\chi(p)}{p}\right) = \frac{4}{\pi} + O\left(\frac{1}{\log z}\right),$$

in which χ is the non-principal character mod 4, the standard Mertens' formula gives

$$2 \prod_{\substack{p<z \\ p\equiv 3, \bmod 4}} \left(1 + \frac{1}{p}\right) \bigg/ \left(1 - \frac{1}{p}\right) = \frac{4e^\gamma}{\pi} \log z + O(1),$$

the factor 2 on the left arising from the prime $p = 2$. Thus

$$2 \prod_{\substack{p<z \\ p\equiv 3, \bmod 4}} \left(1 - \frac{1}{p}\right)^{-2} = \left(\frac{4e^\gamma}{\pi} \log z + O(1)\right) \prod_{\substack{p<z \\ p\equiv 3, \bmod 4}} \left(1 - \frac{1}{p^2}\right)^{-1},$$

so that

$$\prod_{\substack{p<z \\ p\equiv 3, \bmod 4}} \left(1 - \frac{1}{p}\right) = \sqrt{\frac{\pi}{2e^\gamma \log z}} \prod_{\substack{p\equiv 3, \bmod 4}} \left(1 - \frac{1}{p^2}\right)^{1/2} + O\left(\frac{1}{\log^{3/2} z}\right),$$

that is to say

$$V(P(z)) = c\sqrt{\frac{\pi}{e^\gamma \log z}} \left(1 + O\left(\frac{1}{\log z}\right)\right), \tag{2.8}$$

where c is the constant in (2.6).

Because of (2.2) the upper estimate provided by using Proposition 4.4.1 in Theorem 1.3.1 is now

$$S(\mathcal{A}^+(X), P(D^{1/s})) \leq X V(P(D^{1/s}))\Big(F(s) + O(\eta(D,s))\Big) + O(D), \quad (2.9)$$

where $\eta(D,s) \to 0$ as $D \to \infty$ as in Theorems 4.4.1 or 4.4.2, and F is as determined in Lemma 4.2.5(ii):

$$F(s) = 2\sqrt{\frac{e^\gamma}{\pi s}}, \quad f(s) = 0 \quad \text{if} \quad 1 \leq s \leq 2. \quad (2.10)$$

Since $V(P(z))$ is as in (2.8) this gives

$$V(P(D^{1/s}))F(s) = \frac{2c}{\sqrt{\log D}} + O\left(\frac{1}{\log D}\right). \quad (2.11)$$

We may take $D = X/\log X$ in (2.9). Then, the distinction between D and X being unimportant as in Sect. 4.5.1, (2.7) shows that the entry $F(s)$ given in (2.9) cannot be improved.

Proof of Theorem 2. We have only to check that the conditions of Lemma 1 are satisfied, with $\delta = \frac{1}{2}$, when ρ and ψ are as specified by (2.4) and (2.5).

First, the requirement (1.11) for the product P in (2.3) was supplied in (2.8). Next, we need to check that ψ_r, as now defined, satisfies the Buchstab identity (1.10). As in Sect. 3.1.3 start from

$$\sum_{\substack{d|a \\ d|P(z)}} \mu(d) = 1 - \sum_{\substack{p|a \\ p|P(z)}} \sum_{d|(a/p,P(p))} \mu(d)$$

to obtain when $w < z$

$$S(a, P(w)) = S(a, P(z)) + \sum_{\substack{w \leq p < z \\ p \equiv 3, \bmod 4}} S(a/p, P(p)).$$

On summing over a in $\mathcal{A}^\pm$ this gives that when $s < \sigma$

$$\psi_r(X, \sigma) = \psi_r(X, s) + \sum_{\substack{X^{1/\sigma} \leq p < X^{1/s} \\ p \equiv 3, \bmod 4}} \psi_{r+1}\left(\frac{X}{p}, \frac{\log X}{\log p} - 1\right), \quad (2.12)$$

which is (1.10), with ρ as in (2.4).

It remains to check that (1.12) is valid when $1 \leq s < 2$ and $\delta = \frac{1}{2}$. When r is even this follows using (2.7) and (2.11). When r is odd and $1 \leq s < 2$ we have $\phi_r(s) = 0$ as in (2.10). But

$$\psi_1(X, s) = S(\mathcal{A}^-(X), P(X^{1/s})),$$

in which the members n of $\mathcal{A}^-(X)$ satisfy $n \equiv 3$, mod 4 and $n \leq X$. Thus the only numbers counted by $\psi_1(X, s)$ for these s are primes, $\ll X/\log X$ in number, so that (1.9) again holds, and the proof of Theorem 2 is complete.

4.6 Notes on Chapter 4

Sect. 4.1.3. This section illustrates a point that will reappear elsewhere, that in inductive arguments of the type considered here it is possible to incur an unpleasant accumulation of error terms as the inductive parameter n increases.

Sect. 4.2.1. In the special case $\kappa = 1$ the functions F and f considered here have long been implicit in the literature. In fact

$$F(s) = e^\gamma \left(\omega(s) + \frac{\rho(s-1)}{s} \right), \quad f(s) = e^\gamma \left(\omega(s) - \frac{\rho(s-1)}{s} \right),$$

where $\omega(s)$ and $\rho(s)$ are well-known functions introduced in [Buchstab (1937)] and [Dickman (1930)] respectively. As may be inferred from the text, ω is continuous on $[1, \infty)$ and satisfies

$$\frac{d}{ds}(s\omega(s)) = \omega(s-1) \quad \text{if} \quad s > 2, \qquad s\omega(s) = 1 \quad \text{if} \quad 1 < s < 2.$$

It arises in the problem of estimating the number of numbers not exceeding x that have no prime factor smaller than $x^{1/s}$. In a related way the companion function ρ occurs in the problem of counting numbers all of whose prime factors are small. The reader may consult [Tenenbaum (1990)] for a recent account. Rather than quote results from the literature, we have given a self-contained account in the text, which is not confined to the case $\kappa = 1$.

Sometimes in the sieve literature $\rho(s)$ is written in place of the more standard notation $\rho(s-1)$ used here.

Sect. 4.2.2. The concept of the adjoint equation appears in [de Bruijn (1950)], in particular in a short purely analytic proof that $\omega(s) \to e^{-\gamma}$ as $s \to \infty$. It application to questions arising in the theory of sieves was developed in several papers by H. Iwaniec and others, in particular in [Iwaniec (1980a)], on which most of this chapter is based.

Sect. 4.2.4. Calculations carried out by F. Romani and later by J. van de Lune and H. J. J. te Riele lead to the following tabulation of the sifting limit β described by (4.10).

Table 1. Sifting Limits for Rosser's Sieve

κ	0.5	0.55	0.6	0.65	0.7	0.75	0.8	0.85	0.9	0.95	1
β_κ	1	1.0340	1.1042	1.1922	1.2912	1.3981	1.5107	1.6279	1.7489	1.8731	2

We have not tabulated β for larger values of κ (and the calculations quoted were not performed for large κ) because when $\kappa > 1$ other methods incorporating Selberg's idea from Chap. 2 lead to better results. Note, for future reference, that $\beta_\kappa < 2\kappa$ if $\frac{1}{2} < \kappa < 1$.

In Lemma 9, the function u is allowed to have discontinuities of the specified type because of an application that will be made in Chap. 7.

Sect. 4.2.5. Theorem 1 and the related Corollary 1.1 are taken from [Iwaniec 1980a], except that in this source the result was given with a weaker error term $O(\kappa^{2/3})$. The O-term stated is the right answer, in the sense that there is an asymptotic expansion, determined in [Tsang (1989)],

$$\beta = c\kappa + c_1 \kappa^{1/3} + \cdots + c_{n-2}\kappa^{(4-n)/3} + O(\kappa^{1-n/3}),$$

in which the O-constant depends upon n. In particular this implies that the inequality of Corollary 1.1 can be replaced by an equality, and that Theorem 1 is best possible in the sense that there is a constant c' such that $q(s) < 0$ when $s < c\kappa + c'\kappa^{1/3}$.

Sect. 4.3. The method described here for settling the convergence question raised by Sect. 4.2 is the best of those known, in that it leads to the good value for the exponent of $\log D$ appearing in Theorem 4.4.2. It also has the merit of applying whenever $\kappa > \frac{1}{2}$. Other approaches are possible, as indicated below.

Sect. 4.4. There are several ways in which the theorems of this section may be approached. In the first place, if we are not particular about the size of the exponent of $\log D$ in Theorem 4.4.2, and if we are concerned only with the cases $\kappa = \frac{1}{2}$ or $\kappa = 1$, then it is possible to proceed in the following style.

One approach is that mentioned already in the text, in which uses a function of some suitable simple form in place of the functions H and h of this section. For a treatment of the case $\kappa = \frac{1}{2}$ along these lines the reader may consult [Iwaniec (1976). The case $\kappa = 1$ can be treated in an analogous style, though the author knows of no published account of this type.

Rather different treatments of the case $\kappa = 1$ was given in [Motohashi (1981)] and in [Halberstam (1982)]. These papers deal with the estimations of the relevant error terms in the rather efficient style used in [Jurkat and Richert (1965)]. This paper dealt with the case $\kappa = 1$ of the problem considered in this chapter in a somewhat different way, which used Selberg's upper bound to estimate certain terms from above, as well as using the device employed in the combinatorial method of estimating certain other terms from below, by zero.

The method of Jurkat and Richert is described in a more general context, akin to that of Theorem 4.2, in [Halberstam and Richert (1974)]. It gave (when $\kappa = 1$) the same main term as that in Theorem 4.2 but a weaker error term. In the context of the weighted sieve of Chap. 5, however, we will observe in the notes that the Jurkat-Richert method has something to offer that the combinatorial method (at least in the present state of the art) does not.

The case $\kappa = \frac{1}{2}$ of Proposition 4.4.1 has a number of interesting applications in the literature. One of the first occurs in [Iwaniec (1972)], where a lower bound of the presumably correct order of magnitude $X/\log^{3/2} X$ is obtained for the number of primes p not exceeding X of the form $x^2 + y^2 + 1$, or more generally of the form stated in the title of the paper. A further generalisation version is in [Iwaniec (1974)]. The central idea is to show (with the aid of the Bombieri-Vinogradov Theorem) that $p - 1$, when restricted to be $\equiv 1$, mod 4, can be guaranteed to take values not divisible by any sifting prime $\varpi \equiv 3$, mod 4 not exceeding a suitable z. It is not possible to work with $z = \sqrt{X}$, as one might like, so a separate "switching" argument involving Selberg's upper bound is necessary to complete the argument.

A similar pattern of argument was used in [Greaves (1976)] to show that each large N, not congruent to 0, 1 or 5 mod 8, is representable as a sum of two squares and two squares of primes. In this case the necessary tool from prime number theory was the mean value theorem of Barban, Davenport and Halberstam.

Other applications of the sieve with $\kappa = \frac{1}{2}$ are to be found in [Iwaniec (1976)].

Sect. 4.4.1. The weakness of the induction from Sect. 4.1.1, on which Theorem 1 depends, is illustrated by the rather poor error term described in Corollary 1.1. On the other hand the method used is relatively straightforward, and already the error term is as good as will be useful in Chap. 6. The main argument from Sect. 4.1.1 must not be used for large s because it is for these values that the error term in question would become too large. For these s, where the sifting primes involved are small, the "fundamental lemma" from Chap. 3 is used.

The expert reader may feel that the method used in the text to weld these two estimates together, via the method of composition of two sieves supplied in Sect. 1.3.7, is a rather idiosyncratic one. It is more usual to apply the main argument to a sequence that has already been subjected to a pre-sieving via the fundamental lemma. The argument in the text has been selected for ease of exposition, since it has the feature that when considering the main part of the argument we do not need to be concerned with any error terms that arise from the use of the fundamental lemma.

Sect. 4.4.3. The argument here, and that in Sect. 4.3, is from [Iwaniec (1980a)]. The induction procedure is rather delicate, but leads to the rather good exponent $1 - \Delta$ of the entry $\log D$ in Theorem 2. In [Iwaniec (1980a)] this exponent is given as $\frac{1}{3}$, presumably because this is the simplest rational fraction that is valid whenever $\kappa \geq \frac{1}{2}$.

In the important case $\kappa = 1$ it is shown in [Iwaniec (1971)] that it is possible to take $\Delta = 0$, and an application is discussed in which the value of Δ is important. This is to the maximum length $C_0(r)$ of a sequence of consecutive integers each divisible by one of the first r primes. It is shown that $C_0(r) \ll r^2 \log^2 r$. In [Iwaniec (1978c)] this result was extended to the number $C(r)$, for which the r forbidden prime factors may be from any suitably chosen set.

On the other hand it seems unlikely that negative values of Δ can be attained, since this would lead to conclusions (about the number of primes in an interval, for example) that would imply the non-existence of Siegel zeros of Dirichlet L-functions. A weaker edition of this remark is discussed in [Siebert (1983)].

Sect. 4.5.1. The example used here was pointed out in [Selberg (1952a)], in the first place in connection with the "special situation" of the text. Theorem 1 is to the effect that when $\kappa = 1$ the functions F and f are optimal throughout their range. The reader will note that the method of proof is by straightforward induction, and that this produces an adverse effect on the error term. The entry $s^2 \log s$ is related to that arising for similar reasons in another situation that will be mentioned in a note on Sect. 7.1. In the context arising in Sect. 7.1 a much better estimate is obtained in the text, and it seems likely that with due effort a similarly improved error term ought to be obtainable in Theorem 1. On the other hand, in the accounts in [Motohashi (1981)] and [Selberg (1991)] no explicit bound for the error term is obtained, it being left as $o(1)$ as $X \to \infty$, not uniformly in s.

The example (1.1) can be used to shed some further light on Rosser's construction than is given by the remarks in Sect. 3.3.1. The optimal nature of the main terms in the theorems of Sect. 4.4 indicates that the cast-out terms in the fundamental identity in Lemma 3.3.2, viz.

$$\sum_{a \in \mathcal{A}} \sum_{t | (a, P(z))} \mu(t) \bar{\chi}(t) S\left(\frac{a}{t}, P(q(t))\right),$$

(see (3.1.2.15)) should be negligibly small when $\mathcal{A}$ is the example (1.1) with r of appropriate parity. Let us look at the case of the lower bound sieve, so that $r \equiv 1$, mod 2. Then the first two cast-out terms are

$$\Sigma_2 = \sum_{\substack{p_1 p_2 | a \\ p_1 p_2^3 \geq D}} S\left(\frac{a}{p_1 p_2}, P(p_2)\right),$$

$$\Sigma_4 = \sum_{\substack{p_1 p_2 p_3 p_4 | a \\ p_1 p_2^3 < D \leq p_1 p_2 p_3 p_4^3}} S\left(\frac{a}{p_1 p_2 p_3 p_4}, P(p_4)\right),$$

in which $z > p_1 > p_2 > \cdots$ as usual. For the purpose of this discussion let us oversimplify the example from (1.1) to one where $2 | \Omega(a)$ and a is "about D" in size. Then, in Σ_4, the condition $p_1 p_2^3 < D$ forces $p_1 p_2 p_3 p_4 < D$, and the condition $p_1p_2p_3p_4^3 \geq D$ forces $p_1 \ldots p_6 > D$ when $(a, P(p_4)) = 1$. Thus $\Sigma_4 = 0$, except in special cases in which two of the p_i are nearly equal. Similarly when $z < \sqrt{D}$ we would find $\Sigma_2 = 0$ for such a. Observe that this line of reasoning would fail if the exponent 3 in the conditions of summation were replaced by any other number.

This reasoning appears in a slightly less heuristic style in [Greaves (1989)]. We shall refer to it again in connection with the weighted sieve of Chap. 5.

The example (1.1) when r is odd, which demonstrates the optimal nature of the sifting limit 2 when $\kappa = 1$, is often referred to as the "parity obstacle" which prevents the sieve method from revealing the existence of primes in a suitable set. On the other hand if the sieve is used in conjunction with suitable extra information then there is no fundamental reason why the parity obstacle can not be overcome. It is of course essential that this extra information should be sufficient to exclude the possibility that one might be talking about the example (1.1).

An example of an argument answering to this description occurs in [Iwaniec and Jutila (1979)], where the sieve (in the form discussed in Chap. 6) is used in conjunction with analytic information to improve the known bound for g such that an interval $[X^g - X, X^g]$ with X large is guaranteed to contain a prime. Recent results of this type have been obtained by a somewhat different method, still using sieve ideas. See [Baker and Harman (1996)] for a result that is reported to be the subject of a forthcoming improvement by the same authors.

A somewhat different parity-breaking argument was given relatively recently in [Friedlander and Iwaniec (1998b)]. This paper introduces an extra parity-breaking hypothesis, and produces a sieve capable of detecting primes. On the other hand the extra hypothesis is of such a nature that it seems extremely difficult to show that it is satisfied. The only case to date in which this has been done is in the application for which the new sieve seems to have been created. In a paper [Friedlander and Iwaniec (1998a)] in the same volume it is shown that the polynomial $x^2 + y^4$ represents infinitely many primes.

A related result is that of [Fouvry and Iwaniec (1997)], where a similar result is proved for $x^2 + p^2$, in which p is prime. In the same category is a result of D. R. Heath-Brown, yet to appear in the literature. Heath-Brown uses a different parity-breaking argument in a proof that the binary form $x^3 + 2y^3$ represents infinitely many primes.

Sect. 4.5.2. The example discussed here was given in [Selberg (1972)], with a detailed discussion in [Selberg (1991)].

The formula (2.6) was proved in [Landau (1909)] by the now usual methods of analytic number theory initially created to prove the Prime Number Theorem. There is a more recent exposition in [LeVeque (1956)]. An elementary method is described in [Selberg (1991)], which is very much shorter and simpler than the analogous situation of an elementary proof of the prime number theorem. A third method is in [Iwaniec (1976)], which is of interest in this context in that the method used is the sieve with $\kappa = \frac{1}{2}$, the optimal nature of which is the topic currently under discussion.

Some limiting examples relating to the situation where $\kappa > 1$ are given in [Selberg (1991)], but these do not have the decisive character of the examples supplied for the cases $\kappa = 1$ and $\kappa = \frac{1}{2}$.

5. The Sieve with Weights

A characteristic of the methods described in previous chapters is that, although they may have originally been created in the hope of finding primes in suitable sequences of interest, in practice one has to be satisfied with less. For example, let $f(n)$ denote a polynomial of degree g, irreducible over the integers and not having any fixed prime factor. Then it is an easy inference from Proposition 4.4.1 that $\gg X/\log X$ of the numbers $f(n)$ with $1 \leq n < X$ have no prime factors smaller than $X^{1/s}$, when $s > 2$ and X is large enough. Since $f(n) \ll X^g$ this implies (if s is chosen sufficiently close to 2) that these numbers $f(n)$ do not have more than $2g$ distinct prime factors, all of them exceeding $X^{1/s}$. Furthermore, the word "distinct" can easily be removed from this statement, since the number of $f(n)$ divisible by the square of a prime p with $p \geq X^{1/s}$ is $\ll X^{1-1/s}$.

Numbers of this type, the number of whose prime factors does not exceed a specified bound, are usually referred to as being *almost-prime*. The symbol P_r is frequently used to denote a number having at most r prime factors. Typically, it is numbers of this type that are delivered by applying a lower bound sieve to a suitable sequence.

In this situation it is reasonable to revise one's objective. Instead of asking that the prime factors of the survivors of a sifting process should not be small, ask that the number of these prime factors should not be too large. In this case one would not object if one of these prime factors was rather small. The addition of weights to the sieve will provide an approach to this revised objective. Indeed, while the process described in Sect. 5.1 is a combination of a rather simple weighting device to the (unweighted) sieve of Chap. 4, it is our objective to describe a combination of sieving and weighting that functions as a unified whole. This objective will be achieved, at least in part, in Sect. 5.3.

The simplest procedure that we describe appears in Sect. 5.1. In many situations this will already be rather successful. For example, in connection with the polynomial $f(n)$ of degree g alluded to above, it will be a straightforward inference from Theorem 5.1.1 that there are $\gg X/\log X$ values of n with $1 \leq n < X$ for which $f(n)$ has at most $g+1$ prime factors. The bound $g+1$ is a significant reduction on the value $2g$ achieved by the sieve without weights.

Theorem 5.1.1 is more general than the particular application just described, and this and the other results of this chapter may be summarised as follows. As usual, $\mathcal{A}$, ρ and $r_{\mathcal{A}}$ are to satisfy (1.3.1.1). For the most part we confine our attention to the situation in which the function ρ has sifting density 1. Being more specific than in Sect. 1.3.1, say that the sequence $\mathcal{A}$ has a level of distribution D if the remainder $r_{\mathcal{A}}$ satisfies

$$\sum_{d<D} |r_{\mathcal{A}}(d)| \leq X/\log^2 X \ .$$

Then all the results in this chapter imply corollaries of the following type, for various values of a number Λ_R.

Proposition 1. *Let R be an integer with $R \geq 2$. Suppose that $\mathcal{A}$ has level of distribution D and degree g in the sense that*

$$a \in \mathcal{A} \Rightarrow a < D^g \ .$$

Suppose $g < \Lambda_R = R - \delta_R$. Then $\mathcal{A}$ contains at least $\gg X/\log X$ numbers having at most R distinct prime factors.

The number Λ_R is sometimes referred to as "the sifting limit for the weighted sieve", although the analogy with the sifting limit of previous chapters is not a close one. The number Λ_R has not been identified as the zero of any function analogous to the expression f defined in Sect. 4.2, for example.

Write $\Lambda_R = R - \delta_R$, as in Proposition 1. In Theorem 5.1.1 the number Λ_R is described by an explicit elementary expression, which implies that we may take

$$\delta_R = \frac{\log 4}{\log 3} - 1 = 0 \cdot 261859 \ldots \text{ for all } R \ , \quad \delta_2 = \frac{\log 3 \cdot 6}{\log 3} - 1 = 0 \cdot 165956 \ldots \ .$$

In Sect. 5.2 we outline a method that originated with Buchstab. In this case the evaluation of the corresponding Λ_R depends upon a numerical integration, which shows that we may take

$$\delta_R = 0 \cdot 144001 \ldots \quad \text{for all } R \ , \quad \delta_2 = 0 \cdot 108014 \ldots \ .$$

The method described in Sect. 5.3 involves integrating the weighting procedure of Sect. 5.1 with the mechanism of Rosser's sieve, as opposed to using them separately. In this instance the resulting computation (and the construction that leads to it) is rather more formidable, but leads to the admissible choice

$$\delta_R = 0 \cdot 124820 \ldots \text{ for all } R \ , \quad \delta_2 = 0 \cdot 063734 \ldots \ .$$

The method of Sect. 5.3 is not quite the best that is known to date, but can be improved by a suitable injection of Selberg's λ^2 idea. We do not give a full account of this development in this volume, but a brief sketch and a detailed reference is supplied in the notes to this chapter. In this way improved admissible values have been found when $R \le 4$:

$$\delta_2 = 0{\cdot}044560\,, \quad \delta_3 = 0{\cdot}074267\,, \quad \delta_4 = 0{\cdot}103974\,.$$

There is one more remark that can be made about the theorems of this chapter. The "almost-prime" numbers delivered by Theorem 5.1.1 also have the property that all their prime factors exceed $X^{1/4}$. Thus the original feature of an unweighted sifting argument has been reduced but not eliminated. This remark applies also to in the more elaborate weighted sieves described in subsequent sections. Thus the methods of this chapter still refuse to make an effective consideration of some of the numbers P_r that they were designed to find, namely those for which one of their prime factors is distinctly small. One may form an opinion that this refusal is a residual weakness that may be associated with the failure of these methods to prove quite all the results for which one might hope.

An Extremal Example for the Weighted Sieve. It is conjectured that Proposition 1 ought to be true, in a suitable quantitative form, even when $\delta_R = 0$. Such a result, if it could be proved, would correspond to the case $\kappa = 1$ of Proposition 4.4.1 on Rosser's sieve, in that it would be best possible. Moreover, the example which we use to show that these two propositions are best possible are very closely related. This remark makes it more tantalising that the Proposition in Chap. 4 was made into a theorem, but that with $\delta_R = 0$ Proposition 1 is still a conjecture.

The following example is a straightforward modification of the construction from Sect. 4.5.1. For given sufficiently large X fix distinct primes p_1, $p_2, \ldots, p_{R-1}$, to lie in the interval $\left(\frac{1}{2}X, X\right)$, this particular choice not being crucial. Then take

$$\mathcal{A} = \mathcal{A}(X) = \left\{a = bp_1\ldots p_{R-1} \,.\, 2X \le b < 4X,\ \Omega(a) \equiv 0,\ \mathrm{mod}\,2\right\}.$$

Then when $d < \frac{1}{2}X$ we find

$$\mathcal{A}_d = \frac{X}{d} + r_{\mathcal{A}}(d) \quad \text{with} \quad \sum_{d \le X^{1-\varepsilon}} |r_{\mathcal{A}}(d)| \ll \frac{X}{\log^2 X}\,,$$

exactly as in (4.5.1.4), if we again make $\varepsilon = A(\log \log X)^6/\log X$. Thus $\mathcal{A}$ has level of distribution $D = X^{1-\varepsilon}$, in the sense used in Proposition 1. Furthermore

$$a \in \mathcal{A} \Rightarrow a \le X^R = D^{R/(1-\varepsilon)}\,,$$

so that $\mathcal{A}$ has degree g for each $g > R$, when X is sufficiently large. But every member of the set $\mathcal{A}$ has at least $R + 1$ prime factors. Thus Proposition 1 cannot hold for any $\delta_R < 0$.

5.1 Simpler Weighting Devices

The constructions in this section count the numbers a in the sequence $\mathcal{A}$ to be sifted with a certain additive weight attached to the prime factors of the number a. There are advantages to be gained by adding suitable weights to certain of the composite factors of a, but we will see that the methods of this section are already rather effective, as well as being (provided the sieve of Chap. 4 is taken for granted) rather straightforward.

5.1.1 Logarithmic Weights

The machinery described below counts the number of prime factors p of a number a in a slightly special way, according to multiplicity when $p \geq D$ but counting each p at most once when $p < D$. Let $\nu(D, a)$ denote this number of prime factors,

$$\nu(D,a) = \sum_{\substack{p<D \\ p|a}} 1 + \sum_{\substack{p,\alpha \\ p^\alpha|a;\, p\geq D}} 1 \, . \tag{1.1}$$

The following lemma describes an idea that is basic to much of what follows.

Lemma 1. *Let $R \geq 2$ be an integer, and suppose $a \leq D^g$, where $g \leq R$. Define*

$$m_1(a) = 1 - \sum_{p|a;\, p<D} \left(1 - \frac{\log p}{\log D}\right) . \tag{1.2}$$

Then $\nu(D, a) \leq R$ whenever $m_1(a) > 0$.

If $m_1(a) > 0$ then

$$0 < 1 - \sum_{\substack{p<D \\ p|a}} \left(1 - \frac{\log p}{\log D}\right) - \sum_{\substack{p,\alpha \\ p^\alpha|a;\, p\geq D}} \left(1 - \frac{\log p}{\log D}\right) ,$$

where the sum has been extended to one over all prime factors of a, counted as in (1.1) (the additional summands being negative). Here the terms $\log p/\log D$ contribute at most $\log D^g/\log D$. Hence

$$0 < 1 - \nu(D,a) + g \leq R + 1 - \nu(D,a) \, ,$$

so that $\nu(D, a) < R + 1$, whence $\nu(D, a) \leq R$ as stated.

Lemma 1 is of little use on its own. One might hope to show that $m_1(a) > 0$ for some a in a suitable set $\mathcal{A}$. However, if we invoked the usual expression (1.3.1.1) in the simplest case $\rho(p) = 1$, and ignored the effect of error terms $r_A(d)$, we should obtain only the negative estimate

$$\sum_{a\in A} m_1(a) = X\left(1 - \sum_{p<D}\frac{1}{p}\left(1 - \frac{\log p}{\log D}\right)\right)$$
$$\sim -X\log\log D \quad\text{as}\quad D\to\infty. \tag{1.3}$$

This approach, as it stands, is far too simplistic to succeed, corresponding as it does to ignoring all d except $d = 1$ and $d = p$ in Legendre's construction (1.1.2.5) (or in Brun's construction introduced in Lemma 3.1.1).

We will consider a certain weighted analogue of Brun's construction in Sect. 5.3. Meantime, observe that it is the smallest primes that contribute most to the negative factor $-\log\log D$ in (1.3). Accordingly we may seek to exclude them by applying (1.2) only to those numbers a that are free of small prime factors, using the usual sieve method (without "weights") for this purpose.

In this approach, the simple construction (1.2) thus appears as a "weighting" adjunct to the conventional sieve method, which is used to obtain as estimate of the type appearing in Lemma 2 below. For simplicity, the account is restricted to the context of sifting density 1, but we allow a more general weighting function than (1.2).

Lemma 2. *Suppose $W(1) > 0$ and that $W(x)$ is positive, increasing and bounded over $1/s \leq x < 1/u$, where $1 < u < s$. Define*

$$\mathcal{W}(\mathcal{A}) = \sum_{\substack{a\in\mathcal{A}\\(a,P(D^{1/s}))=1}} \left\{W(1) - \sum_{\substack{p|a\\D^{1/s}\leq p<D^{1/u}}}\left(W(1) - W\left(\frac{\log p}{\log D}\right)\right)\right\}. \tag{1.4}$$

Suppose that the function ρ has sifting density 1. Then

$$\mathcal{W}(\mathcal{A}) \geq XV(P(D^{1/s}))\left(M(W) + O\left(\frac{e^{\sqrt{L}}}{\log^{1-\Delta}D}\right)\right) - W(1)\sum_{d<D}|r_A(d)|,$$

where $\Delta > 0$ is arbitrary, the implied constant may depend upon u, s, Δ and the function ρ, and

$$M(W) = W(1)f(s) - \int_{1/s}^{1/u}(W(1) - W(x))F(s(1-x))\frac{dx}{x}, \tag{1.5}$$

with F and f as in Proposition 4.4.1.

The error term quoted follows from Theorem 4.4.2. A much weaker estimate that suffices for most purposes follows using Theorem 4.4.1.

5. The Sieve with Weights

Rewrite the sum (1.4) as

$$W(\mathcal{A}) = W(1) \sum_{\substack{a \in \mathcal{A} \\ (a, P(D^{1/s}))=1}} 1 \\
- \sum_{D^{1/s} \leq p < D^{1/u}} \left(W(1) - W\left(\frac{\log p}{\log D}\right) \right) \sum_{\substack{a \in \mathcal{A}_p \\ (a, P(D^{1/s}))=1}} 1, \quad (1.6)$$

where, as usual, $a \in \mathcal{A}_p$ means $a \in \mathcal{A}$ and $a \equiv 0$, mod p. The first sum is dealt with directly by using Proposition 4.4.1:

$$\sum_{\substack{a \in \mathcal{A} \\ (a, P(D^{1/s}))=1}} 1 \geq X V(P(D^{1/s}))\Big(f(s) - \eta(D,s)\Big) - \sum_{\substack{d < D \\ d | P(D^{1/s})}} |r_\mathcal{A}(d)|. \quad (1.7)$$

For the second sum in (1.6) note that when $p > D^{1/s}$ and $d | P(D^{1/s})$ the set $\mathcal{A}_p$ satisfies

$$\sum_{n \in \mathcal{A}_{dp}} 1 = \frac{X \rho(p)}{p} \cdot \frac{\rho(d)}{d} + r_\mathcal{A}(dp).$$

This follows directly from Definition 1.3.4, since $(d,p) = 1$ in the current situation. Apply Theorem 1.3.1 to the set $\mathcal{A}_p$, again using Proposition 4.4.1, but with level $D_p = D/p$. Since $D^{1/s} = (D/p)^{1/s_p}$ with $s_p = s(1 - \log p/\log D)$, this gives

$$\sum_{\substack{a \in \mathcal{A}_p \\ (a, P(D^{1/s}))=1}} 1 \leq \frac{X\rho(p)}{p} V(P(D^{1/s}))\Big(F(s_p) + O(\eta(D_p, s_p))\Big) \\
+ \sum_{d_p \leq D/p} |r_\mathcal{A}(pd_p)|. \quad (1.8)$$

Now sum over p. For the error terms η in (1.7) and (1.8), Theorem 4.4.2 gives

$$\eta(D,s) + \sum_{D^{1/s} \leq p < D^{1/u}} \frac{\rho(p)}{p} \eta(D_p, s_p) \\
\ll \frac{1}{\log^{1-\Delta} D} + \sum_{D^{1/s} \leq p < D^{1/u}} \frac{\rho(p)}{p (\log D/p)^{1-\Delta}} \ll \frac{1}{\log^{1-\Delta} D}, \quad (1.9)$$

using the observation $\log(D/p) \geq (1 - 1/u)\log D$ and then Lemma 1.3.5. The contribution to the main term from (1.8) is estimated, using partial summation as in Lemma 1.3.6(ii), by

$$\sum_{D^{1/s} \leq p < D^{1/u}} \frac{\rho(p)}{p} \left(W(1) - W\left(\frac{\log p}{\log D}\right) \right) F\left(s\left(1 - \frac{\log p}{\log D}\right) \right) \\
\leq \int_{1/s}^{1/u} (W(1) - W(x)) F(s(1-x)) \, d\log x + O\left(\frac{L}{\log D^{1/s}}\right). \quad (1.10)$$

5.1 Simpler Weighting Devices

Lemma 2 as stated is an assembly of (1.6), (1.7), (1.8), (1.9) and (1.10).

If we were to apply Lemma 2 to the weight function $m_1(a)$ described in Lemma 1, we would take $W(1) = 1$, $W(p) = \log p/\log D$. We need $M(W) > 0$ if the result is to be non-trivial. The expression (1.5) for $M(W)$ is then particularly simple when $s \leq 4$, so that F and f are as in (4.2.4.16) and

$$M(W) = \frac{2e^\gamma}{s}\left(\log(s-1) - \int_{1/s}^{1/u} \frac{dx}{x}\right) = \frac{2e^\gamma}{s}\left(\log\frac{s-1}{s} + \log u\right). \quad (1.11)$$

The best choice here is $s = 4$, in which case $M(W) > 0$ provided $u > \frac{4}{3}$. The function $m_1(a)$ was not constructed with the situation $u > 1$ in mind, however, and it is more efficient to proceed as in the following section.

5.1.2 Modified Logarithmic Weights

The weighting function described below is a modification of the possibly more natural logarithmic weights from Sect. 5.1.1. This modification is more successful when Lemma 2 is applied.

In the following, the number U plays the part performed by $1/u$ in Lemma 2. Choose numbers $V < U \leq 1$, where V may be negative. Define

$$W(1) = U - V, \qquad W(x) = \begin{cases} x - V & \text{if } V \leq x < U \\ U - V & \text{if } U \leq x \\ 0 & \text{if } 0 \leq x < V. \end{cases} \quad (2.1)$$

Then specify

$$m_2(a) = W(1) - \sum_{\substack{p < D^U \\ p \mid a}} \left(W(1) - W\left(\frac{\log p}{\log D}\right)\right), \quad (2.2)$$

where the condition $p < D^U$ is actually redundant because of the definition of $W(t)$. Further, denote

$$w(p) = W\left(\frac{\log p}{\log D}\right). \quad (2.3)$$

Then Lemma 1 generalises as follows.

Lemma 3. *Let $\nu(D, a)$ be as in (1.1). Suppose g chosen so that $a \leq D^g$, and that $RU + V \geq g$. Let q_a denote the smallest prime factor of a. Suppose also that $m_2(a) > 0$ in (2.2). Then $\nu(D^U, a) \leq R$ and $m_2(a) \leq w(q_a)$.*

The proof follows the lines of Lemma 1. When $m_2(a) > 0$ we find

$$0 < U - V - \sum_{\substack{p<D \\ p|a}} \left(U - \frac{\log p}{\log D}\right) - \sum_{\substack{p,\alpha \\ p^\alpha | a; \, p \geq D}} \left(U - \frac{\log p}{\log D}\right),$$

where the sum has been extended over all prime factors of a, with multiple prime factors exceeding D^U being counted multiply. Thus the number of these primes is just $\nu(D^U, a)$, defined as in (1.1). Hence

$$0 < U - V - U\nu(D^U, a) + g \leq U(R + 1 - \nu(D^U, a)),$$

because $g \leq RU + V$. Consequently $\nu(D^U, a) < R + 1$, whence $\nu(D^U, a) \leq R$ as asserted. Lastly, since $W(t) \leq W(1)$ in (2.1), it follows from (2.2) that

$$m_2(a) \leq W\left(\frac{\log q_a}{\log D}\right) = w(q_a).$$

In fact $m_2(a) = W(1)$ if $(a, P(D^U)) = 1$, $m_2(a) = w(p)$ if $(a, P(D^U)) = p$, in which case $p = q_a$, and $m_2(a) \leq w(q_a)$ otherwise because of extra summands $W(1) - w(p)$ that this argument casts aside.

The last step in this proof of Lemma 3 is not always a sharp one, a point that will be taken up in Sect. 5.3.

The principal result of this section follows from Lemmas 2 and 3. The argument is no deeper than that leading to (1.11), but there are some matters of detail to be attended to.

Theorem 1. *Suppose that ρ has sifting density 1 and that $a \leq D^g$ for all a in $\mathcal{A}$. Suppose that $g = \Lambda_R - \eta$, where $\eta > 0$, and specify*

$$\Lambda_R = R - \frac{\log(4U/3)}{\log 3}, \qquad U = \frac{1}{1 + 3^{-R}}, \qquad V = g - RU. \qquad (2.4)$$

Take W, w as in (2.1) and (2.3). Then

$$\sum_{\substack{a \in \mathcal{A}; \, q_a \geq D^{1/4} \\ \nu(D^U, a) \leq R}} w(q_a) \geq X \frac{e^\gamma}{4} V(P(D^{1/4})) \left(2\eta \log 3 + O\left(\frac{1}{\log^{1-\Delta} D}\right)\right)$$

$$- W(1) \sum_{d \leq D} |r_\mathcal{A}(d)|, \qquad (2.5)$$

where $\Delta > 0$ as in Lemma 2, and the O-constant depends on R and Δ.

Theorem 1 counts the numbers a with the weight $w(q_a)$ attached to the smallest prime factor q_a of the number a. It is easy to deduce a result that counts them with the constant weight 1. The factor $e^\gamma V(P)$ can also be expressed in a rather different way, as follows.

5.1 Simpler Weighting Devices

Corollary 1.1. *The number of a in $\mathcal{A}$ counted in Theorem 1 satisfies*

$$\sum_{\substack{a\in\mathcal{A};q_a\geq D^{1/4}\\ \nu(D^U,a)\leq R}} 1 \geq \frac{X}{\log D}\left(\prod_{p<D^{1/4}}\frac{p-\rho(p)}{p-1}\right)\left(\frac{2\eta\log 3}{\frac{\log 4}{\log 3}+\eta}+O\!\left(\frac{1}{\log^{1-\Delta}D}\right)\right)$$
$$-\sum_{d\leq D}|r_\mathcal{A}(d)|\,. \tag{2.6}$$

The ratio $2\eta\log 3/(\log 4/\log 3 + \eta)$ appearing in Corollary 1.1 could be replaced by $\eta\log 3$, for example, if η is not too large. This is legitimate if $\eta < \frac{14}{19}$, for instance, because $2^{19} = 524\,288 < 531\,441 = 3^{12}$, so that $\log 4/\log 3 < \frac{24}{19}$.

The last inequality implies a convenient rational approximation to the number Λ_R in Theorem 1. Since, in (2.4), $U < 1$ for all R we obtain

$$\Lambda_R > R - \tfrac{5}{19} \quad \text{for all } R\,. \tag{2.7}$$

In a similar way it can be verified that

$$\Lambda_2 > 2 - \tfrac{1}{6}\,,\quad \Lambda_3 > 3 - \tfrac{2}{9}\,,\quad \Lambda_4 > 4 - \tfrac{1}{4}\,. \tag{2.8}$$

Almost-primes. The discussion of Lemma 1 shows why the number of prime factors of a number a is in the first place counted as described by the function $\nu(D^U, a)$. To count them more conventionally proceed as follows.

Say a number a is an *almost-prime of order R*, and write $a = P_R$, if the total number $\Omega(a)$ of prime factors of a satisfies $\Omega(a) \leq R$.

To obtain a result counting prime factors in this way from Theorem 1, we need a suitable bound for the number of a in $\mathcal{A}$ divisible by p^2 when $p \leq D^U$. The hypothesis (2.10), which could be relaxed somewhat, is satisfied in most situations likely to be of interest.

Corollary 1.2. *Suppose D is chosen so that*

$$X^\theta \leq D \leq \frac{X}{\log^2 X}\,,\quad \sum_{d\leq D}|r_\mathcal{A}(d)| \leq \frac{X}{\log^2 X}\,, \tag{2.9}$$

for some $\theta > 0$, and suppose there exists $\alpha > 0$ so that

$$\sum_{D^{1/s}\leq p<D^U}|\mathcal{A}_{p^2}| \ll \frac{X}{D^{1/s}} + D^{1-\alpha(1-U)+\varepsilon}\,, \tag{2.10}$$

for each $\varepsilon > 0$. Then under the conditions of Theorem 1 the number of P_R-numbers in $\mathcal{A}$ satisfies

$$\sum_{a\in\mathcal{A},a=P_R} 1 \geq \frac{X}{\log D}\left(\prod_{p<D^{1/4}}\frac{p-\rho(p)}{p-1}\right)\left(\frac{2\eta\log 3}{\frac{\log 4}{\log 3}+\eta}+O\!\left(\frac{1}{\log^{1-\Delta}D}\right)\right)$$
$$+O\!\left(\frac{X}{\log^2 X}\right). \tag{2.11}$$

5. The Sieve with Weights

The numbers counted in (2.6) which are not of the form P_R are divisible by p^2 for some prime p with $D^{1/4} \leq p < D^U$, so that because of (2.10) and (2.9) their number is $O(X/\log^2 X)$. Thus Corollary 1.2 will follow immediately from Corollary 1.1.

Proof of Theorem 1. Use of Lemma 3 gives

$$\sum_{\substack{a \in \mathcal{A} \\ \nu(D,a) \leq R}} w(q_a) \geq \sum_{\substack{a \in \mathcal{A} \\ \nu(D,a) \leq R}} m_2(a)$$

$$\geq \sum_{\substack{a \in \mathcal{A} \\ m_2(a) > 0}} m_2(a) \geq \sum_{a \in \mathcal{A}} m_2(a) = W(\mathcal{A}) , \qquad (2.12)$$

the expression estimated in Lemma 2. The numbers a counted will satisfy $q_a \geq D^{1/4}$, as asserted, if we take $s = 4$ in Lemma 2. It only remains to choose the parameters U and V so that $M(W) = \frac{1}{4} e^\gamma \cdot 2\eta \log 3$ in Lemma 2 and the constraint $RU + V \geq g$ of Lemma 3 is satisfied. We require to show that this is possible when $g \leq \Lambda_R - \eta < \Lambda_R$, with Λ_R is as in (2.4). Actually this value of Λ_R is optimal under the restriction $s \leq 4$ that keeps the expressions for F and f elementary.

With $RU + V = g$ as in (2.4) take $s \leq 4$. Since W is as in (2.1) the expression in (1.5) is

$$M(W) = (U - V)f(s) - \int_{1/s}^{U} (U - x) F(s(1-x)) \frac{dx}{x}$$

$$= \frac{2e^\gamma}{s} \left((U(R+1) - g) \log(s-1) - \int_{1/s}^{U} \frac{U - x}{1 - x} \frac{dx}{x} \right) . \qquad (2.13)$$

If $U = 1$, then (2.13) gives $M(W) = 2e^\gamma((R+1-g)\log(s-1) - \log s)/s$, so that $M(W) > 0$ if $g < R - \log \frac{4}{3}/\log 3$, the choice $s = 4$ being the best one. Actually $U < 1$ is required to meet the requirements of Lemma 2.

The value for g stated in Theorem 1 follows by selecting U optimally when $s = 4$. For later purposes in Sect. 5.3 we rewrite (2.13) as

$$M(W) = \frac{2e^\gamma}{s} \mathcal{M}(W) = \frac{2e^\gamma}{s} I(U, V)$$

$$I(U,V) = \int_{1/4}^{1/2} \frac{U - V}{1 - x} \frac{dx}{x} - \int_{1/4}^{U} \frac{U - x}{1 - x} \frac{dx}{x} + \alpha - \frac{1}{3}\beta(1 - U) , \qquad (2.14)$$

and α and β are certain constants, currently given by $\alpha = \beta = 0$, but which will be non-zero in other situations (for example if (2.13) were used with $s > 4$.)

5.1 Simpler Weighting Devices

We will show that the choices in (2.4) maximise $\Lambda_R = RU + V_0$ subject to $I(U, V_0) = 0$. Here

$$I(U, V_0) = -\int_{1/2}^{U} \frac{U-x}{1-x} \frac{dx}{x} + \int_{1/4}^{1/2} \frac{x - \Lambda_R + RU}{1-x} \frac{dx}{x} + \alpha - \tfrac{1}{3}\beta(1-U) \,. \quad (2.15)$$

To maximise Λ_R, make $\partial I/\partial U = 0$ in (2.15). Thus we seek

$$R\int_{1/4}^{1/2} \frac{1}{1-x} \frac{dx}{x} = \int_{1/2}^{U} \frac{1}{1-x} \frac{dx}{x} - \tfrac{1}{3}\beta = -\int_{2}^{1/U} \frac{dt}{t-1} - \tfrac{1}{3}\beta, \quad (2.16)$$

so we make $1/U = 1 + 3^{-R}e^{-\beta/3}$, which when $\beta = 0$ is the choice specified in (2.4). Then (2.16) reduces (2.15) to

$$\begin{aligned}
I(U, V_0) &= -\Lambda_R \log 3 + \log \tfrac{3}{4} - \log(1-U) + \alpha - \tfrac{1}{3}\beta \\
&= -\Lambda_R \log 3 - \log \frac{4U}{3} - \log\left(\frac{1}{U} - 1\right) + \alpha - \tfrac{1}{3}\beta, \\
&= -\Lambda_R \log 3 + R\log 3 - \log \frac{4U}{3} + \alpha \,.
\end{aligned}$$

Thus $I(U, V_0) = 0$ when

$$\Lambda_R = R - \frac{\log(4U/3e^\alpha)}{\log 3}, \qquad U = \frac{1}{1 + 3^{-R}e^{-\beta/3}}, \quad (2.17)$$

which at $\alpha = \beta = 0$ is as stated in (2.4).

To obtain Theorem 1, take $g = RU + V$ with $V = V_0 - \eta$, so that (2.14) and (2.15) show

$$I(U, V) = I(U, V_0) + \eta \log 3 = \eta \log 3 \,,$$

and $g = \Lambda_R - \eta$ as stated. Theorem 1 follows using (2.12), Lemma 2, and (2.14).

We should check that $W(1) > 0$, as required in (2.1). Reference to (2.1) and (2.4) shows

$$W(1) = U - V = U(R+1) - g = U(R+1) - \Lambda_R + \eta \,. \quad (2.18)$$

Here $\eta > 0$ and

$$R - \frac{\log 4/3}{\log 3} < \Lambda_R \leq R \,, \quad (2.19)$$

because $U < 1$ in (2.4). Thus

$$W(1) > U(R+1)) - R = 1 - (1-U)(R+1) = 1 - \frac{R+1}{3^R + 1},$$

so $W(1) > 0$ because $R < 3^R$. In a similar way the current choice of U and V gives

$$W(1) < \frac{\log 4}{\log 3} + \eta \,. \quad (2.20)$$

Proof of Corollary 1.1. First replace the factor $\frac{1}{4}e^{\gamma}V(P(D^{1/4}))$ in Theorem 1 by the expression

$$\frac{1}{\log D}\left(\prod_{p<D^{1/4}}\frac{p-\rho(p)}{p-1}\right)\left(1+O\left(\frac{1}{\log D}\right)\right),$$

using the consequence (2.2.2.4) of Mertens' formula. Next, use the inequality $w(q_a) \leq W(1)$, immediate from (2.3) and (2.1), and divide through by $W(1)$. Corollary 1.1 then follows using the proof of Theorem 1, and in particular (2.20). A slightly sharper statement could be made when R is small.

5.1.3 Some Applications

There follow some relatively simple illustrations of the utility of Theorem 1, or of the stronger theorems of the same type to be discussed later.

Almost-primes in Short Intervals. Corollary 1.3 provides an upper bound for the difference between consecutive P_R numbers.

Corollary 1.3. *Suppose $g = \Lambda_R - \eta$, where $\eta > 0$ and Λ_R is as in Theorem 1. Write $D = X/\log^2 X$. Then the number of almost-primes P_R of order R in the interval*

$$\mathcal{A} = \{a : X^g - X \leq a < X^g\} \tag{3.1}$$

satisfies the estimate (2.11) in Corollary 1.2. In particular there are at least two P_R numbers in this interval, provided X is sufficiently large.

The permissible values of g satisfy the inequalities given in (2.7) and (2.8), so that in particular we may take

$$g = R - \tfrac{5}{19} \text{ for all } R, \qquad g = 2 - \tfrac{1}{6} \text{ if } R = 2.$$

We have only to check that the conditions of Corollary 1.2 are satisfied. Of these, (2.9) is immediate since the discussion in Sect. 1.2.2 shows that in the context of (3.1) we obtain $|r_\mathcal{A}(d)| \leq 1$ in (1.3.1.1), also

$$|\mathcal{A}_{p^2}| \leq \frac{X}{p^2} + 1,$$

so that (2.10) holds with $\alpha = 1$. Corollary 1.3 now follows.

Almost-primes in Polynomial Sequences. Corollary 1.4 still represents the state of the art on this topic, in that the value $R = g + 1$ has not been improved, except in the case $g = 2$.

Corollary 1.4. *Let f be a polynomial of degree g irreducible over the integers. Set $R = g + 1$ and $D = X/\log^3 X$. Then the number of almost-primes P_R of order R in*
$$\mathcal{A} = \{f(n) : 1 \leq n \leq X\}$$
satisfies the estimate (2.11) in Corollary 1.2, where the O-constant may now depend on f. In particular, there are infinitely many almost-primes P_{g+1} represented by the polynomial expression $f(n)$.

As in Sect. 1.2.3, take $\rho(d)$ to be the number of roots of the congruence $f(n) \equiv 0, \bmod d$. Then, because of the consequence (2.3.3.3) of the Prime Ideal Theorem, Lemma 1.3.4 shows that ρ has sifting density 1, so that in particular
$$\frac{1}{V(P(D))} \ll \log D ,$$
from (1.3.5.7).

In the current context the remainder term $r_\mathcal{A}(d)$ in Definition 1.3.4 satisfies $|r_\mathcal{A}(d)| \leq \rho(d)$, as in (1.2.3.2). Now use of (1.3.1.12) gives
$$\sum_{d \leq D} |r_\mathcal{A}(d)| \leq \frac{D}{V(P(D))} \leq D \log D \leq \frac{X}{\log^2 X} ,$$
by our choice of D, so that the hypothesis (2.9) in Corollary 1.2 is satisfied. A second use of (1.2.3.2) gives
$$\mathcal{A}_{p^2} \leq \frac{X \rho(p^2)}{p^2} + O\big(\rho(p^2)\big) \ll \frac{X}{p^2} + 1 ,$$
since $\rho(p^2) \leq g\Delta^2$, where Δ denotes the discriminant of f. The hypothesis (2.10) now follows as in Corollary 1.3, so Corollary 1.2 leads to Corollary 1.4.

5.2 More Elaborate Weighted Sieves

Inspection of the logarithmic weight $w(p) = \log p/\log D$ used in Lemma 5.1.1 shows that the expression (5.1.1.5) takes the values
$$m_1(a) = \begin{cases} 1 & \text{if } (a, P(D)) = 1 \\ \log \dfrac{p_1 p_2 \ldots p_r}{D^{r-1}} & \text{if } (a, P(D)) = p_1 p_2 \ldots p_r . \end{cases}$$

5. The Sieve with Weights

Lemma 5.1.1 showed that these values are acceptable for our purposes. Strictly negative values, such as $\log p_1 p_2 / D$ when $p_1 p_2 < D$, are a tolerable feature but an unwelcome one, as they make a negative contribution to a sum $\sum_a m_1(a)$ for which we seek a positive estimate. Some of these negative values may be eradicated by the addition of an extra term

$$\frac{1}{\log D} \sum_{p_1 p_2 < D} S\left(\frac{a}{p_1 p_2}, p_2\right) \log \frac{D}{p_1 p_2},$$

the factor $S(\cdot, \cdot)$ ensuring that this extra term appears only when p_2 is the smallest prime factor of a.

A weighting device incorporating this improvement, in which the the logarithmic weight $w(p)$ is modified as in Sect. 5.1.2, is discussed in Sect. 5.2.2. This discussion has been kept fairly brief, because the device just described is capable of improvement by the addition of a further term. The improved version, whose use will be the subject of Sect. 5.3, is described in a direct way in Sect. 5.2.1.

5.2.1 An Improved Weighting Device

The additional term that makes up the device to be described is the entry (1.6) below. The effect of introducing this term is to repair the obvious residual deficiency that would otherwise remain. This remark is not as significant as it may sound, however, because it is not obvious how the sum just written down should be made the subject of an effective sieve estimate.

The function W will be as in Sect. 5.1, save that in some situations it will be necessary to allow the positive "weight" $W(x)$ to be smaller than was specified in (5.1.2.1). Thus we now require

$$W(1) = U - V, \qquad W(x) \begin{cases} \leq x - V & \text{if } 0 \leq x < U \\ = U - V & \text{if } U \leq x \\ = 0 & \text{if } 0 \leq x < V. \end{cases} \qquad (1.1)$$

Denote

$$w(1) = W(1), \qquad w(p) = W\left(\frac{\log p}{\log D}\right), \qquad (1.2)$$

which is as before, save for the notation $w(1) = W(1)$, which will be convenient in Sect. 5.3.3.

Lemma 1 below, which will be our standard instrument for detecting almost-primes, improves upon Lemma 5.1.3 with respect to the remark made immediately following its proof. The discussion applies in particular to the simplest case $U = 1$, $V = 0$, with $W(x) = x$ for $0 \leq x \leq 1$, so that (1.2) would give $w(p) = \log p / \log D$.

Introduce a notation

$$m(a) = m_2(a) + m_3(a) + m_4(a), \qquad (1.3)$$

where

$$m_2(a) = w(1) - \sum_{p_1 | a}(w(1) - w(p_1)),\qquad(1.4)$$

as in (5.1.2.2), and

$$m_3(a) = \sum_{\substack{p_2 < p_1;\, p_1 p_2 | a \\ w(p_1) + w(p_2) < w(1)}} (w(1) - w(p_1) - w(p_2)) S\left(\frac{a}{p_1 p_2}, P(p_2)\right),\qquad(1.5)$$

$$m_4(a) = \sum_{\substack{p_3 p_2 < p_1;\, p_1 p_2 p_3 | a \\ w(p_1) + w(p_3) < w(1)}} w(p_3) S\left(\frac{a}{p_1 p_2 p_3}, P(p_2)\right).\qquad(1.6)$$

Here, $S(a/p_1 p_2, P(p_2))$ counts those integers a divisible by p_1 and p_2 but by no prime smaller than p_2. Similarly, when $p_3 < p_2 < p_1$ the expression $S(a/p_1 p_2 p_3, P(p_2))$ counts those a for which p_3 and p_2 are respectively the smallest and second smallest prime factors of a. As is observed below, the condition $p_1 < D^U$ is implicit from other aspects of the notation employed.

Lemma 1. *Let w satisfy (1.1) and (1.2). As in Lemma 5.1.3, suppose that $a < D^g$ and $RU + V \geq g$. Let $m(a)$ be defined as in (1.3). Denote $A = (a, P(D^U))$. Then $m(a) = m(A)$. Furthermore, if $m(A) > 0$, then $\nu(D^U, a) \leq R$ and $m(a) \leq w(q_a)$, where q_a denotes the smallest prime factor of a.*

The importance of this lemma is that since $m(a) \geq m_2(a)$, with strict inequality for many integers a, it should *a priori* be possible to derive a better lower estimate for $\sum m(a)$ than the bound for $\sum m_2(a)$ provided by Lemma 3.1.2.

The equality $m(a) = m(A)$ holds because (1.2) gives $0 \leq w(p) \leq w(1)$, so the condition $w(p_1) < w(1)$ is explicit in (1.5) and (1.6) and implicit in (1.4) if the sum is non-empty. Thus $p_i < D^U$ for all the primes that are involved. In this proof we write $m(a)$ rather than $m(A)$.

For some integers a it may happen that $m_3(a) = m_4(a) = 0$ (because the sums defining these quantities happen to be empty). For these a, Lemma 1 follows from Lemma 5.1.3.

The proof is now completed by showing

$$m_3(a) = m_4(a) = 0 \quad \text{whenever} \quad m(a) > 0,\qquad(1.7)$$

and then using Lemma 5.1.3. Thus the improvement in Lemma 1 over Lemma 5.1.3 is that it makes a sharper statement about some of those a for which $m(a) \leq 0$.

First, consider those a for which $m_3(a) > 0$ and their prime factors p_i for which $q_a < p_i$ and $w(q_a) + w(p_i) \leq w(1)$. These primes are $r \geq 1$ in number, say. Every entry $w(1) - w(p_1)$ in (1.5) has $p_1 = p_i$ for some i and is cancelled by a corresponding entry in (1.4), the entry in (1.4) with $p_1 = q_a$ not being used in this process. Hence

$$m_2(a) + m_3(a) \leq w(1) - \big(w(1) - w(q_a)\big) - \sum_{\substack{q_a < p_i;\, p_i q_a \mid a \\ w(p_i) + w(q_a) \leq w(1)}} w(q_a)$$

$$= (1 - r)w(q_a) \leq 0, \qquad (1.8)$$

since $r \geq 1$. This shows that the circumstance $m_3(a) > 0$, $m_4(a) = 0$ does not arise when $m(a) > 0$, for then we would obtain $m(a) = m_2(a) + m_3(a) \leq 0$ from (1.8).

It remains to consider those a for which $m_4(a) > 0$. Then $m_3(a) > 0$, because to any term in (1.6) with $p_1 = p$, $p_3 = q$ there corresponds a positive term in (1.5) with $p_1 = p$, $p_2 = q$. Let $q_1(a) = q_a$ and $q_2(a)$ denote the smallest and second smallest prime factors of a, and let r be as in (1.8). There is a contribution to (1.6) arising whenever p_1 takes any of the r values p_1 other than $q_2(a)$ (the conditions $p_2 = q_2(a)$, $p_3 = q_1(a) = q_a$ being forced). This shows $m_4(a) = (r-1)w(q_a)$, so that (1.8) gives

$$m(a) = m_2(a) + m_3(a) + m_4(a) \leq 0,$$

completing the proof of (1.7) and hence of Lemma 1.

5.2.2 Buchstab's Weights

The simplest improvement on the weight (1.4) that follows from Lemma 1 is that where we retain the term $m_3(a)$ from (1.5) but dismiss the entry (1.6) with the remark that $m_4(a) \geq 0$. Thus the last conclusion of Lemma 1 may be inferred from the strengthened hypothesis $m'(A) > 0$, where $m'(a)$ is as in Lemma 2 below. This lemma provides an alternative form in which this weight may be described.

Lemma 2. *Define $m'(a) = m_2(a) + m_3(a)$, where m_2 and m_3 are as in (1.4) and (1.5) respectively. As in Lemma 1, let q_a denote the smallest prime factor of a. Then*

$$m'(a) = w(1) - \sum_{p_1 \mid a} w'(a), \qquad (2.1)$$

with

$$w'(a) = \begin{cases} w(1) - w(p_1) & \text{if } p_1 = q_a \\ \min\big\{w(1) - w(p_1), w(q_a)\big\} & \text{if } p_1 > q_a. \end{cases} \qquad (2.2)$$

The effect of the entry $S(\cdot)$ in (1.5) is that only those terms where $p_2 = q_a$ (so that $p_1 > q_a$) are counted. Furthermore, when this entry (1.5) is non-zero it has the effect that the terms in $w(1) - w(p_1)$ in (1.4) and (1.5) cancel out. Thus $m'(a)$ is as in (2.1), with $w'(a) = w(q_a)$ precisely when $p_1 > q_a$ and $w(q_a) \leq w(1) - w(p_1)$. This establishes Lemma 2.

5.2 More Elaborate Weighted Sieves

Logarithmic Weights. In the construction outlined in Lemma 2, take the function $w(p)$ to be the weight from Sect. 5.1.2, so that in particular

$$w(1) = U - V, \qquad w(p) = \log p/\log D - V \quad \text{if} \quad D^V \le p < D^U. \qquad (2.3)$$

In this case there is a slightly specialised way of describing the weight $m'(a)$ from Lemma 2, based on the identity in Lemma 3.

Lemma 3. *The function $S(a, P)$ satisfies the identity*

$$\int_{D^x < \xi} S(a, P(D^x))\, dx = S(a, P(\xi)) \frac{\log \xi}{\log D} + \sum_{p|a;\, p<\xi} \frac{\log p}{\log D} S\left(\frac{a}{p}, P(p)\right).$$

In the following treatment, $p = Q(d)$ denotes the largest prime factor of $d = pd_1$. The definition in (1.2.1.2) of S in terms of the Möbius function μ gives

$$\int_{D^x < \xi} S(a, P(D^x))\, dx = \sum_{d|a} \mu(d) \int_{Q(d) < D^x < \xi} dx$$

$$= \sum_{d|a;\, d|P(\xi)} \mu(d) \left(\frac{\log \xi}{\log D} - \frac{\log Q(d)}{\log D} \right).$$

Lemma 3 now follows after writing $Q(d) = p$ and $d = pd_1$, in which $d_1 \,|\, P(p)$ as previously encountered in Sect. 3.1.3.

In the application of the identity of Lemma 3 it is more straightforward, from the current point of view, to ignore the description of $m'(a)$ given in (2.2) and proceed directly from the description supplied by (1.5). We will use the identities in Lemma 4, of which the first follows from Lemma 3 and the second is an instance of Buchstab's identity (3.1.3.2).

Lemma 4. *Let ξ be arbitrary. Then*

$$\sum_{\substack{p_2|a/p_1 \\ p_2 < \xi}} \frac{\log p_2}{\log D} S\left(\frac{a}{p_1 p_2}, P(p_2)\right)$$

$$= \int_{D^x < \xi} S\left(\frac{a}{p_i}, P(D^x)\right) dx - S\left(\frac{a}{p_1}, P(\xi)\right) \frac{\log \xi}{\log D}.$$

Also

$$\sum_{\substack{p_2|a/p_1 \\ p_2 < \xi}} S\left(\frac{a}{p_1 p_2}, P(p_2)\right) = S\left(\frac{a}{p_1}, 1\right) - S\left(\frac{a}{p_1}, P(\xi)\right).$$

5. The Sieve with Weights

Laborde's Weights. Take the expression (1.5) at face value, so that we examine the expression

$$m_3(a) = \sum_{\substack{D^V \le p_2 < p_1 < D^U; p_1 p_2 | a \\ p_1 p_2 < D^{U+V}}} \left(U + V - \frac{\log p_1}{\log D} - \frac{\log p_2}{\log D}\right) S\left(\frac{a}{p_1 p_2}, P(p_2)\right),$$

the contribution from terms where $p_1 \ge D^U$ being zero. As in Sect. 5.1, the construction will be applied only to numbers a having no prime factors smaller than $D^{1/s}$. The number s will satisfy $V < 1/s$, so that the conditions $D^V < p_2$, $D^V < p_1$ will automatically be satisfied. In this situation the weight $m'(a)$ can be written in the following lengthier form. The purpose of Lemma 5 is that an analogue of Lemma 5.1.2 follows from it on summing over $a \in \mathcal{A}$ and again estimating the resulting expressions using Theorem 4.4.2.

The reader who dislikes the apparent complications of Lemma 5 may first consider the case when $V = 0$ and $s \to \infty$, in which the last two sums in Lemma 5 disappear.

Lemma 5. *Let w be as in (2.3). Suppose that the number a has no prime factors smaller than $D^{1/s}$, where $1/s \ge V$. Then the expression*

$$m'(a) = m_2(a) + m_3(a)$$

given by (1.4) and (1.5) can be rewritten as

$$m'(a) = (U - V)S(a, P(D^{1/s}))$$

$$- \sum_{p_1 < D^{(U+V)/2}} \left(U + V - \frac{2 \log p_1}{\log D}\right) S\left(\frac{a}{p_1}, P(p_1)\right)$$

$$- \sum_{D^{1/s} \le p_1 < D^U} \int_{D^x < \min\{p_1, D^{U+V}/p_1\}} S\left(\frac{a}{p_1}, P(D^x)\right) dx \qquad (2.4)$$

$$- \sum_{D^{1/s} \le p_1 < D^{U+V-1/s}} \left(\frac{1}{s} - V\right) S\left(\frac{a}{p_1}, P(D^{1/s})\right)$$

$$- \sum_{D^{U+V-1/s} \le p_1 < D^U} \left(U - \frac{\log p_1}{\log D}\right) S\left(\frac{a}{p_1}, P(D^{1/s})\right).$$

In (1.5), apply the identities from Lemma 4, in which (2.3) indicates we should now take $\xi = \min\{p_1, D^{(U+V)}/p_1\}$. This gives

$$m_3(a) = \sum_{p_1 < D^U} S\left(\frac{a}{p_1}, P(\xi)\right) \frac{\log \xi}{\log D} - \sum_{p_1 < D^U} \int_{D^x < \xi} S\left(\frac{a}{p_1}, P(D^x)\right) dx$$

$$+ \sum_{p_1 < D^U} \left(U + V - \frac{\log p_1}{\log D}\right) \left(S\left(\frac{a}{p_1}, 1\right) - S\left(\frac{a}{p_1}, P(\xi)\right)\right).$$

5.2 More Elaborate Weighted Sieves

But the expression (1.4) for $m_2(a)$ is now

$$m_2(a) = U - V - \sum_{p_1 < D^U} \left(U - \frac{\log p_1}{\log D}\right) S\left(\frac{a}{p_1}, 1\right),$$

so the resulting identity for $m'(a) = m_2(a) + m_3(a)$ is

$$m'(a) = U - V + \sum_{p_1 < D^U} S\left(\frac{a}{p_1}, P(\xi)\right) \frac{\log \xi}{\log D}$$

$$- \sum_{p_1 < D^U} \int_{D^x < \xi} S\left(\frac{a}{p_1}, P(D^x)\right) dx$$

$$- \sum_{p_1 < D^U} \left(U + V - \frac{\log p_1}{\log D}\right) S\left(\frac{a}{p_1}, P(\xi)\right) + V \sum_{p_1 < D^U} S\left(\frac{a}{p_1}, 1\right).$$

If $p_1 \geq D^{(U+V)/2}$ then $\xi = D^{U+V}/p_1$, so the coefficient of $S(a/p_1, P(\xi))$ is zero for these p_1, while for smaller p_1 it is $U + V - 2\log p_1/\log D$. Furthermore $\xi = p_1$ for these p_1. Thus

$$m'(a) = U - V - \sum_{p_1 < D^{(U+V)/2}} \left(U + V - 2\frac{\log p_1}{\log D}\right) S\left(\frac{a}{p_1}, P(p_1)\right)$$

$$+ V \sum_{p_1 < D^U} S\left(\frac{a}{p_1}, 1\right) - \sum_{p_1 < D^U} \int_{D^x < \xi} S\left(\frac{a}{p_1}, P(D^x)\right) dx. \quad (2.5)$$

When this identity is applied to numbers a having no prime factor smaller than $D^{1/s}$ there are some simplifications to be made. When $x < 1/s$ the integrand reduces to $S(a/p_1, P(D^{1/s}))$. The conditions of integration are now $x < 1/s$ if $p_1 < D^{U+V-1/s}$ and $x < U + V - \log p_1/\log D$ for larger p_1. Thus the contribution to the last sum in (2.5) from these x is

$$\sum_{D^{1/s} \leq p_1 < D^U} \int_{x < \min\{1/s, \log \xi/\log D\}} dx$$

$$= \sum_{D^{1/s} \leq p_1 < D^{U+V-1/s}} \frac{1}{s} + \sum_{D^{U+V-1/s} \leq p_1 < D^U} \left(U + V - \frac{\log p_1}{\log D}\right).$$

On the other hand the contribution from $x > 1/s$ is

$$\int_{1/s < x} \sum_{D^x < p_1 < D^{U+V-x}} S\left(\frac{a}{p_1}, P(D^x)\right) dx,$$

so that $x < \frac{1}{2}(U + V)$ in the integral.

Applying (2.5) when $(a, P(D^{1/s})) = 1$, after these substitutions have been made, gives the identity in Lemma 5.

5. The Sieve with Weights

Theorem 1. *Let $m'(a)$ be as in Lemma 5, and suppose that ρ has sifting density 1. Then*

$$\sum_{\substack{a \in \mathcal{A} \\ (a, P(D^{1/s}))=1}} m'(a) \geq \frac{Xe^{-\gamma}}{\log D} \left(\prod_{p<D} \frac{p - \rho(p)}{p-1} \right)$$

$$\times \left((U-V)f(s) - M(W) + O\left(\frac{1}{\log^{1-\Delta} D}\right) \right)$$

$$- O\left(\sum_{d \leq D} |r_{\mathcal{A}}(d)| \right),$$

where Δ, f, F, and the O-constant are as described in Lemma 5.1.2, and

$$M(W) = \int_{\frac{2}{U+V} < t < s} \left(U + V - \frac{2}{t} \right) F(t-1) \frac{dt}{t} + \int_{t<s} \int_{\substack{x<1/t \\ x<U+V-1/t}} F\left(\frac{t-1}{x}\right) \frac{dx}{x} \frac{dt}{t}$$

$$+ \int_{\frac{1}{s} < \frac{1}{t} < U+V - \frac{1}{s}} (1-Vs) F(t-1) \frac{dt}{t} + \int_{U+V-\frac{1}{s} < \frac{1}{t} < U} s\left(U - \frac{1}{t} \right) f(t-1) \frac{dt}{t}.$$

This lemma follows in a straightforward way by applying Theorem 4.4.2 to the expression on the right of (2.4). Accordingly we shall be brief and consider only the second one in any detail. After summation over those $a \in \mathcal{A}$ indicated in Lemma 5 the contribution of this term to the expression to be estimated in Theorem 1 is

$$\sum_{D^{1/s} < p < D^{(U+V)/2}} \left(U + V - \frac{2 \log p}{\log D} \right) \sum_{a \in \mathcal{A}} S\left(\frac{a}{p}, P(p) \right)$$

$$= \sum_{D^{1/s} < p < D^{(U+V)/2}} \left(U + V - \frac{2 \log p}{\log D} \right) \qquad (2.6)$$

$$\times \left(\frac{X \rho(p)}{p} V(P(p)) \left(F\left(\frac{\log D/p}{\log p}\right) + O\left(\frac{1}{\log^{1-\Delta} D/p}\right) \right) - \sum_{\substack{d_p \leq D/p \\ d | P(p)}} |r_{\mathcal{A}}(pd_p)| \right).$$

The contribution from the $r_{\mathcal{A}}$ term is absorbed by the corresponding term in Theorem 1. The term involving F can be estimated by partial summation as in Lemma 1.3.6(ii). Using also (1.3.5.2) with $\kappa = 1$, we find

$$\sum_{D^{1/s} < p < D^{(U+V)/2}} \left(U + V - \frac{2 \log p}{\log D} \right) \frac{\rho(p)}{p} \frac{V(P(p))}{V(P(D))} F\left(\frac{\log D/p}{\log p}\right)$$

$$\leq \sum_{D^{1/s} < p < D^{(U+V)/2}} \left(U + V - \frac{2\log p}{\log D}\right) \frac{\rho(p)}{p} \frac{\log D}{\log p} F\left(\frac{\log D/p}{\log p}\right)$$

$$\leq \int_{2/(U+V)}^{s} \left(U + V - \frac{2}{t}\right) F(t-1) \frac{dt}{t} + O\left(\frac{1}{\log D}\right), \qquad (2.7)$$

where the O-term is as stated because s is bounded. Since $U + V < 2$ (otherwise the sum over p would be empty) the contribution of the O-term in (2.6) is absorbed that in (2.7).

On re-expressing $V(P(D))$ using Mertens' formula in the style already used in Corollary 5.1.1 this leads to the first contribution to the expression $M(W)$ in Theorem 1, and the other three entries are obtained in a similar way.

5.3 A Weighted Sieve Following Rosser

The sieves with weights described thus far depend on a relatively simple weighting device such as that in Lemma 5.1.3 or (in a simpler version) in Lemma 5.1.1. As was noted at (5.1.1.3), these constructions are too simplistic to be used in isolation. To obtain an effective instrument the approach used hitherto has been to combine the weighting construction with the machinery of the sieve (without weights) in the fashion described in Lemma 5.1.2.

A genuine weighted sieve, on the other hand, might be viewed as a construction in which the weighting construction is introduced at the outset. This section describes such a construction, in which the weighting device is integrated with the procedures used to set up the sieves described in Chapters 3 and 4. This will raise some technical questions that are rather more formidable than those appearing earlier in this chapter. In particular, the manipulations in Sects. 5.3.2 and 5.3.3 are necessary to transform the construction defined in Sect. 5.3.1 into one where the involvement of a weight function w is of the type

$$\sum_p w(p) \Psi(a, p) ,$$

is which $\Psi(a, p)$ is such that the resulting sum over $a \in \mathcal{A}$ can be estimated using the results developed in Chap. 4.

The principal result of this section (Theorem 1) is stated in Sect. 5.3.1. With a fairly significant amount of numerical computation, of which we give only the result, this leads to the values of the numbers δ_R that were described in the introduction to this chapter.

5.3.1 Combining Sieving and Weighting

The construction described in this section is of the form

$$\Sigma(A, \chi, \xi) = \sum_{d|A} \mu(d)\chi(d)\psi(d) ,\qquad(1.1)$$

with ψ of the shape

$$\psi(d) = \xi(1) - \sum_{p|d} \xi(p) .\qquad(1.2)$$

We will use (1.1) with

$$\chi = \chi_D^- , \quad \xi = w ,$$

where w is a weight function of the type described in Sect. 5.2.1 and χ_D^- is the characteristic function of Rosser's sieve from (3.3.1.1). When the function ρ has sifting density 1 (the only case considered in detail in the sequel) we will take $\beta = 2$ as in Chapter 4.

The connection between the object $\Sigma(A, \chi^-, \xi)$ defined in (1.1) and the weighting device from Lemma 5.2.1 is as described in Lemma 1 below. For this purpose it is preferable to make a slight adjustment to the definitions (5.2.1.5) and (5.2.1.6). Set

$$m_2(a) = w(1) - \sum_{p_1|a}\bigl(w(1) - w(p_1)\bigr) ,\qquad(1.3)$$

as in (5.2.1.4), but specify

$$m_3'(a) = \sum_{\substack{p_2<p_1;\, p_1p_2|a \\ p_1p_2^{\beta+1}<D}} \bigl(w(1) - w(p_1) - w(p_2)\bigr) S\!\left(\frac{a}{p_1p_2}, p_2\right) ,\qquad(1.4)$$

$$m_4'(a) = \sum_{\substack{p_3<p_2<p_1;\, p_1p_2p_3|a \\ p_1p_2^{\beta+1}<D}} w(p_3) S\!\left(\frac{a}{p_1p_2p_3}, p_2\right) .\qquad(1.5)$$

From this point, we will normally rewrite the parameter U from Sect. 5.2 as $1/u$, so that $u > 1$.

Lemma 1, the proof of which appears later, describes an additional condition which the function w must satisfy.

Lemma 1. *Let χ_D^- denote Rosser's function as specified in (3.3.1.1). Suppose w satisfies the constraint*

$$\sum_{i=1}^{2j} w(p_i) \leq w(1) \quad \text{if} \quad j \geq 2,\; p_1 p_2^{\beta+1} \leq D,\; p_1 \ldots p_{2j-3} p_{2j-2}^{\beta+1} \leq D ,\qquad(1.6)$$

where $p_1 > p_2 > \cdots > p_{2j}$ as usual. Write $A = (a, P(D^{1/u}))$, as in Lemma 5.2.1. Then the sum (1.1) satisfies

$$\Sigma(A, \chi_D^-, w) \leq m_2(A) + m_3'(A) + m_4'(A) \leq m(A), \qquad (1.7)$$

where $m(a)$ is as in Lemma 5.2.1.

In Lemma 5.2.1 and its precursors in Sect. 5.1 the quotient $w(p)/w(1)$ is required to be $\log p/\log D$ or a close relative thereof. Thus the condition (1.6) will be most naturally satisfied when $\beta = 2$, as arises in the context $\kappa = 1$ in Chapter 4.

When Lemma 1 is used by applying (1.3.1.1) in the usual way we are led to an estimate

$$\sum_{a \in \mathcal{A}} m(a) \geq X \, \Theta(P(D^{1/u})) - E(w),$$

where $E(w)$ is as in (1.13) below and

$$\Theta(P(D^{1/u})) = \sum_{d | P(D^{1/u})} \frac{\mu(d)\chi_D^-(d)\rho(d)}{d} \left(w(1) - \sum_{p|d} w(p) \right). \qquad (1.8)$$

The estimation of $\Theta(P(D^{1/u}))$ in Lemma 8 will involve the following expression $h(x,s)$. For $x > 0$, $s \geq 1$ and integers $n \geq 2$ let

$$h_1(x,s) = \int_{\substack{x < x_1 < 1/s \\ x + \beta x_1 < 1 < (\beta+1)x_1}} \frac{1}{(1 - x - x_1)^\kappa} \frac{\kappa \, dx_1}{x_1}, \qquad (1.9)$$

$$h_n(x,s) = \int_{\substack{s < t \\ t < \beta+1 \text{ if } 2 \nmid n}} h_{n-1}\left(\frac{tx}{t-1}, t-1 \right) \frac{dt^\kappa}{(t-1)^\kappa}, \qquad (1.10)$$

$$h(x,s) = \sum_{r \geq 1} h_{2r}(x,s). \qquad (1.11)$$

Here the integration over t and the summation over r are finite because $h_n(x,s) = 0$ when $\beta + n > 1/x$ or when $s > \beta + n$.

The recursive nature of the definition of h is similar to that of the functions F and f from Sect. 4.2. These functions do not appear explicitly in Theorem 1, but the analysis leading to the occurrence of the function h depends heavily upon them. The discussion in Sect. 5.3.4 will require that the quantity $V^+(D, P(z))$ is estimated from below (in terms of F) as in Theorem 4.4.3, rather than from above. Consequently the two-sided sifting density property from Definition 1.3.8 is needed in Theorem 1.

Theorem 1 has been constructed with application to the context $\kappa = 1$ in mind, via Corollary 1. Since only a little extra work is involved, the result is given for the context $\kappa \geq \frac{1}{2}$. There is, however, less need for such a result when $\kappa = \frac{1}{2}$, in view of the relative success of the sieve (without weights) in this case. An analogous result could be derived when $\kappa < \frac{1}{2}$, but the form of the result would then differ (because Lemma 7 uses $f(\beta) = 0$, but when $\kappa < \frac{1}{2}$ better than this is true).

Theorem 1. *Assume that ρ has a two-sided sifting density $\kappa \geq \frac{1}{2}$. Let $\beta = \beta(\kappa)$ be as in Proposition 4.2.1. Suppose w satisfies the condition (1.6) from Lemma 1, and assume $w(p) = 0$ if $p < D^T$, where $T > 0$. Then the expression $m(a)$ appearing in Lemma 1 satisfies*

$$\sum_{a \in \mathcal{A}} m(a) \geq CXV(P(D))\left(\mathcal{M}(W) + O\left(\frac{L'e^{\sqrt{L}}}{\log^{1-\Delta}D}\right)\right) + E(w). \quad (1.12)$$

Here $C > 0$ is the constant from Lemma 4.2.5, Δ is as in Theorem 4.4.2, and

$$E(w) = \sum_{d \mid P(D^{1/u})} \mu(d)\chi_D^-(d)\left(w(1) - \sum_{p \mid d} w(p)\right) r_{\mathcal{A}}(d). \quad (1.13)$$

The O-constant depends on $u > 1$ and $T > 0$, and

$$\mathcal{M}(W) = -\int_{1/2}^{1/u} \frac{W(1)-W(x)}{(1-x)^\kappa}\frac{\kappa\,dx}{x} + \int_T^{1/2} W(x)\left(\frac{1}{(1-x)^\kappa} - h(x,1)\right)\frac{\kappa\,dx}{x},$$

where h is as in (1.11), and $W(x) = w(p)$ with $x = \log p / \log D$, as in (5.2.1.2).

If, in addition, w satisfies the conditions of Lemma 5.2.1 then the expression (1.12) is a lower bound for

$$\sum_{a \in \mathcal{A}: \nu(D^{1/u}, a) \leq R} w(q_a),$$

in which $\nu(D^{1/u}, a)$ counts the number of prime factors of a in the way described in Lemma 5.2.1.

In Theorem 1, the non-elementary component $h(x, 1)$ in the result is supported on $x < 1/(\beta+2)$. This follows because in (1.9) $h_1(x, s)$ is supported where $sx < 1$ and $(\beta+1)x < 1$, so that in (1.10) $h_2(x, s)$ is supported where $tx < 1$ and $(\beta+1)x < 1 - 1/t$. Inspection of Lemma 5.1.2 shows that (with reference to the case $\kappa = 1$) the result in Theorem 1 coincides with that in Sect. 5.1 in the "elementary" range $T \geq \frac{1}{4}$.

Theorem 1 will be proved in the following way. Lemmas 7 and 8 (and Lemma 9 in the case $\kappa = \frac{1}{2}$) will show that the right side of (1.12) is a valid lower estimate for the quantity on the left side of (1.7). Then (1.12) will follow using Lemma 1. The last statement in Theorem 1 is merely a reiteration of Lemma 5.2.1.

5.3 A Weighted Sieve Following Rosser

The primes p between $D^{1/u}$ and D are not relevant to the sum estimated by Theorem 1, but it is simpler to state the result in terms of $V(P(D))$. However, our control of the O-term in Theorem 1 will assume that $w(p)$ is supported only on an interval $[D^T, D^{1/u}]$, where $T > 0$. The requirement $p < D^{1/u}$ is necessary for reasons already encountered in Lemma 5.1.2. The condition $T > 0$ is not an inconvenience, since the shape of the main term $\mathcal{M}(W)$ in Theorem 1 indicates that no advantage is to be gained by taking $W(x)$ non-zero for x so small that $(1-x)^\kappa h(x,1) > 1$.

Inspection of the proof shows that the entry $h(x,1)$ in Theorem 1 arises as the possibly smaller number $h(x,u)$, but this is actually constant for u close to 1, so that in practice no loss will have been incurred.

To transform Theorem 1 into a useful instrument we will adopt the natural choice of the function w that satisfies the conditions of Lemma 1 and of Lemma 5.2.1. In the following corollary the condition $\kappa = 1$ gives $\beta = 2$. Consequently (1.11) implies that the factor $h(x,1)$ in Theorem 1 is supported on the interval $[T, \frac{1}{4}]$. In this interval the function w will be linear, so that the directly relevant properties of h are the numbers

$$\alpha = \int_T^{1/4} \phi(x) \kappa \, dx, \quad \beta = \int_T^{1/4} \phi(x) \frac{\kappa \, dx}{x}, \tag{1.14}$$

where $\phi(x) = 1/(1-x)^\kappa - h(x,1)$.

In the light of Theorem 1 the natural choice of T is the solution of

$$\frac{1}{(1-T)^\kappa} = h(T,1) \, .$$

In the case $\kappa = 1$ then numerical calculations, of which an outline is given in the notes, indicate the values

$$T = 0{\cdot}074308\ldots, \quad \alpha = 0{\cdot}150552\ldots, \quad \beta = 0{\cdot}87695\ldots .$$

Corollary 1.1 gives an "applicable" consequence of Theorem 1. It is part (i) that is generally more useful, but part (ii) is better in the context $R = 2$. This leads to the values of δ_R stated in the introduction to this chapter.

There are situations, arising in connection with the bilinear error term discussed in Chap. 6, where a larger choice of T might be appropriate, closer to $\frac{1}{6}$. The point $T = \frac{1}{6}$ has the attraction that the necessary computations reduce to one-dimensional numerical integrations. For this value of T such a calculation, again with $\kappa = 1$, shows

$$\alpha(\tfrac{1}{6}) = 0{\cdot}098580\ldots, \quad \beta(\tfrac{1}{6}) = 0{\cdot}474533\ldots .$$

Corollary 1.1. *Suppose ρ has sifting density $\kappa = 1$, and that $a \leq D^g$ if $a \in \mathcal{A}$. Assume $g \leq RU + T$, and write $\Lambda_R = R - \delta_R$. Let $\alpha, \overline{\beta}$ be as in (1.14).*

(i) *Let*
$$U = \frac{1}{1+3^{-R}e^{-\beta/3}}, \qquad \delta_R = \frac{\log(4U/3e^\alpha)}{\log 3}. \qquad (1.15)$$

Suppose $g \le \Lambda_R - \eta$, *where*
$$\eta > 0, \qquad \eta \ge (1-U)(R - \tfrac{1}{3}) - \delta_R. \qquad (1.16)$$

Then in Theorem 1 we may take
$$\mathcal{M}(W) = \eta \log 3. \qquad (1.17)$$

(ii) *In place of* (1.15) *suppose*
$$U \log\left(\frac{1}{U} - 1\right) + \log\frac{3}{4(1-U)} + \alpha - \frac{1-U}{3}\log 3e^\beta = 0, \qquad (1.18)$$
$$\delta_R = (1-U)(R - \tfrac{1}{3}). \qquad (1.19)$$

Then Theorem 1 again applies, now with $\mathcal{M}(W)$ *as in* (1.17).

Specify $w(1) = W(1) = U - V$, $RU + V = g$, and
$$W(x) = \begin{cases} x - V & \text{if } \tfrac{1}{4} \le x \le U, \\ \min\left\{x - V, x - \tfrac{1-U}{3}\right\} & \text{if } T \le x < \tfrac{1}{4}, \end{cases} \qquad (1.20)$$

with $W(x) = 0$ if $0 \le x < T$. This satisfies the conditions of Lemma 5.2.1, because the condition $g \le RU + T$ guarantees $V \le T$, so that $W(x) = 0$ when $x \le V$ as required.

To use Theorem 1 we also need to check the condition (1.6) of Lemma 1. First suppose $U + 3V \ge 1$, so that $W(x) = x - V$ when $T \le x \le U$. The conditions granted in (1.6) imply $p_1 p_2 \ldots p_{2j} \le D$ when $j \ge 2$, because $\beta = 2$ when $\kappa = 1$. Thus
$$\sum_{i=1}^{2j} w(p_i) \le \sum_{i=1}^{2j}\left(\frac{\log p_i}{\log D} - V\right) \le 1 - 2jV \le 1 - 4V \le U - V = w(1).$$

If $U + 3V \le 1$ then the condition $p_1 p_2^{\beta+1} < D$ in (1.6) gives $p_2 < D^{1/4}$, because $\beta = 2$. Thus
$$w(p_1) = \frac{\log p_1}{\log D} - V, \qquad w(p_i) \le \frac{\log p_i}{\log D} - \frac{1-U}{3} \quad \text{if } i \ge 2,$$

so when $2j - 1 \ge 3$
$$\sum_{i=1}^{2j} w(p_i) \le 1 - V - (2j-1)\frac{1-U}{3} \le U - V = w(1).$$

This verifies (1.6).

5.3 A Weighted Sieve Following Rosser

It remains to show that $\mathcal{M}(W)$ can be estimated as stated in (1.17). When $U + 3V \leq 1$ the specification (1.20) gives $\mathcal{M}(W) = I(U, V)$, with

$$I(U,V) = -\int_{1/2}^{U} \frac{U-x}{1-x}\frac{dx}{x} + \int_{1/4}^{1/2} \frac{x-V}{1-x}\frac{dx}{x} + \alpha - \tfrac{1}{3}\beta(1-U), \quad (1.21)$$

this notation being as in (5.1.2.14). We showed in (5.1.2.17) that the choice (1.15) gives $I(U, V_0) = 0$ with $\Lambda_R = RU + V_0$. Now make $V = V_0 - \eta$ and $g = \Lambda_R - \eta$. Then $g = RU + V$ and $\mathcal{M}(W) = I(U, V) = \eta \log 3$.

It is necessary to ensure that this choice does in fact satisfy $U + 3V \leq 1$. In fact

$$U + 3V = U + 3V_0 - 3\eta = U + 3(\Lambda_R - RU - \eta),$$

and (because $\Lambda_R = R - \delta_R$)

$$U + 3V_0 - 1 = U - 1 + 3(R - \delta_R - RU) = (1-U)(3R-1) - 3\delta_R.$$

Thus $U + 3V \leq 1$ when (1.16) is satisfied. This establishes part (i).

The constraint (1.16) becomes unimportant as soon as R is at all large (actually when $R \geq 3$), because (1.15) shows that $(1-U)(3R-1)$ tends to 0 rather strongly as $R \to \infty$. When $R = 2$, however, numerical work shows that it is better not to seek an optimum within the region $U + 3V < 1$, but to make $U + 3V_0 = 1$. Then the condition (1.18) says that $I(U, V_0) = 0$ in (1.21). With this U make $V = V_0 - \eta$ with $\eta > 0$, so that $U + 3V < 1$. Then (1.21) again gives $I(U, V) = \eta \log 3$. Also, this U, V give $RU + V = \Lambda_R - \eta$, where

$$\Lambda_R = RU + V_0 = R - (R - \tfrac{1}{3})(1-U) = R - \delta_R,$$

with δ_R as in (1.19). Thus (1.18) and (1.19) also lead to (1.17), as stated in part (ii).

One might also seek an optimum value of Λ_R from the region $U + 3V > 1$. If this were done we would encounter a constraint of the form

$$0 < \eta \leq (1-U)(R - \tfrac{1}{3}) - \delta_R$$

in place of (1.16), and some numerical work indicates values of U and δ_R for which this constraint cannot be satisfied.

The proof of Lemma 1 uses an instance of the "Fundamental Identity" from Lemma 3.1.2. In Lemma 2, χ and the related object $\bar{\chi}(d)$ may be as described in Lemma 3.1.1. When the particular choice $\chi = \chi_D^-$ is made, as in Lemma 1, the factor $\mu(t)$ in (1.24) may be dropped, since then $\bar{\chi}(t) = \bar{\chi}_D^-(t)$, which is supported only where $\mu(t) = 1$.

Lemma 2. *When $t|B$ and B is squarefree write*

$$B_t = \left(\frac{B}{t}, P(q(t))\right), \quad (1.22)$$

5. The Sieve with Weights

in which $q(t)$ is the smallest prime factor of t. When ξ is supported at 1 and the primes, as in (1.2), write

$$\xi(B,t) = \begin{cases} \xi(1) - \sum_{p|t} \xi(p) & \text{if } B_t = 1 \\ \xi(p) & \text{if } B_t = p. \end{cases} \quad (1.23)$$

Then the quantity in (1.1) satisfies the identity

$$\Sigma(B,\chi,\xi) = \xi(B) - \sum_{t|B} \mu(t)\bar{\chi}(t)\xi(B,t). \quad (1.24)$$

Use Lemma 3.1.2 in the form (3.1.2.16). This gives

$$\Sigma(B,\chi,\xi) = \sum_{d|B} \mu(d)\psi(d) - \sum_{t|B} \mu(t)\bar{\chi}(t) \sum_{\substack{f|B/t \\ f|P(q(t))}} \mu(f)\psi(ft).$$

Because ψ is as in (1.2) and B is squarefree, this gives

$$\sum_{d|B} \mu(d)\psi(d) = \sum_{f|B} \mu(f) \sum_{h|f} \mu(h)\xi(h) = \sum_{h|B} \mu(h)\xi(h) \sum_{h|f;f|B} \mu(f) = \xi(B).$$

Similarly, since in (1.22) B_t is also squarefree,

$$\sum_{\substack{f|B/t \\ f|P(q(t))}} \mu(f)\psi(ft) = \sum_{\substack{f|B/t \\ f|P(q(t))}} \mu(f) \sum_{h|ft} \mu(h)\xi(h)$$

$$= \sum_{h|B} \mu(h)\xi(h) \sum_{\frac{h}{(t,h)}|f;f|B_t} \mu(f)$$

$$= \sum_{\frac{h}{(t,h)}=B_t} \mu(h)\xi(h)\mu\left(\frac{h}{(t,h)}\right) = \sum_{k|t} \mu(k)\xi(kB_t),$$

because $h = (t,h)B_t$ is determined by t and B_t when $k = (t,h)$ is given. Here the last summand is non-zero only when B_t (and kB_t) is 1 or a prime, so the sum over k is exactly $\xi(B,t)$, as in (1.23). Now (1.24) follows.

Proof of Lemma 1. If Lemma 2 were applied to the expression (1.1) in the most direct way then we would encounter terms for which $\bar{\chi}_D^-(p_1 p_2) = 1$, which allows $p_1 p_2 > D$. This is avoided by first separating the terms with $d = 1$ and $d = p_1$ in the expression (1.1). When $d = p_1 p_2 d_2$ with $p_1 > p_2$, $d_2 | P(p_2)$ (our usual notation) the specification of χ_D^- gives

$$\chi_D^-(d) = \chi_{D/p_1 p_2}^-(d_2) \quad \text{when} \quad p_1 p_2^{\beta+1} < D, \quad (1.25)$$

5.3 A Weighted Sieve Following Rosser

as also follows from the recurrences noted at (4.1.1.5). Then (1.1) becomes

$$\Sigma(A, \chi_{\bar{D}}, w) = w(1) - \sum_{p_1 | A}(w(1) - w(p_1))$$
$$+ \sum_{\substack{p_2 < p_1; \, p_1 p_2 | A \\ p_1 p_2^{\beta+1} < D}} \Sigma(A_2, \chi_{\bar{D}/p_1 p_2}^{-}, \xi), \qquad (1.26)$$

where
$$A_2 = (A/p_1 p_2, P(p_2)),$$
$$\xi(1) = w(1) - w(p_1) - w(p_2), \quad \xi(p) = w(p). \qquad (1.27)$$

In (1.26), apply Lemma 2 to $\Sigma(A_2, \chi^-, \xi)$, with χ^- as in (1.25). When $\bar{\chi}_{\bar{D}/p_1 p_2}^{-}(t) \neq 0$ in the inner sum in (1.24), the description of $\bar{\chi}$ given by Lemma 3.1.2 shows that we may write $t = q_1 \ldots q_{2j}$, where

$$p_2 > q_1 > \cdots > q_{2j}, \quad q_1 \ldots q_{2j-3} q_{2j-2}^{\beta+1} \leq \frac{D}{p_1 p_2} \quad \text{and} \quad j \geq 2.$$

In the last sum in (1.26), $p_1 p_2^{\beta} = 1$, and $\xi(t)$ and $\xi(B, t)$ are as in (1.27) and (1.23). Thus the hypothesis (1.6) in Lemma 1 now gives

$$w(1) - w(p_1) - w(p_2) - \sum_{p | t} w(p) \geq 0,$$

so that (1.23) shows $\xi(A_2, t) \geq 0$. Now (1.24) gives

$$\Sigma(A_2, \chi_{\bar{D}/p_1 p_2}^{-}, \xi) \leq \xi(A_2). \qquad (1.28)$$

It follows from (1.27) that

$$\xi(A_2) = \begin{cases} w(1) - w(p_1) - w(p_2) & \text{if } A_2 = 1 \\ w(p) & \text{if } A_2 = p \end{cases}$$
$$= (w(1) - w(p_1) - w(p_2)) S\left(\frac{A}{p_1 p_2}, P(p_2)\right)$$
$$+ \sum_{p < p_2} w(p) S\left(\frac{A}{p_1 p_2 p}, P(p_2)\right).$$

Now (1.26) and (1.28) give $\Sigma(A, \chi_{\bar{D}}, w) \leq m_2(A) + m_3'(A) + m_4'(A)$ in the notation of (1.4) and (1.5), as required in Lemma 1. Lastly, the hypothesis (1.6) in Lemma 1 shows that $w(p_1) + w(p_2) \leq w(1)$ in (1.26), in which $p_1 p_2^{\beta+1} < D$. Consequently, comparison of (5.2.1.5) with (1.4) shows $m_3'(A) \leq m_3(A)$, and $m_4'(A) \leq m_4(A)$ follows similarly. The last inequality claimed in Lemma 1 now follows.

5.3.2 The Reduction Identities

Subsequent progress will rest on the identities in Lemma 4 for the quantities

$$V^\pm(D, P(D^{1/s})) = \sum_{d | P(D^{1/s})} \frac{\mu(d)\chi_D^\pm(d)\rho(d)}{d} \tag{2.1}$$

studied in Chapter 4. Actually these identities apply to

$$V^\pm(D, A) = \sum_{d|A} \mu(d)\chi_D^\pm(d)\psi(d) \tag{2.2}$$

for any squarefree number A and multiplicative function ψ. Here χ_D is as specified in (3.3.1.1).

In Sect. 5.3.3 we deduce a corresponding identity for

$$\sum_{d|A} \mu(d)\chi_D^\pm(d)\psi(d)\left(w(1) - \sum_{p|d} w(p)\right).$$

When $A = (a, P(D^{1/u}))$ and $\psi(d) = 1$ this gives an identity for the sum $\Sigma(A, \chi_D^-, w)$ that appears in Lemma 1. Our ultimate concern will however be with the sum (1.8) when ρ has sifting density κ. The identities in Lemma 4 are formulated in a way that is motivated entirely by considerations relating to this context.

When ρ has sifting density κ and the parameter β is appropriately chosen, the sum (2.1) was estimated when $s \geq \beta$, in terms of the function $f(s)$. In the current application the condition $s \geq \beta$ will not always hold (at one point we need to take $s = u$, possibly arbitarily close to 1), but the matter is resolved by using the identity (2.6) of Buchstab's type in Lemma 3.

For brevity denote

$$v_D^\pm(A, s) = V^\pm\left(D, (A, P(D^{1/s}))\right), \tag{2.3}$$

and define s_1 in terms of p_1 by either of the equivalent relations

$$s_1 = \frac{\log D}{\log p_1} - 1, \qquad p_1 = \left(\frac{D}{p_1}\right)^{1/s_1}, \tag{2.4}$$

so that

$$v_{D/p_1}^\pm\left(\frac{A}{p_1}, s_1\right) = \sum_{d|(A/p_1, P(p_1))} \mu(d)\chi_{D/p_1}^\pm(d)\psi(d).$$

Recall from (3.3.1.1) that $\chi_D^+(d) = 0$ when $p_1 > D^{1/(\beta+1)}$. Consequently

$$v_D^+(A, s) = v_D^+(A, \beta+1) \quad \text{when} \quad 0 < s \leq \beta+1. \tag{2.5}$$

The object of Lemmas 3 and 4 is to express all values of $v_D^\pm(A, s)$ in terms of the special values $v_D^+(A, \beta+1)$ and $v_D^-(A, \beta)$, for which the corresponding functions from Sect. 4.2.1 take the "elementary" values $F(\beta+1) = C/(\beta+1)^\kappa$ and $f(\beta) = 0$.

5.3 A Weighted Sieve Following Rosser

Lemma 3. *Let v_D be as in (2.3) and s_1 as in (2.4). Suppose ψ is multiplicative, and that the parameter β implicit in (2.2) satisfies $\beta \geq 1$.*

(i) *If $1 \leq s \leq \beta$, then*

$$v_D^-(A, s) = v_D^-(A, \beta) - \sum_{\substack{D^{1/\beta} \leq p_1 < D^{1/s} \\ p_1 | A}} \psi(p_1) v_{D/p_1}^+ \left(\frac{A}{p_1}, s_1\right). \qquad (2.6)$$

(ii) *If $s \geq \beta$, then*

$$v_D^-(A, s) = v_D^-(A, \beta) + \sum_{\substack{D^{1/s} \leq p_1 < D^{1/\beta} \\ p_1 | A}} \psi(p_1) v_{D/p_1}^+ \left(\frac{A}{p_1}, s_1\right). \qquad (2.7)$$

(iii) *If $s \geq \beta - 1$, then*

$$v_D^+(A, s) = v_D^+(A, \beta + 1) + \sum_{\substack{D^{1/s} \leq p_1 < D^{1/(\beta+1)} \\ p_1 | A}} \psi(p_1) v_{D/p_1}^- \left(\frac{A}{p_1}, s_1\right). \qquad (2.8)$$

In the terms with $d > 1$ in (2.2) set $d = p_1 d_1$, where p_1 is the greatest prime factor of d. Proceed as with the prototype Buchstab Identity (3.1.3.1), but also note $\chi_D^-(d) = \chi_{D/p_1}^+(d_1)$, from (4.1.1.6). Thus

$$v_D^-(A, t) = 1 - \sum_{p_1 < D^{1/t}; p_1 | A} \psi(p_1) \sum_{d_1 | (A/p_1; P(p_1))} \mu(d_1) \chi_{D/p_1}^+(d_1) \psi(d_1)$$

$$= 1 - \sum_{p_1 < D^{1/t}; p_1 | A} \psi(p_1) v_{D/p_1}^+ \left(\frac{A}{p_1}, s_1\right).$$

To obtain (2.6), use this with $t = s$ and with $t = \beta$ and subtract. In this instance $s \leq \beta$, so (2.4) shows $s_1 \leq \beta - 1$. Then the entry s_1 can be replaced by $\beta + 1$ because of (2.5). The identity (2.7) is derived similarly, but with the rôles of s and β interchanged.

The proof of (2.8) is similar when $s \geq \beta + 1$. When $s \leq \beta + 1$ the sum in (2.8) is empty, so (2.8) is also valid as stated because of (2.5), and in fact holds when $s > 0$.

Repeated use of Lemma 3 gives the decomposition of $v^\pm$ in Lemma 4. Here, the constraints on the permissible values of s are those inherited from Lemma 3. Let $q(f)$ and $Q(f)$ denote respectively the least and greatest prime factors of f, as in Sect. 5.3.1, and let $\nu(f)$ be the total number of these prime factors. Adopt the convention

$$Q(1) = 1, \quad q(1) = \infty, \quad \nu(1) = 0,$$

so that a summand with $f = 1$ always appears in Lemma 4.

5. The Sieve with Weights

In Lemma 4 and the ensuing argument we adopt the abbreviations

$$\sigma_D^+(A) = v_D^+(A, \beta+1), \quad \sigma_D^-(A) = v_D^-(A, \beta) \tag{2.9}$$

for these special values of the expressions (2.3).

The significant property of these special values is that Theorem 4.4.2 estimates them in terms of elementary expressions. The expansion in Lemma 4 expresses (2.3) for larger s in terms of these elementary special values.

Lemma 4. *Assume the conditions of Lemma 3. Suppose $s \geq \beta$ when r is odd and $s \geq \beta - 1$ when r is even. Then the expression (2.3) can be expanded as*

$$v_D^{(-)^r}(A, s) = \sum_{\substack{f \mid A;\, \nu(f) \equiv r,\, \mathrm{mod}\, 2 \\ D^{1/s} \leq q(f);\, fQ^{\beta-1}(f) < D}} \psi(f) \sigma_{D/f}^+(A/f)$$

$$+ \sum_{\substack{f \mid A;\, \nu(f) \equiv r+1,\, \mathrm{mod}\, 2 \\ D^{1/s} \leq q(f);\, fQ^{\beta}(f) < D}} \psi(f) \sigma_{D/f}^-(A/f) . \tag{2.10}$$

Note that $Q(f) \geq q(f)$ and $f \geq q(f)$ when $\nu(f) \geq 1$, while $f \geq Q(f)q(f)$ when $\nu(f) \geq 2$. This gives the implications

$$\begin{aligned} fQ^{\beta-1}(f) < D,\ \nu(f) \geq 2 &\Longrightarrow q(f) < D^{1/(\beta+1)} \\ fQ^{\beta}(f) < D,\ \nu(f) \geq 1 &\Longrightarrow q(f) < D^{1/(\beta+1)} , \end{aligned} \tag{2.11}$$

as apply to the terms in (2.10) with $f > 1$ when $2 \mid r$.

We establish Lemma 4 when $s \leq \beta + k$ by induction on k. Suppose first that $s \leq \beta + 1$. Then (2.10) reduces to

$$v_D^-(A, s) = \sigma_D^-(A) + \sum_{\substack{p_1 \mid A \\ D^{1/s} \leq p_1 < D^{1/\beta}}} \psi(p_1) \sigma_{D/p_1}^+(A/p_1) , \tag{2.12}$$

$$v_D^+(A, s) = \sigma_D^+(A) , \tag{2.13}$$

no other terms from (2.10) appearing because (2.11) applies to them, so the condition $D^{1/s} \leq q(f)$ gives $s > \beta + 1$, which is excluded. Here (2.13) is just (2.5), in the notation (2.9). Also, $s_1 \leq s - 1 \leq \beta$ in (2.7) because of (2.4). Therefore s_1 may be replaced by $\beta + 1$ because of (2.5). Now (2.12) follows. This verifies Lemma 4 when $s \leq \beta + 1$.

Suppose now that Lemma 4 has been established when $s \leq \beta + K$. We require it when $s \leq \beta + K + 1$, in which case the entries s_1 in (2.7) and (2.8) satisfy $s_1 \leq \beta + K$. Therefore we may apply (2.10) with the substitutions

$$D \mapsto D_1 = D/p_1 , \quad A \mapsto A_1 = A/p_1 , \quad s \mapsto s_1 , \quad f \mapsto f_1 .$$

Then $D_1^{1/s_1} = p_1$ from (2.4), so that the conditions of summation in (2.10) give $p_1 \leq q(f_1)$, and actually $p_1 < q(f_1)$ since $f_1 \mid A_1 = A/p_1$ and A is squarefree, so that $(f_1, p_1) = 1$. Now write $p_1 f_1 = f$, so that $q(f) = p_1$ and $Q(f) = Q(f_1)$.

First consider the case of (2.10) when r is odd. The sum in (2.7) has become

$$\sum_{\substack{p_1 \mid A \\ D^{1/s} \leq p_1 < D^{1/\beta}}} \psi(p_1) \sum_{\substack{1 < f \mid A;\, \nu(f) \equiv 1,\, \text{mod } 2 \\ q(f) = p_1;\, fQ^{\beta-1}(f) < D}} \psi(f_1) \sigma_{D/f}^+(A/f)$$

$$+ \sum_{\substack{p_1 \mid A \\ D^{1/s} \leq p_1 < D^{1/\beta}}} \psi(p_1) \sum_{\substack{1 < f \mid A;\, \nu(f) \equiv 0,\, \text{mod } 2 \\ q(f) = p_1;\, fQ^{\beta}(f) < D}} \psi(f_1) \sigma_{D/f}^-(A/f) \,,$$

in which the conditions $\nu(f) \equiv i$ arose as $\nu(f_1) \equiv i+1$. Here the condition $D^{1/s} < p_1$ says just $D^{1/s} < q(f)$. Furthermore the condition $p_1 < D^{1/\beta}$ in the outer sum is redundant, because in the inner sums

$$D > Q^{\beta-1}(f) q(f) > q^\beta(f) = p_1^\beta \,.$$

This gives the terms with $f > 1$ in (2.10), that with $f = 1$ having appeared separately in (2.7).

The treatment is similar when r is even, except that under either of the sets of conditions

$$\{f > 1,\ \nu(f) \equiv 0,\, \text{mod } 2,\ q(f) = p_1,\ fQ^{\beta-1}(f) < D\}$$
$$\{\nu(f) \equiv 1,\, \text{mod } 2,\ q(f) = p_1,\ fQ^\beta(f) < D\}$$

we need to infer $p_1 < D^{1/(\beta+1)}$. This follows from (2.11), so the treatment of Lemma 4 is complete.

5.3.3 An Identity for the Main Term

The treatment in Lemma 6 of the main term Θ, given in (3.9), depends on a transformation

$$\sum_{d \mid A} \mu(d) \chi_D^-(d) \psi(d) \sum_{p \mid d} w(p) = - \sum_{p \mid d} w(p) \Psi_D\left(\frac{A}{p}, p\right),$$

so that $w(p)$ becomes the "subject" of the resulting expresson. The initial description in (3.14) of Ψ_D will be simplified by using the following partial cancellation arising from the occurrence of the Möbius function μ.

Part (i) of Lemma 5 leads to the entry $h(x, 1)$ in the main term of Theorem 1. Part (ii) deals with quantities that ultimately contribute only to the error term, and its use could be avoided. It is however essential if we wish the identity in Lemma 6 to be in fact an identity, rather than an approximate asymptotic formula.

In the important case $n = 1$ the sum in Lemma 5(i) reduces to 1 whenever $p < D$, since $Q(1) = 1$. Lemma 5 deals with the case $n > 1$.

Lemma 5. *Suppose $p < D$. Let q and Q be as in Lemma 4. Assume n is squarefree, $n > 1$ and that the prime p satisfies $p < q(n)$, the least prime factor of n.*

(i) *If $\nu(n)$, the number of distinct prime factors of n, is even then*

$$\sum_{\substack{fg=n \\ fgQ^{\beta-1}(f)<D/p}} \mu(g)\chi_D^-(gp) = \begin{cases} -\bar{\chi}_D^-(n) & \text{if } q^{\beta-1}(n) < D/np \\ 0 & \text{otherwise,} \end{cases} \qquad (3.1)$$

where $\bar{\chi}^-$ is as in (3.1.2.10).

(ii) *If $\nu(n)$ is odd then*

$$\sum_{\substack{fg=n \\ fgQ^{\beta}(f)<D/p}} \mu(g)\chi_D^-(gp) = \begin{cases} -\chi_D^-(np) & \text{if } q^{\beta}(n) \geq D/np \\ 0 & \text{otherwise,} \end{cases} \qquad (3.2)$$

where χ^- is as in (3.3.2.1).

In part (i), the term with $g = n$ (so $f = Q(f) = 1$) contributes exactly $\chi_D^-(np)$, since this is already 0 if $pn \geq D$.

For $g < n$ (so $f > 1$) it follows that $f \geq Q(f) \geq p$, because $p < q(n)$ (so that p is the smallest prime factor of $pn = pfg$). In (3.1) the condition $D > pgfQ^{\beta-1}(f)$ now gives $p^{\beta+1}g < D$. So for these g the entry $\chi_D^-(gp)$ can be replaced by $\chi_D^-(g)$.

Next, set $\varpi = Q(n/g)$, the largest prime factor of $f = n/g$, and write $n = n_1\varpi n_2$, where $q(n_1) > \varpi > Q(n_2)$. Thus n_2 is the product of the prime factors of n smaller than ϖ, and $f | \varpi n_2$, so that $g = n_1 t$ for some t with $t | n_2$.

In this situation observe

$$\chi_D^-(gp) = \chi_D^-(g) = \chi_D^-(n_1 t) = \chi_D^-(n_1)\chi_{D/n_1}^{(-)^{\nu(n_1)+1}}(t), \qquad (3.3)$$

from the specification of χ_D^- given in (3.3.1.1). Moreover, when

$$\varpi^{\beta-1} < D/pn = D/pn_1\varpi n_2, \qquad (3.4)$$

as follows from the conditions in (3.1), it follows that $\chi_{D/n_1}^{\pm}(t) = 1$ whenever $t | n_2$. These facts are easily checked: write

$$n_1 = p_1 p_2 \ldots p_{j-1}, \quad t = p_j p_{j+1} \ldots p_k,$$

where $p_1 > p_2 > \cdots > p_{j-1} > \varpi = p_j > \cdots$. Then $\chi_D^-(n_1 t) = 1$ precisely when $p_1 p_2 \ldots p_{2i}^{\beta+1} < D$ for $2i \leq k$, which says that both factors on the right of (3.3) take the value 1, and for $2i \geq j$ these inequalities follow from (3.4), since it gives $n_1 n_2 \varpi_1^{\beta} < D$ for any prime factor ϖ_1 of n_2.

5.3 A Weighted Sieve Following Rosser

These observations show that the contribution to (3.1) from $g < n$ is

$$\sum_{\varpi^{\beta-1} < D/pn} \mu(n_1)\chi_{\bar{D}}(n_1) \sum_{t|n_2} \mu(t)\chi_{\bar{D}/n_1}^{(-)^{\nu(n_1)+1}}(t) \qquad (3.5)$$

$$= \sum_{\varpi^{\beta-1} < D/pn} \mu(n_1)\chi_{\bar{D}}(n_1) \sum_{t|n_2} \mu(t) .$$

Here the inner sum over t counts precisely those terms where $n_2 = 1$, so that $\varpi = q(n)$, the smallest prime factor of the number $n = \varpi n_1$. Thus the expression (3.5) reduces, because $\nu(n)$ is even in part (i), to

$$\begin{cases} -\chi_{\bar{D}}(n/q(n)) & \text{if } nq^{\beta-1}(n) < D/p \\ 0 & \text{otherwise.} \end{cases}$$

This is to be added to the contribution $\chi_{\bar{D}}(np)$ from the term with $g = n$. When $nq^{\beta-1}(n) < D/p$ the total is

$$\chi_{\bar{D}}(n) - \chi_{\bar{D}}(n/q(n)) = \bar{\chi}_{\bar{D}}(n) ,$$

since $\bar{\chi}$ is as in (3.1.2.12). When $nq^{\beta-1}(n) \geq D/p$ it follows that $nq^{\beta}(n) \geq D$ because it is given that $p < q(n)$. Also n has at least two prime factors, of which the two largest now satisfy $p_1 p_2^{\beta+1} \geq D$. Hence $\chi_{\bar{D}}(n) = 0$, and the total contribution to (3.5) reduces to zero. This establishes part (i) of Lemma 5.

The treatment of the less important part (ii) is along similar lines, but some details differ, not only that the exponents of q and Q have increased to β. The term with $g = n$ (i.e. $f = 1$) contributes $-\chi_{\bar{D}}(np)$. In the other terms, where $f > 1$, argue as before to find that they sum to

$$\begin{cases} -\chi_D(n/q(n)) & \text{if } nq^{\beta}(n) < D/p \\ 0 & \text{otherwise.} \end{cases}$$

Now add these two contributions together. When $nq^{\beta}(n) < D/p$ it follows that $p^{\beta+1}n < D$, whence $\chi_{\bar{D}}(np) = \chi_{\bar{D}}(n/q(n))$, and the total contribution reduces to zero. When $nq^{\beta}(n) \geq D/p$ the total contribution is just $-\chi_{\bar{D}}(np)$, so part (ii) of the lemma follows.

When $\psi(d) = 1$ the expression $\Theta(A)$ appearing in part (ii) of Lemma 6 is the quantity $\Sigma(A, \chi_{\bar{D}}, w)$ defined in (1.1). For the proof of Theorem 1 we need the case $\psi(d) = \rho(d)/d$.

Lemma 6. *Suppose that A is squarefree, and that ψ is multiplicative as in Lemmas 3 and 4.*
(i) *When $p \nmid A$ define*

$$\Psi_D(A, p) = \sum_{d_1 | A/p} \mu(d_1)\chi_{\bar{D}}(pd_1)\psi(d_1) . \qquad (3.6)$$

5. The Sieve with Weights

Then

$$\Psi_D(A/p,p) = \sigma^+_{D/p}(A/p) + \Psi^+_{D,3}(A/p,p) + \Psi^-_{D,4}(A/p,p), \quad (3.7)$$

where

$$\Psi^+_{D,3}(A/p,p) = -\sum_{\substack{1<n|A/p; 2|\nu(n) \\ p<q(n); nq^{\beta-1}(n)<D/p}} \psi(n)\sigma^+_{D/pn}\left(\frac{A}{pn}\right)\bar{\chi}_D(n) \quad (3.8)$$

$$\Psi^-_{D,4}(A/p,p) = \sum_{\substack{1<n|A/p; 2\nmid \nu(n) \\ p<q(n); nq^{\beta}(n) \geq D/p}} \psi(n)\sigma^-_{D/pn}\left(\frac{A}{pn}\right)\chi^-_D(np),$$

in the notation used in Lemmas 3 and 4.

(ii) *The expression*

$$\Theta(A) = \sum_{d|A}\left(w(1) - \sum_{p|d}w(p)\right)\mu(d)\,\chi^-_D(d)\psi(d) \quad (3.9)$$

can be expanded as $\Theta_1(A) + \Theta_2(A) + \Theta_3(A) + \Theta_4(A)$, *where*

$$\Theta_1(A) = \sum_{p|A; D^{1/\beta}\leq p}(w(p)-w(1))\sigma^+_{D/p}(A/p)\psi(p), \quad (3.10)$$

$$\Theta_2(A) = \sum_{p|A; p<D^{1/\beta}}w(p)\sigma^+_{D/p}(A/p)\psi(p), \quad (3.11)$$

$$\Theta_3(A) = \sum_{p|A}w(p)\Psi^+_{D,3}(A/p,p)\psi(p), \quad (3.12)$$

$$\Theta_4(A) = w(1)\sigma^-_D(A) + \sum_{p|A}w(p)\Psi^-_{D,4}(A/p,p)\psi(p), \quad (3.13)$$

in which $\Psi^+_{D,3}$ *and* $\Psi^-_{D,4}$ *are as in part* (i) *and* σ^-_D *is as in* (2.9).

Write $d = pd_1 = gpd_2$, where $Q(d_2) < p < q(g)$, in the style employed already in the treatment of Lemma 5. Then

$$\chi^-_D(pd_1) = \chi^-_D(d) = \chi^-_D(gp)\chi^{(-)^{\nu(g)}}_{D/gp}(d_2),$$

rather as with (3.3). Then the expression in (3.6) is

$$\Psi_D(A/p,p) = \sum_{\substack{g|A/p \\ q(g)>p}}\mu(g)\chi^-_D(gp)\psi(g)\sum_{\substack{d_2|A/gp \\ d_2|P(p)}}\mu(d_2)\chi^{(-)^{\nu(g)}}_{D/gp}(d_2)\psi(d_2). \quad (3.14)$$

5.3 A Weighted Sieve Following Rosser

Here, the inner sum over d_2 may be expanded by Lemma 4, into

$$\sum_{\substack{f|A/gp;\, \nu(f)\equiv\nu(g),\,\text{mod } 2 \\ p<q(f);\, fQ^{\beta-1}(f)<D/gp}} \psi(f)\sigma^+_{D/gpf}(A/gpf)$$

$$+ \sum_{\substack{f|A/gp;\, \nu(f)\not\equiv\nu(g),\,\text{mod } 2 \\ p<q(f);\, fQ^{\beta}(f)<D/gp}} \psi(f)\sigma^-_{D/gpf}(A/gpf) \,.$$

where we can write $p < q(f)$ rather than $p \leq q(f)$ because A is squarefree.

Write $gf = n$, for the purpose of collecting together all the terms $\sigma^{\pm}$ with the same arguments. The expression (3.14) now becomes

$$\sum_{\substack{n|A/p \\ p<q(n);\, 2|\nu(n)}} \psi(n)\sigma^+_{D/pn}(A/pn) \sum_{\substack{fg=n \\ Q^{\beta-1}(f)<D/np}} \mu(g)\chi^-_D(gp)$$

$$+ \sum_{\substack{n|A/p \\ p<q(n);\, 2\nmid\nu(n)}} \psi(n)\sigma^-_{D/pn}(A/pn) \sum_{\substack{fg=n \\ Q^{\beta}(f)<D/np}} \mu(g)\chi^-_D(gp) \,. \quad (3.15)$$

There are now several contributions to be distinguished. First, the term with $n = 1$ gives

$$\sigma^+_{D/p}(A/p) \,,$$

which supplies the first entry on the right of (3.7). Next, consider the terms with $2|\nu(n)$ and $n > 1$. Because of Lemma 5(i) their contribution to (3.15) is

$$- \sum_{\substack{1<n|A/p;\, 2|\nu(n) \\ p<q(n);\, q^{\beta-1}(n)<D/np}} \psi(n)\sigma^+_{D/pn}(A/pn)\bar{\chi}^-_D(n) \,, \quad (3.16)$$

so that the entry $\Psi_{D,3}(A/p,p)$ in (3.7) is correct. In a similar way the terms where $\nu(n)$ is odd give

$$- \sum_{\substack{n|A/p;\, 2\nmid\nu(n) \\ p<q(n);\, q^{\beta}(n)\geq D/np}} \psi(m)\sigma^-_{D/pn}(A/pn)\chi^-_D(np) \,,$$

so that the last remaining entry $\Psi_{D,4}(A/p,p)$ in (3.7) is also correct.

Part (ii) of Lemma 6 is now immediate. When $p|A$ the coefficient of $w(p)$ in (3.9) is

$$- \sum_{\substack{d|A \\ d\equiv 0,\,\text{mod } p}} \mu(d)\chi^-_D(d)\psi(d) = \psi(p)\Psi_D(A,p) \,,$$

where Ψ_D is as defined in (3.6), as follows at once on writing $d = pd_1$. Part (i) now gives that the coefficient of $w(p)$ in (3.9) is as described by (3.10)–(3.13).

As for the coefficient of $w(1)$, this is also correct because if we write $D^{1/s}$ for the largest prime factor of A then

$$\sum_{d|A} \mu(d)\chi_D^-(d)\psi(d) = v_D^-(A,s),$$

the expression defined in (2.2). When $s \leq \beta$ this is as given in Lemma 3(i), in which the entry $v_D^-(A,\beta)$ is the term $\sigma_D^-(A)$ in (3.13). This completes the proof of Lemma 6.

5.3.4 The Estimate for the Main Term

Following Lemma 1 we require an estimate (from below) of

$$\sum_{a \in A} \Sigma\Big((a, P(D^{1/u})), \chi_D^-, w\Big),$$

using the fact that the summand is precisely the case $\psi(d) = 1$ of the expression (3.9) considered in Lemma 6. Here, the occurrence of the entry (3.11) raises the technical point that when $p < D^{1/\beta}$ the quantity $\sigma_{D/p}^+$ needs to be estimated from below, rather than from above as usual. Consequently it will be necessary to invoke the estimates from Sect. 4.4.4, for which the two-sided version of the sifting density hypothesis is required.

The arguments used for Lemma 7 would not work for arbitrarily small p, a feature seen earlier in connection with Lemma 4.4.3, for example. In this case we need not hesitate to assume (4.1), because in Theorem 1 we find we actively want to take $T > 0$. Consequently we may adopt a much more simplistic treatment of the error terms than would have been permissible in Sect. 4.4.

Lemma 7. *Suppose that ρ has a two-sided sifting density κ, as in Theorem 4.4.3, and assume $\kappa > \frac{1}{2}$. Further, suppose there exists $T > 0$ so that*

$$w(p) = 0 \quad \text{if} \quad p < D^T. \tag{4.1}$$

Denote $M(w) = M_1(w) + M_2(w) + M_3(w)$, with

$$M_1(w) = -\sum_{D^{1/\beta} \leq p < D^{1/u}} \frac{(w(1) - w(p))\rho(p)}{p} \left(\frac{\log D}{\log D/p}\right)^\kappa,$$

$$M_2(w) = \sum_{p < D^{1/\beta}} \frac{w(p)\rho(p)}{p} \left(\frac{\log D}{\log D/p}\right)^\kappa,$$

$$M_3(w) = -\sum_{p < D^{1/(\beta+2)}} \frac{w(p)\rho(p)}{p} \sum_{\substack{1 < n | P(D^{1/u})/p; 2|\nu(n) \\ p < q(n);\, nq^{\beta-1}(n) < D/p}} \frac{\rho(n)\bar{\chi}_D^-(n)}{n} \left(\frac{\log D}{\log(D/pn)}\right)^\kappa,$$

5.3 A Weighted Sieve Following Rosser

Then the quantity Σ appearing in Lemma 2 satisfies the estimate

$$\sum_{a \in \mathcal{A}} \Sigma\Big((a, P(D^{1/u})), \chi_D^-, w\Big)$$
$$\geq X\, M(w) C\, V(P(D))\left(1 + O\left(\frac{L' e^{\sqrt{L}}}{\log^{1-\Delta} D}\right)\right) + E(w),$$

where F is the function from Proposition 4.2.1, $C = (\beta+1)^\kappa F(\beta+1)$ is the associated constant determined in Lemma 4.2.5, Δ is as in Theorem 4.4.2, and $E(w)$ is as stated in Theorem 1.

We will use Lemma 6 with

$$\psi(d) = \rho(d)/d, \qquad A = P(D^{1/u}). \tag{4.2}$$

On invoking (1.3.1.1) in the standard way we obtain from (1.1)

$$\sum_{a \in \mathcal{A}} \Sigma(A, \chi_D^-, w)$$
$$= \sum_{d \mid P(D^{1/u})} \mu(d)\chi_D^-(d)\left(w(1) - \sum_{p \mid d} w(p)\right)\left(\frac{X \rho(d)}{d} + r_A(d)\right)$$
$$= X \sum_{i=1}^{4} \Theta_i(P(D^{1/u})) + E(w), \tag{4.3}$$

where $E(w)$ is as in (1.13) as required, and $\Theta = \sum \Theta_i$ is as in Lemma 6, now with the choices (4.2).

The quantity Θ_4 contributes only to the error term $E(w)$. In (3.13) there occurs the entry $\sigma_D^-(A) = V^-(D, P(D^{1/\beta}))$, as defined via (2.9) and (2.3). Thus Theorem 4.4.2 gives

$$\sigma_D^-(A) \geq -V(P(D^{1/\beta}))\frac{c_1 e^{\sqrt{L}}}{\log^{1-\Delta} D},$$

for some constant $c_1 > 0$, since $f(\beta) = 0$ when $\kappa \geq \frac{1}{2}$ (as noted at (4.2.1.7)). In a similar way the remaining contribution to Θ_4 in (3.13) is

$$\geq - \sum_{D^T \leq p < D^{1/u}} \frac{w(p)\rho(p)}{p} \sum_{\substack{n \mid P(D^{1/u})/p \\ p < q(n); 2 \nmid \nu(n)}} V\Big(P((D/pn)^{1/\beta})\Big)\frac{\chi_D^-(np)\rho(n)c_2 e^{\sqrt{L}}}{n\big(\log(D/pn)\big)^{1-\Delta}}.$$

Here we can use $\log D/pn \geq \log p \geq T \log D$, because p is the smallest prime factor of pn, $\nu(pn)$ is even, and $\chi_D^-(np) = 1$, so that $np^{\beta+1} < D$ from the specification (3.3.1.1). But $\beta \geq 1$, so $D/pn > p$.

Next, the sum $\sum \rho(n)/n$ is bounded by a product $\prod(1+\rho(p_i)/p_i)$ in which there are at most $1/T$ factors, each of which is $O(1)$ because $\rho(p_i) \le p_i$. Thus, for some constant A,

$$\sum_{p<q(n);n|P(D)} \frac{\rho(n)}{n} \le A^{\log D/\log p} \le A^{1/T} = O(1) \,. \tag{4.4}$$

Actually $(1-\rho(p_i)/p_i)^{-1} = O(1)$ (see (1.3.5.5)), so the V-factor may be replaced by $V(P(D))$, at the cost of adjusting the constant c_2. Hence

$$\Theta_4(P(D^{1/u})) \ge -\frac{c_3 e^{\sqrt{L}} V(P(D))}{\log^{1-\Delta} D}\,.$$

The expression Θ_3 is handled on similar lines. Theorem 4.4.2 gives

$$\sigma^+_{D/pn}(A/pn) \le V\left(P\left((D/pn)^{\frac{1}{\beta+1}}\right)\right)\left(F(\beta+1) + O\left(\frac{e^{\sqrt{L}}}{\log^{1-\Delta}(D/pn)}\right)\right)$$

$$\le C V(P(D))\left(\frac{\log D}{\log D/pn}\right)^\kappa \left(1+O\left(\frac{e^{\sqrt{L}}}{\log^{1-\Delta}(D/pn)}\right)\right). \tag{4.5}$$

Here we re-expressed the V-factor in terms of $V(P(D))$ by the sifting density property (1.3.5.3), and rewrote the resulting product $(\beta+1)^\kappa F(\beta+1)$ as C, as in Sect. 4.2.

In Lemma 6, the expression on the left of (4.5) appears with values of n for which

$$2|\nu(n), \quad \bar{\chi}_D(n)=1, \quad p<q(n), \quad nq^{\beta-1}(n) < D/p\,.$$

Currently $p > D^T$ and $\beta > 1$, so that $D/np > p^{\beta-1} > D^{(\beta-1)T}$. Thus the O-term can be replaced by $O(e^{\sqrt{L}}/\log^{1-\Delta} D)$ and summed using the considerations already applied to Θ_4. This shows

$$\Theta_3(P(D^{1/u})) \ge -V(P(D)) C\, M_3(w)\left(1+O\left(\frac{e^{\sqrt{L}}}{\log^{1-\Delta} D}\right)\right), \tag{4.6}$$

where $M_3(w)$ and $C = (\beta+1)^\kappa F(\beta+1)$ are as stated in Lemma 7.

Rather more easily, the same upper bound for σ^+ (now with $n=1$) gives

$$\Theta_1(P(D^{1/u})) \ge -C V(P(D)) M_1(w)\left(1+O\left(\frac{e^{\sqrt{L}}}{\log^{1-\Delta} D}\right)\right).$$

In a similar way the lower estimate for σ^+ from Theorem 4.4.3 leads to the requisite lower bound

$$\Theta_2(P(D^{1/u})) \ge C V(P(D)) M_2(w)\left(1+O\left(\frac{L' e^{\sqrt{L}}}{\log^{1-\Delta} D}\right)\right)$$

Lemma 7 as stated follows from (4.3) and these estimates of Θ_i for $i \le 4$.

5.3 A Weighted Sieve Following Rosser

The last step in the proof of Theorem 1 is an application of partial summation, as described in Sect. 1.3.

Lemma 8. *Suppose that $D \geq e^L$, that ρ has a two-sided sifting density $\kappa > \frac{1}{2}$, as used in Theorem 4.4.3, and that $\kappa > \frac{1}{2}$. Then the sum $M(w)$ appearing in Lemma 7 satisfies the estimate*

$$M(w) \geq \mathcal{M}(W) + O\left(\frac{L'}{\log D}\right),$$

where $w(p) = W(\log p / \log D)$ and $w(1) = W(1)$ as in (5.2.1.2), and $\mathcal{M}(W)$ is as in Theorem 1.

The inequality $D \geq e^L$ ensures $(L/\log D)^n \leq L'/\log D$ whenever $n \geq 1$, since Definition 1.3.8 specified $L' \geq L$ in the two-sided density condition.

In Lemma 7, the expression $M(w)$ appeared as $M_1(w) + M_2(w) + M_3(w)$. The estimate for M_1 follows directly from Lemma 1.3.6:

$$M_1(w) \geq -\int_{1/\beta}^{1/u} \frac{W(1) - W(x)}{(1-x)^\kappa} \frac{\kappa \, dx}{x} - O\left(\frac{L}{\log D}\right). \tag{4.7}$$

In an exactly analogous way the estimate

$$M_2(w) \geq \int_T^{1/\beta} \frac{W(x)}{(1-x)^\kappa} \frac{\kappa \, dx}{x} - O\left(\frac{L'}{\log D}\right) \tag{4.8}$$

is obtained starting from (1.3.5.6).

In the expression for $M_3(w)$ in Lemma 7, the contribution to the sum over n from those n having exactly $r \geq 1$ prime factors is expressible in terms of

$$\Sigma_r(D, s, p) = \sum_{\substack{n \mid P(D^{1/s})/p;\, \nu(n)=r \\ p < q(n);\, nq^{\beta-1}(n) < D/p}} \frac{\rho(n)\tilde{\chi}_D^{(-)^{r+1}}(n)}{n} \left(\frac{\log D}{\log(D/pn)}\right)^\kappa. \tag{4.9}$$

As above, write $x = \log p / \log D$. An induction will establish

$$\Sigma_r(D, s, p) \leq h_r(x, s) + O\left(\frac{L}{\log p}\right) \quad \text{when} \quad r \geq 1, \tag{4.10}$$

where the r-dependence of the O-constant can be ignored since r is bounded by $\log D/\log p < 1/T$. Since (1.9) and (1.10) show that $h_r(x,s)$ is bounded (in (1.9) $x + x_1$ can approach 1 only when $\beta = 1$, which arises only if $\kappa \leq \frac{1}{2}$) the induction will simultaneously show that $\Sigma_r(D, s, p)$ is bounded by some constant depending only on T.

5. The Sieve with Weights

In the first place (4.9) and Lemma 1.3.6 give

$$\Sigma_1(D,s,p) = \sum_{\substack{D^{1/(\beta+1)} \leq p_1 < D^{1/s} \\ p \leq p_1;\, pp_1^\beta < D}} \frac{\rho(p_1)}{p_1} \left(\frac{\log D}{\log(D/pp_1)}\right)^\kappa \leq h_1(x,s) + O\left(\frac{L}{\log p}\right),$$

since $h_1(x,s)$ is as given in (1.9). Here, the condition $p_1 \geq D^{1/(\beta+1)}$ arises from the description of $\tilde{\chi}$ given in Sect. 4.1.

For the inductive step observe that (4.1.1.5) gives that if p_1 is the largest prime factor of $p_1 m$ and $r > 1$ then

$$\tilde{\chi}_D^{(-)^{r+1}}(p_1 m) = \chi_D^{(-)^{r+1}}(p_1)\tilde{\chi}_D^{(-)^r}(m),$$

the factor χ_D requiring $p_1^{\beta+1} < D$ when $r \geq 2$ is even. Thus Σ_r satisfies the recursion

$$\Sigma_r(D,s,p) = \sum_{\substack{p_1 < D^{1/s};\, p_1^{\beta+1} < D \text{ if } 2|r \\ m|P(p_1);\, \nu(m)=r-1 \\ p<q(m);\, mq^{\beta-1}(m)<D/pp_1}} \frac{\rho(p_1)}{p_1} \frac{\rho(m)\tilde{\chi}_D^{(-)^r}(m)}{m} \left(\frac{\log D}{\log(D/pp_1 m)}\right)^\kappa$$

$$= \sum_{\substack{p_1 < D^{1/s} \\ p_1^{\beta+1} < D \text{ if } 2|r}} \frac{\rho(p_1)}{p_1} \left(\frac{\log D}{\log D/p_1}\right)^\kappa \Sigma_{r-1}\left(\frac{D}{p_1}, \frac{\log D}{\log p_1} - 1, p\right),$$

since $p_1 = (D/p_1)^{1/s_1}$ with $s_1 = \log D/\log p_1 - 1$.

Now use the inductive step (4.10) for Σ_{r-1}, and Lemma 1.3.6, so that $\log p_1/\log D$ will be replaced by a continuous variable x_1. Since Σ_{r-1} and $\log D/\log(D/p_1)$ are bounded (the latter by $\log D/\log p < 1/T$) this gives

$$\Sigma_r(D,s,p) \leq \int_{\substack{x_1 < 1/s \\ x_1 < 1/(\beta+1) \text{ if } 2\nmid r}} \frac{1}{(1-x_1)^\kappa} h_{r-1}\left(\frac{x}{1-x_1}, \frac{1}{x_1} - 1\right) \frac{\kappa\, dx_1}{x_1}$$

$$+ O\left(\frac{L}{\log p}\right),$$

which gives the inductive step (4.10) for Σ_r because the substitution $x_1 = 1/t$ yields the expression (1.10) for $h_r(x,s)$, as required.

After a (finite) summation over r the estimate (4.10) for the quantity (4.9) shows that the sum M_3 from Lemma 7 satisfies

$$M_3(w) \geq -\sum_{p<D^{1/(\beta+2)}} \frac{w(p)\rho(p)}{p}\left(h\left(\frac{\log p}{\log D}, u\right) + O\left(\frac{L}{\log D}\right)\right)$$

$$\geq -\int_T^{1/(\beta+2)} W(x)h(x,u) \frac{\kappa\, dx}{x} - O\left(\frac{L}{\log D}\right), \qquad (4.11)$$

5.3 A Weighted Sieve Following Rosser

after a further appeal to Lemma 1.3.6, because $h(x,u)$ is as in (1.11), and the O-terms are as stated when $p > D^T$. The upper limit of integration may be replaced by $\frac{1}{2}$ since (1.9), (1.10) and (1.11) gave that $h(x,s) = 0$ when $x > 1/(\beta + 2)$.

Lemma 8 now follows from (4.7), (4.8) and (4.11).

The case $\kappa > \frac{1}{2}$ of Theorem 1 now follows from Lemmas 1, 7 and 8 as previously described, since $\beta > 1$ for these κ. In fact the only point at which this hypothesis was used was in deriving the estimate for Θ_3 in Lemma 7. The proof of Theorem 1 when $\kappa = \frac{1}{2}$ is completed by replacing (4.6) and (4.11) by the following estimate.

Lemma 9. *If $\kappa = \frac{1}{2}$ then the expression M_3 in Lemma 7 satisfies*

$$M_3(w) \geq -\int_T^{1/3} W(x)h(x,u)\frac{dx}{x} - O\left(\frac{e^{\sqrt{L}}}{\log^{1-\Delta} D}\right),$$

where $0 < \Delta < \frac{1}{2}$ as in Theorem 4.4.2.

The difficulty that arises is that when $\beta = 1$ the entry D/pn in the O-term in (4.5) may be close to 1. In this case the estimate (4.5) becomes trivial, but may be replaced by

$$\sigma^+_{D/pn}(A/pn) \ll 1 \quad \text{if} \quad pn < D,$$

as follows by direct reference to (4.1.1.2) when D is so small that Theorem 4.4.2 is insufficient. Hence

$$\sigma^+_{D/pn}(A/pn) \ll V(P(D))\sqrt{\log D},$$

where the sifting density estimate (1.3.5.2) has been used. This shows that when $\psi(n) = \rho(n)/n$ in (3.12) the contribution to Θ_3 from those pairs n,p with $np \geq D/3$ is, because of (3.8),

$$\ll V(P(D))\sqrt{\log D} \sum_{D^T \leq q(n); n|P(D)} \frac{\rho(n)}{n} \sum_{\substack{D/3n \leq p < D/n \\ p \geq D^T}} \frac{w(p)\rho(p)}{p}.$$

The fact that $p \geq D^T$ in Lemma 7 gives that the inner sum is empty unless $D/n \geq D^T$. Since $w(p) \leq w(1)$, Lemma 1.3.5 now shows that the inner sum is

$$\leq \log\frac{\log D/n}{\log D/3n} + \frac{L}{\log D/3n} \ll \frac{L}{\log D},$$

while the outer sum is $O(1)$ as in (4.4). This contribution is absorbed by the O-term in Lemma 9.

216   5. The Sieve with Weights

For the terms with $np < D/3$ the main term in Lemma 9 is derived as when $\kappa > \frac{1}{2}$, but the effect of the O-term in (4.5) must be reconsidered. Its contribution to the O-term in Lemma 9 is

$$\ll \sum_{n} \sum_{D^T \leq p < D/3n} \left(\frac{\log D}{\log D/pn}\right)^{1/2} \frac{e^{\sqrt{L}}}{\log^{1-\Delta}(D/pn)}$$

$$\ll \sum_{D^T \leq q(n): n|P(D)} \frac{e^{\sqrt{L}}}{\log^{1-\Delta} D} \int_T^{1-\frac{\log 4n}{\log D}} \frac{1}{\left(1 - \frac{\log n}{\log D} - t\right)^{\frac{3}{2}-\Delta}} \frac{dt}{t} + O\left(\frac{1}{\log D}\right),$$

in which the integral is $\ll \log 1/T = O(1)$, and the resulting sum over n is $O(1)$ as before. This leads to Lemma 9 as stated, so that the proof of Theorem 1 is now complete.

5.4 Notes on Chapter 5

Sect. 5.1.1. A weighting device simpler than that described here appeared in [Kuhn (1941)] and [Kuhn (1954)]. In Kuhn's work, the place occupied by $w(p)$ was taken by a function that is constant on a certain interval. In the language of Lemma 5.1.1, one may consider

$$m(a) = 1 - \frac{1}{b+1} \sum_{p|a;\, p<D^U} 1 \, .$$

If this can be shown to be positive for some a, by averaging over a with $(a, P(D^{1/s})) = 1$ as in Lemma 5.1.2, then these a have at most b distinct prime factors not exceeding D^U, perhaps as small as $D^{1/s}$. There might be $R - b$ prime factors exceeding D^U, provided $(R-b)u + b/s < g$, where $a < D^g$ for all a, as in the text. This is the basis of Kuhn's device.

This device is most likely to be effective with $b = 1$. In this form it played an important rôle in the celebrated theorem of J.-R. Chen [Chen (1973)] that every sufficiently large even number N is expressible as a sum of a prime and a P_2 number. In this work, the sequence $N - p$ is sifted, in which p is a prime not exceeding N. Some numbers of the form $p_1 p_2 p_3$ survive the weighted sifting procedure, but the number of these survivors is majorised by a separate sifting procedure. This procedure required a mean value theorem of Chen's own, related to but different from the Bombieri-Vinogradov theorem that was an adjunct to the main sifting procedure. There is an account of this work in [Halberstam and Richert (1974)].

The logarithmic weight introduced in Sect. 5.1.1 appears to have been first used in [Ankeny and Onishi (1964)] in connection with a proof that there are infinitely many pairs P_2, P_3 in which the almost-primes concerned differ by 2.

Sect. 5.1.2. This departure from the possibly more natural weight introduced in Sect. 5.1.1 appears in [Richert (1969)], and is also discussed in [Halberstam and Richert (1974)]. As is clear from the text, this "warping" of the simple logarithmic weights was more effective in the then current state of the art. This is still the case today, although we will argue in a note on Sect. 5.3 that it is to be hoped that this state of affairs will not persist for ever.

Sect. 5.1.3. The applications described here are from [Richert (1969)]. This paper also describes how the Bombieri-Vinogradov theorem enables the methods of this chapter to show that every large even N is representable in the form $p + P_3$. This is of course attained without the special device of Chen alluded to above.

Another application of Theorem 1 that might be mentioned is the author's proof [Greaves (1971)] that a suitable irreducible binary cubic form, such as $m^3 + 2n^3$, represents infinitely many P_2 numbers. This result has been recently overtaken by the "parity-breaking" work of D. R. Heath-Brown referred to in the notes on Chap. 4.

Not mentioned in the text, but nevertheless important, is the question of applying a weighting device in situations where $\kappa > 1$. In [Diamond and Halberstam (1997)] there is a description of the results obtainable by combining the currently best known sieve for moderate values of κ with the device from Sect. 5.1. As these authors remark, their results should be improvable by using a better weighting device. The device from Sect. 5.2 should be usable in this context, as (possibly with more trouble) should that from Sect. 5.3.

Sect. 5.2.1 The improved weighting device described here appeared in [Halberstam, Heath-Brown and Richert (1981)] and in [Greaves (1982a)], although it was used in a rather different way in each case. The first-named authors applied the construction to a particular application, and treated the sums arising from the term $m_4(a)$ in isolation from the other aspects of the problem, using the lower bound sieve from Chap. 4. The estimate obtained in this way is non-trivial only when $p_1 p_2^3 p_3 < D$. One may observe that in this process one is working on the wrong side of the parity phenomenon, in that in the extremal example relating to the lower bound sieve (the case $r = 1$ in Sect. 4.5.1) the estimation now being performed is not a sharp one. It should also be pointed out that this work was used in conjunction with the bilinear form of the remainder which we discuss in Chap. 6. In [Greaves (1982a)] the construction was used in the context of the general sifting situation of sifting density $\kappa = 1$, and the treatment of the term $m_4(a)$ was, as far as possible, combined with other aspects of the problem. This is the method that is discussed in Sect. 5.3, in which account is taken of the entry $m_4(a)$ whenever $p_1 p_2^3 < D$. A comparison of these two methods with that of Sect. 5.2.2 was the subject of the survey in [Greaves (1985)].

218   5. The Sieve with Weights

Sect. 5.2.2. The weighting device described here has a long history, which we briefly summarise. The somewhat complicated paper [Buchstab (1965)] achieved results approximately as good (if not slightly better) than those in [Richert (1969)], is spite of using a slightly weaker sifting input; instead of using results equivalent to those of Chap. 4, Buchstab used a sieve derived from Brun's method by a finite number of iterations of the Buchstab transform described in Sect. 7.2. Moreover, instead of using a continuous weight $w(p)$, of the type appearing in Sect. 5.1, Buchstab used a function which was a constant on each of many intervals, in this way being an elaboration of Kuhn's device.

Buchstab's paper was studied by M. Laborde, who produced a continuous version of Buchstab's weights in [Laborde (1979)], which (apart from unimportant changes of notation) is as described in Theorem 5.2.1.

H. Iwaniec, in an important paper [Iwaniec (1978a)] devoted to showing that $n^2 + 1$ is infinitely often a P_2 number, referred to "a new weighted sum of Richert", in which the description of the weighting device was a slight variant of that appearing in Lemma 5.2.2. The author understands, from a conversation with the late H.-E. Richert, that Richert's student R. Rotermund had similarly studied Buchstab's paper and had been led to similar conclusions about the nature of the underlying ideas.

The relationship of these devices with the sum $m_2(a)$ appearing in Sect. 5.2.1 is a later observation from [Greaves (1985)].

The treatment of the Buchstab–Laborde weighting device in Sect. 5.2.2 is in a style rather different from that used in either Sect. 5.1 or Sect. 5.3. The reason for this departure is a wish to make Theorem 5.2.1 visibly the same, apart from minor changes in notation, as that in [Laborde (1979)].

A development of the Buchstab-Laborde weighting device was described in [Buchstab 1985], for which see also [Vakhitova (1999)] or [Vakhitova (1995)]. It seems to the author that this development takes some account of the term $m_4(a)$ appearing in Sect. 5.2.1. However, it appears that these "new Buchstab weights" take account only of the contributions to (1.6) from the terms with $p_1^3 p_2 < D$, a more severe restriction than arises in the treatments described in Sect. 5.3 or referred to in the note to Sect. 5.2.1.

Sect. 5.3. The device described in this section is that from [Greaves (1982a)], but the exposition contains some modifications based on [Halberstam and Richert (1985a)].

Before we get involved with the technicalities, there are some general remarks to be made. In the first place it is instructive to examine the details of the construction used in the light of Selberg's extremal example. This is, nearly enough for our purposes, the same as in Sect. 4.5, and may be examined in the same way as in the notes to Sect. 4.5.1. In this case we find we have cast out terms like

$$\sum_{\substack{p_1p_2p_3|a \\ p_1p_2 \leq D; p_1p_2^3 \geq D}} S\left(\frac{a}{p_1p_2p_3}, P(p_2)\right).$$

On this occasion significant contributions occur when $\Omega(a)$ is even and a is about D in size, with $p > p_2$, $pp_3 < p_2^2$, and $a = p_1pp_2p_3 < p_1p_2^3$. Consequently we may at once infer that the construction used is not going to establish the standard "$\Lambda_R = R$" conjecture, which essentially says that there exists a weighted lower bound construction that is sharp with respect to Selberg's extremal example.

A fuller edition of this discussion appears in [Greaves (1989)], where is is also pointed out (following an examination of the construction in a way slightly different from that of the text) that any significant modification of the weight function w from the logarithmic form of Sect. 5.1.1 is similarly going to have a disastrous effect on any attack on the standard conjecture.

Unfortunately it appears to be very much harder to find a construction that does not suffer from this sort of defect than it is to point out the defects in existing methods.

Naturally one hopes that the conjecture will some day be proved. It is for these reasons that our note on Sect. 5.1.2 expressed the hope that the warping of the weight function made in Sect. 5.1.2 will not prove to be part of the long-term future of the subject.

An analogous upper bound for the sieve with weights could be discussed, but has not been provided in this account. However, a heuristic discussion is sufficient to indicate, with the weight function $w(p) = p$ that we would like to use, that a better upper bound can be obtained by use of Selberg's method. One may now consider a weighted analogue of the construction of [Jurkat and Richert (1965)], also not discussed in any detail in this volume, in which Selberg's upper bound is used as an adjunct to a combinatorial construction. In contrast to the unweighted situation, we now expect to obtain improved results in this way. This remark was followed up in [Greaves (1986)], and used to obtain the improved values for $\delta_2(= 0.04456...)$, δ_3 and δ_4 given in the introduction to this chapter.

Naturally, an examination of the method in the last paper referred to again shows where the argument is not sharp with respect to Selberg's limiting example.

Sect. 5.3.1. The computations of the numbers α and β referred to immediately preceding Corollary 1.1 have been conducted by two entirely different methods, which are in excellent agreement. A method initiated in [Greaves (1982a)] uses now-familiar ideas relating to the adjoint function, in this case obtained by replacing $f(s)$ in Sect. 4.2 by the moments $\int_0^1 x^n h(x,s)\,dx$ for integers n, with a related replacement for $F(s)$. This led to two numerical problems. The first was the computation of values $J^{(n)}(1)$ of the derivatives of such an adjoint function. These derivatives were computed simultaneously in a parallel process described in [Greaves (1982c)]. The second was the inversion of the moment map to recover the numbers α and β, for which an algorithm was described in [Greaves (1982b)].

The second method departed from the familiar arguments of sieve theory, but treated the problem in a numerically more direct way. The principle used in [Grupp and Richert (1986)] was to represent the functions to be computed by a sequence of Taylor expansions valid in a series of circles.

Both these methods seem more attractive than direct computation of the multiple integrals resulting from following the recurrences (1.10). An exception to this remark is in the computation of the numbers $\alpha(\frac{1}{6})$ and $\beta(\frac{1}{6})$, in which the computation in this way reduces to a straightforward 1-dimensional numerical quadrature.

Sect. 5.3.2. The argument in [Greaves (1982a)], where only the case $\kappa = 1$ was discussed, resorted to approximation of the quantities to be estimated using continuous functions, namely F and f from Sect. 4.2, at the earliest possible stage. To make progress, a certain expansion of these functions was introduced, as follows. For $r \geq 1$ let

$$s\, I_r(s) = \int \cdots \int\limits_{\substack{1/s < v_r < \cdots < v_1 \\ v_r + \cdots + v_1 = 1}} \frac{dv_1 \ldots dv_{r-1}}{v_1 \ldots v_{r-1} v_r},$$

so that in particular $s\, I_1(s) = 1$ when $s \geq 1$. Then it is not hard to verify that $\bigl(s\, I_r(s)\bigr)' = I_{r-1}(s-1)$ when $s > 1$, in which $'$ denotes differentiation with respect to s. Then we claim that we can redefine F and f by

$$F(s) = 2e^\gamma \sum_{r \text{ odd}} I_r(s), \quad f(s) = 2e^\gamma \sum_{r \text{ even}} I_r(s).$$

For $I_r(s) = 0$ when $r \geq s$, so this new definition gives $F(s) = 2e^\gamma/s$ if $1 \leq s \leq 3$ and $f(s) = 0$ if $1 \leq s \leq 2$. Also F and f are continuous in $s \geq 1$ and satisfy $\bigl(s\, F(s)\bigr)' = f(s-1)$ if $s > 1$ and $\bigl(s\, f(s)\bigr)' = F(s-1)$ if $s > 2$. This is sufficient to show that F and f are the functions described in the case $\kappa = 1$ of Sect. 4.2. Note that this description differs in several respects from that in Proposition 4.2.1.

These expansions are implicit in the paper [Bombieri (1976)] on the asymptotic sieve: if the reader tries to recover the theorem of Sect. 4.4 (in the "local" context considered in Bombieri's paper) he will find that the functions F and f are delivered in this expanded form. These expansions also appeared in [Siebert (1983)].

One may now observe that an analogous expansion can be performed at the arithmetical level, before any approximations by continuous functions have been made. This line of thought leads to the reduction identities of the text, which are an arithmetical analogue of relations such as

$$s^\kappa f(s) = \beta^\kappa f(\beta) + \int_\beta^s F(t-1)\,dt^\kappa \quad \text{if} \quad s > \beta,$$

an immediate consequence of the equations of Sect. 4.2.1. The identity of Lemma 4 is similarly an analogue of a corresponding identity involving the functions F and f.

Sect. 5.3.3. The text carries the arithmeticisation of the argument one stage further than [Halberstam and Richert (1985a)], in that it postpones, until Sect. 5.3.4, the stage at which the expressions $v_D^-(A,\beta)$ are estimated by O-terms. In this way Lemma 6 is still an exact identity.

6. The Remainder Term in the Linear Sieve

In the applications of sieve ideas to number-theoretic situations considered earlier in this book, the remainder terms $R^{\pm}$ appearing in Theorem 1.3.1 have been dealt with in the so-called trivial way used to infer Corollary 1.3.1.1, in which we said simply

$$|R(D,P)| = \left|\sum_{d|P} \lambda(d) r_A(d)\right| \leq \sum_d |r_A(d)|,$$

when $|\lambda(d)|$ for each d. On the other hand, in many situations there is much more information available about $r_A(d)$ than is conveyed by bounds of its absolute value, as the following illustration indicates.

Consider the case when $\mathcal{A}$ consists of the integers in an interval $[Y-X,Y)$. As in Sect. 1.2.2, let $[\cdot]$ and $\{\cdot\}$ denote the integer and fractional part functions, but introduce

$$\psi(t) = \begin{cases} \frac{1}{2} - \{t\} & \text{if } t \text{ is not an integer} \\ 0 & \text{if } t \text{ is an integer.} \end{cases}$$

Suppose meantime that X and $Y-X$ are not integers, so that in particular they are not divisible by d. Beginning as in Sect. 1.2.2 observe

$$|\mathcal{A}_d| = [Y/d] - [Y-X/d] = X/d + \{(Y-X)/d\} - \{Y/d\}$$
$$= X/d + \psi(Y/d) - \psi((Y-X)/d) .$$

Here (and throughout this chapter) denote $e(x) = e^{2\pi i x}$. The function ψ is given by the sum of its Fourier series

$$\psi(t) = \frac{1}{\pi} \sum_{h=1}^{\infty} \frac{\sin 2\pi h t}{h} = \frac{1}{2\pi i} \sum_{\substack{h=-\infty \\ h \neq 0}}^{\infty} \frac{e(ht)}{h},$$

in which the last summation over h is to be read as $\lim_{H \to \infty} \sum_{0 < |h| < H}$. It is desirable to make some use of this analytic information, but at first sight the resulting expression for $R(D,P)$, as a difference of two expressions of the shape

$$\frac{1}{2\pi i} \sum_{d|P} \lambda(d) \sum_{\substack{h=-\infty \\ h \neq 0}}^{\infty} \frac{e(hz/d)}{h},$$

appears intractable, in view of the detailed structure of the expression $\lambda(d)$ in the constructions of Brun and Rosser, and of Selberg.

In Selberg's upper bound method it is relatively straightforward to see that the construction has a certain bilinear structure

$$\lambda^+(d) = \sum_{\substack{m<M \ n<N \\ mn=D}} a_m b_n \,,$$

nearly enough for present purposes with $a_m = \lambda(m)$, $b_n = \lambda(n)$, so that we would need to take $M = N = \sqrt{D}$. The details of this structure are not given in this book (but see the notes and references at the end of this chapter). We will however show that in the context of the linear sieve Rosser's upper and lower bound methods also possess such a bilinear structure, actually without the constraint $M = N$. This gives an added flexibility which is very useful in applications.

The consequence is that the remainder term inherits a bilinear structure, in terms of an expression

$$\sum_{\substack{m<M \ n<N \\ mn|P}} a_m b_n r_A(mn) \,,$$

where $MN = D$, and we hope to make some use of analytic input about the entries $r_A(mn)$. The detailed statement of the result we derive in given in Theorem 6.1.1. The factors a_m, b_n will be no less complicated than the expressions $\lambda^\pm$ from which they were obtained, but in some situations these complications can be swept aside by judicious use of Cauchy's inequality. We will see that we do not need to know much more about these numbers than that they satisfy $|a_m| \leq 1$, $|b_n| \leq 1$. An example of these processes in action will be given in detail in Sect. 6.2, but an outline of the salient points follows almost at once.

One important requirement of the use of Cauchy's inequality is the numbers a_m, b_n will need to be independent, in that b_n does not, for example, depend on m. This means that the remainder term $R(D, P)$ needs to be written not as above (where the independence is violated by the occurrence of the condition $mn|P$) but in terms of a modified expression

$$\sum_{\substack{m<M \ n<N \\ m|P \ n|P}} a_m b_n r_A(mn) \,,$$

in which the product mn is not necessarily squarefree. A result of this type is given in Theorem 6.1.2, but for details of the proof of this modified version the reader will be referred to the literature. This result possesses the novel feature of referring to values of $r_A(d)$ for values of $d = mn$ which are not squarefree, in marked contrast to the sieve method from which it was obtained.

6. The Remainder Term in the Linear Sieve

To illustrate the need for this feature, there follows an initial and somewhat heuristic discussion of the application of these ideas to the short interval problem from Sect. 1.2.2. The preceding discussion indicates how, in this context, the remainder $R(D, P)$ can be expressed in terms of certain trigonometrical sums

$$E = \sum_{m<M} \sum_{n<N} a_m b_n \sum_{h \neq 0} c_h e(h\, f(mn)) ,$$

in the first place with $c_h = 1/h$ and $f(mn) = t/mn$. In this case the sum E would have the serious disadvantage that it is not absolutely convergent, a difficulty which will have to be overcome by the introduction of one of a range of possible smoothing devices. Suppose that this has been done.

In such a situation the sum E could be estimated by

$$|E| \leq \sum_{h \neq 0} c_h \left| \sum_{m<M} \sum_{n<N} e(h\, f(mn)) \right| .$$

The inner sum can be estimated via Cauchy's inequality via

$$\left| \sum_{m<M} a_m \left\{ \sum_{n<N} b_n e(h\, f(m,n)) \right\} \right|^2$$

$$\leq \left\{ \sum_{m<M} |a_m|^2 \right\} \sum_{m<M} \left| \sum_{n<N} b_n e(h\, f(m,n)) \right|^2$$

$$\ll M \sum_{m<M} \sum_{n_1<N} \sum_{n_2<N} b_{n_1} b_{n_2} e(h\, f(mn_1) - h\, f(mn_2)) .$$

Here the "diagonal" terms with $n_1 = n_2$ give, even after an absolutely convergent summation over h, an amount

$$\ll M \sum_{m<M} \sum_{n<N} 1 \ll M^2 N ,$$

giving a contribution $\ll M\sqrt{N}$ to E. Since $MN = D$, such an estimate will be more than satisfactory. Observe how the somewhat recondite a_m, b_n have been dealt with simply via bounds for their absolute values.

The remaining off-diagonal terms can be handled using suitable methods from analytic number theory. It will become clear that these methods require that the range of summation for n_1 and n_2 should be precisely an interval, unencumbered by side-conditions such as $(n_i, m) = 1$. It is for such reasons that Theorem 6.1.2 is an indispensable adjunct to Theorem 6.1.1.

In general an effective application of Theorem 6.1.1 to specific problems in Number Theory requires an input relating to Dirichlet polynomials, the dispersion method, or the like, which is beyond the scope of this book. The notes to this chapter refer to some of the relevant literature. On the other

226   6. The Remainder Term in the Linear Sieve

hand the theorems of Sect. 6.1 might seem sterile if no indication were given of how they might be used. Accordingly, a specimen application is provided in Sect. 6.2, giving a bound for the number of primes in a "reasonably short" interval. This improves on the inequality of Theorem 2.1.4, for a certain range of the parameters involved. The account is self-contained, and uses only fairly elementary properties of trigonometrical sums.

6.1 The Bilinear Nature of Rosser's Construction

In this section we establish that in the case $\kappa = 1$ of Rosser's sieve (so that $\beta = 2$ in the construction of Sect. 3.3.1) the remainder term from Theorem 1.3.1 can be expressed in a bilinear form, as just sketched, and as stated in detail in Theorem 1 below. A central idea in the argument is that when $MN = D$ and $\chi_D(d) = 1$, in the notation of Sect. 3.3.1, the number d can be factored as $d = mn$ with $m \leq M$ and $n \leq N$. This fact, which is established in Sect. 6.1.1, does not of itself yield the required bilinear form because it leads only to an expression

$$\sum_{d|P} \mu(d)\chi_D(d)r_A(d) = \sum_m \sum_{\substack{n \\ mn|P}} \mu(mn)\chi_D(mn)r_A(mn)$$

for the remainder $R(D, P)$ in Theorem 1.3.1. Here, apart from the question of the squarefree nature of mn already referred to, the variables m, n are very far from being independent because of the occurrence of the factor $\chi_D(mn)$.

This difficulty is overcome by introducing a certain discretisation, or condensation, of Rosser's sieve. The idea of condensation is an old one, going back at least to Cauchy. In this instance it is applied in a moderately elaborate way. We will replace $\chi_D(d)$ by an approximation which takes constant values when when the prime factors p_j of d lie in certain specified intervals I_j. For each such set of intervals I_j the number $\chi_D(d) = \chi_D(mn)$ is now a constant, so he obstruction it raised to the construction of the desired bilinear form has now been removed. There will remain a summation over all possible sets of intervals I_j which will have to be performed afterwards.

The effect of this discretisation on the main term $\sum \mu(d)\chi_D(d)\rho(d)/d$ in Theorem 1.3.1 has to be estimated, and balanced against the effect it has on the corresponding remainder term. Actually it is the discussion of the effect on the main term that is more troublesome. The treatment of the bilinear expression of the remainder term will then proceed comparatively painlessly.

6.1.1 The Factorisation of χ_D

The principal result of this chapter, on Rosser's sieve with a bilinear form of the remainder term, is enunciated as Theorem 1 below. This theorem retains the structure that in most natural in a sieve argument, that the remainder term $r_A(d)$ appears only for squarefree values of $d = mn$.

For the reasons outlined in the introduction to this chapter, what is required for the purpose of applications in number theory is Theorem 2, in which the variables m and n are independent. This involves reference to $r_A(mn)$ for non-squarefree values of the product mn, and consequently involves a (mild) assumption on the corresponding values of $\rho(d)$ for non-squarefree d. For the proof of the extension of Theorem 1 to Theorem 2, the reader is directed to the original references given in the notes.

The property of Rosser's functions $\chi_D^\pm$ that permits the subsequent construction of the bilinear remainder term is embodied in Lemma 1.

Theorem 1. *Suppose that the function ρ has sifting density $\kappa = 1$ and that $L \leq (\log \log D)^{1/50}$ in (1.2.5.3). Suppose $s \geq 1$. Then, in Theorem 1.3.1, there is a choice of the sifting functions $\lambda^\pm(d)$ with the following properties.*

  (i) The main terms are given by

$$V^+(D,P(D^{1/s})) = V(P(D^{1/s}))F(s)(1 + \eta(D,s))$$
$$V^-(D,P(D^{1/s})) = V(P(D^{1/s}))f(s)(1 - \eta(D,s)),$$

where

$$\eta(D,s) \ll \frac{L^{10}}{(\log \log D)^{1/5}},$$

and the functions F, f are those described in Sect. 4.2 (now with $\kappa = 1$).

  (ii) Suppose $M \geq 1$, $N \geq 1$ and $MN \leq D$. Then for each of the remainder terms R^+, R^- there exist real numbers a_m, b_n with $|a_m| \leq 1$, $|b_n| \leq 1$ such that

$$R^\pm(D, P(D^{1/s})) \ll \log^{1/3} D \sum_{\substack{mn | P(D^{1/s}) \\ 1 \leq m \leq M, 1 \leq n \leq N}} a_m b_n r_A(mn). \qquad (1.1)$$

Theorem 2. *Assume that*

$$\sum_{w \leq p < z} \frac{\rho(p^2)}{p^2} \ll \frac{1}{\log \log w} \quad \text{when} \quad 2 \leq w < z.$$

Then, in Theorem 1, the sum over m and n in (1.1) may be replaced by

$$\sum_{m | P(D^{1/s})} \sum_{n | P(D^{1/s})} a_m b_n r_A(mn).$$

228   6. The Remainder Term in the Linear Sieve

In Lemma 1 below, the numbers $c_1, c_2, \ldots, c_r$ could be taken as the prime factors of a number d. In this case the lemma reduces to a decomposition of those numbers d for which $\chi_D^\pm(d) = 1$, where χ_D^+ and χ_D^- are as described in Sect. 3.3.1. Such a decomposition would not be directly useful, however, and it is to rather different sets of numbers c_i that Lemma 1 will be applied.

It will be convenient to extend the notation χ to such sets of numbers c_i. Let
$$\mathbf{C} = \{c_i : 1 \le i \le r\} \quad \text{when} \quad c_1 > c_2 > \cdots > c_r > 1 \, . \tag{1.2}$$
Define
$$\chi_D^+(\mathbf{C}) = 1 \quad \text{if} \quad c_1 c_2 \ldots c_{2j} c_{2j+1}^{\beta+1} < D \quad \text{when} \quad 2j+1 \le n \tag{1.3}$$
$$\chi_D^-(\mathbf{C}) = 1 \quad \text{if} \quad c_1 c_2 \ldots c_{2j-1} c_{2j}^{\beta+1} < D \quad \text{when} \quad 2j \le n \, , \tag{1.4}$$
with $\chi_D^\pm(\emptyset) = 1$, and $\chi_D^\pm(\mathbf{C}) = 0$ otherwise. Thus in the case when the numbers c_i are primes p_i these expressions $\chi_D^\pm(\mathbf{C})$ reduce to the corresponding $\chi_D^\pm(d)$ described by (3.3.1.1), in which d was the product of the primes p_i. In general, when (1.2) applies denote the product of the numbers c_i by
$$\|\mathbf{C}\| = \prod_{1 \le i \le n} c_i \, , \tag{1.5}$$
with the usual convention $\|\emptyset\| = 1$ for the empty product.

When $\beta \ge 2$ the inequalities in either of (1.3) or (1.4) imply
$$c_1 c_2 \ldots c_{j-1} c_j^2 < D \quad \text{when} \quad j \ge 2 \, . \tag{1.6}$$
As in Sect. 3.3.3, it is only this weakening of (1.3) or (1.4) that is essential for the next lemma, in which the conclusion is expressed using the notation (1.5).

In Lemma 1, the inequality (1.6) is not assumed when $j = 1$, but is replaced by the hypothesis $c_1 \le Y_1$, where Y_1 may be the larger of Y_1, Y_2.

Lemma 1. *Let $\mathbf{C}$ be as in (1.2). Suppose that (1.6) holds and*
$$D = Y_1 Y_2 \quad \text{with} \quad Y_1 \ge c_1 = \max_{c_i \in \mathbf{C}} c_i \, , \quad Y_2 \ge 1 \, .$$
Then the set $\mathbf{C}$ can be partitioned as
$$\mathbf{C} = \mathbf{C}_1 \cup \mathbf{C}_2 \, , \quad \text{where} \quad \|\mathbf{C}_1\| \le Y_1 \, , \ \|\mathbf{C}_2\| \le Y_2 \, .$$

Lemma 1 is proved by induction on n, the cardinality of $\mathbf{C}$. When $n = 1$ take $\mathbf{C}_1 = \{c_1\}$, $\mathbf{C}_2 = \emptyset$. When $n = 2$ the inequality $c_1 c_2^2 \le D = Y_1 Y_2$ gives either $c_1 c_2 < Y_1$ or $c_2 < Y_2$. In the first case take $\mathbf{C}_1 = \{c_1, c_2\}$, $\mathbf{C}_2 = \emptyset$. In the second case take $\mathbf{C}_1 = \{c_1\}$ and $\mathbf{C}_2 = \{c_2\}$.

6.1 The Bilinear Nature of Rosser's Construction

The proof of the general inductive step is similar. Suppose the lemma is satisfied when $n = N$, by sets $\mathbf{C}_1 = \mathbf{C}_1(N)$, $\mathbf{C}_2 = \mathbf{C}_2(N)$, say. Now the inequality $c_1 \ldots c_N c_{N+1}^2 < Y_1 Y_2$ shows that at least one of the inequalities

$$\|\mathbf{C}_1 \cup \{c_{N+1}\}\| < Y_1, \qquad \|\mathbf{C}_2 \cup \{c_{N+1}\}\| < Y_2$$

is satisfied. Thus adjoining c_{N+1} to the appropriate one of $\mathbf{C}_1(N)$ or $\mathbf{C}_2(N)$ gives new sets satisfying $\|\mathbf{C}_1(N+1)\| < Y_1$ and $\|\mathbf{C}_2(N+1)\| < Y_2$, as required.

6.1.2 Discretisations of Rosser's Sieve

To transform the device considered in Sect. 6.1 into a useful instrument, an interval $[w, z)$ (in a context appearing in Sect. 4.4.1) will be dissected into disjoint intervals

$$C_j = [c_{j+1}, c_j), \qquad z = c_1 > c_2 > \cdots > c_J = w. \tag{2.1}$$

The principle underlying the following discussion is that the functions $\chi_D^{\pm}$ describing Rosser's construction should be replaced by modified functions that depend only on the intervals C_j in which the prime factors of d might lie.

The structure of the discretisation is sufficiently elaborate to make it worthwhile to introduce the following terminology. First, the notation χ_D is extended slightly further than in Sect. 6.1.1. Let

$$\mathcal{E} = I_1 I_2 \ldots I_r$$

denote a set-theoretic direct product of (not necessarily distinct) intervals $I_j = C_{\nu_j}$ from (2.1). Write $c_{\nu_j} = i_j$, the larger end-point of the interval I_j. We may suppose the intervals taken so that the intervals with the larger suffices contain the smaller numbers:

$$I_1 \geq I_2 \geq \cdots \geq I_r. \tag{2.2}$$

Here we use a natural notation for inequalities between intervals (or even between direct products of intervals) in which $A > B$ means $a > b$ whenever a, b lie respectively in intervals A, B (or in intervals that are direct factors of A, B). Thus

$$\mathcal{E} = \bigotimes_{j=1}^{r} I_j, \qquad I_j = [h_j, i_j), \qquad z \geq i_1 \geq i_2 \geq \cdots \geq i_r, \tag{2.3}$$

where actually $h_j = c_{\nu_j+1}$, and the inequality $z \geq i_1$ is inherited from (2.1). The dissection $\mathcal{C}$ will be defined as

$$\mathcal{C} = \bigotimes_{j=1}^{J} I_j, \tag{2.4}$$

the case when $\mathcal{C}$ is the product of all the intervals C_j from (2.1).

When (2.3) holds define a Möbius function by $\mu(\emptyset) = 1$ and

$$\mu(\mathcal{E}) = \begin{cases} (-1)^r & \text{if } i_1 > i_2 > \cdots > i_r \\ 0 & \text{otherwise}. \end{cases} \qquad (2.5)$$

Thus $\mu(\mathcal{E}) \neq 0$ precisely when $\mathcal{E}$ is "squarefree" in the sense that the intervals I_r in (2.3) are all distinct. In this case set

$$\nu(\mathcal{E}) = r.$$

When $\mu(\mathcal{E}) \neq 0$ in (2.5), extend the notation χ_D from sets of real numbers, as given in (1.3) and (1.4), to the products $\mathcal{E}$ given in (2.3) by defining

$$\chi_D^-(\mathcal{E}) = \chi_D^-(\mathbf{I}), \quad \chi_D^+(\mathcal{E}) = \chi_D^+(\mathbf{I}), \quad \text{where} \quad \mathbf{I} = \{i_1, \ldots i_r\}, \qquad (2.6)$$

with $\chi_D^\pm(\emptyset) = 1$, and set $\chi_D^\pm(\mathcal{E}) = 0$ otherwise. Similarly, extend the notation (1.5) to

$$\|I_j\| = i_j, \quad \|\mathcal{E}\| = \prod_{j=1}^r \|I_j\|, \quad \|\emptyset\| = 1$$

when (2.3) holds. Thus $\|\mathcal{E}_1 \mathcal{E}_2\| = \|\mathcal{E}_1\| \cdot \|\mathcal{E}_2\|$. Furthermore,

$$\chi_D^\pm(\mathcal{E}) \neq 0 \Longrightarrow \|\mathcal{E}\| < D, \qquad (2.7)$$

so that the usual significance of the suffix D is preserved.

Next, introduce the corresponding functions $\bar{\chi}_D^\pm$ analogous to those introduced in Lemma 3.1.2, most concisely by following the expression

$$\bar{\chi}(d) = \chi(d/q(d)) - \chi(d) \qquad (2.8)$$

given in (3.1.2.12). The place of the least prime factor $q(d)$ will be taken by the interval I_r, the labelling being as in (2.2). Define $\bar{\chi}(\emptyset) = 0$ and

$$\bar{\chi}_D^\pm(\mathcal{E}) = \chi_D^\pm(I_1 I_2 \cdots I_{r-1}) - \chi_D^\pm(I_1 I_2 \cdots I_r) \quad \text{when} \quad r \geq 1. \qquad (2.9)$$

As in Sect. 3.1.2 this expression can then be described at length as

$$\bar{\chi}_D^-(I_1 \ldots I_{2n}) = 1 \quad \text{if} \quad \begin{cases} B_{2n} \leq i_{2n} \\ i_{2j} < B_{2j} \text{ when } j < n, \end{cases}$$

$$\bar{\chi}_D^+(I_1 \ldots I_{2n+1}) = 1 \quad \text{if} \quad \begin{cases} B_{2n+1} \leq i_{2n+1} \\ i_{2j+1} < B_{2j+1} \text{ when } j < n, \end{cases}$$

with $\bar{\chi}_D^\pm(\mathcal{E}) = 0$ otherwise, where the numbers B_j are now given by

$$i_j < B_j \iff i_1 i_2 \ldots i_{j-1} i_j^{\beta+1} < D.$$

6.1 The Bilinear Nature of Rosser's Construction

Say that a squarefree number d is *well-separated* by the dissection $\mathcal{C}$ given in (2.4) if distinct prime factors of d lie in distinct intervals $I_j \in \mathcal{C}$. Further, if all these I_j are in $\mathcal{E}$ and $\mu(\mathcal{E}) \neq 0$, then write

$$d \in \mathcal{E}, \tag{2.10}$$

in a non-standard use of the symbol $\in$. Observe

$$d \in \mathcal{E}, \ \chi_D^{\pm}(\mathcal{E}) = 1 \implies \chi_D^{\pm}(d) = 1. \tag{2.11}$$

Lemma 2, which follows fairly immediately from Lemma 1, refers also to certain products of intervals which are (marginally) not squarefree, lying in a set

$$\mathcal{C}^* = \left\{ I_{\nu_1} \ldots I_{\nu_r} I_{\nu_{r+1}}^2 : \nu_1 > \cdots > \nu_r > \nu_{r+1}, \ r \geq 0 \right\}. \tag{2.12}$$

For future reference observe

$$|\mathcal{C}^*| \leq |\mathcal{C}| = 2^J, \tag{2.13}$$

since there are J intervals in (2.1) from which I_{ν_j} is to be chosen.

Lemma 2. *Let $\mathcal{E}$ be as in (2.3), so that $x < z$ whenever $x \in I_j \in \mathcal{E}$, and suppose $D = Y_1 Y_2$ with*

$$Y_1 \geq z, \quad Y_2 \geq 1.$$

(i) *Suppose that either $\chi_D^+(\mathcal{E}) = 1$ or $\chi_D^-(\mathcal{E}) = 1$. Then $\mathcal{E}$ can be factored as*

$$\mathcal{E} = \mathcal{E}_1 \mathcal{E}_2, \quad \text{with} \quad \|\mathcal{E}_1\| \leq Y_1, \quad \|\mathcal{E}_2\| \leq Y_2, \quad \mathcal{E}_i \in \mathcal{C}. \tag{2.14}$$

(ii) *Suppose that $\mathcal{E} = I^2 \mathcal{G}$, with $\mathcal{G} \in \mathcal{C}$ and $I < G$ for each $G \in \mathcal{G}$ (possibly with $\mathcal{G} = \emptyset$), and that*

$$\chi_D^{(-)^n}(I\mathcal{G}) = 1, \quad \text{where} \quad \nu(\mathcal{G}) \equiv n, \bmod 2.$$

Then $\mathcal{E}$ can again be decomposed as in (2.14), but with $\mathcal{E}_i \in \mathcal{C} \cup \mathcal{C}^*$, where $\mathcal{C}^*$ is as in (2.12).

To obtain part (i), take **C** in Lemma 1 to be the set **I** appearing in (2.6).

Part (ii) is slightly more involved, in that it would not follow from a consequence such as (1.6) from the hypothesis involving χ_D. Note from part (i) that $I\mathcal{G}$ can be decomposed as

$$I\mathcal{G} = \mathcal{K}_1 \mathcal{K}_2, \quad \text{where} \quad \|\mathcal{K}_1\| < Y_1, \ \|\mathcal{K}_2\| < Y_2. \tag{2.15}$$

In the current situation it also follows that

$$\|I^3 \mathcal{G}\| < Y_1 Y_2. \tag{2.16}$$

6. The Remainder Term in the Linear Sieve

To see this, write (consistently with the condition $\nu(\mathcal{G}) \equiv n$, mod 2)

$$\mathcal{G} = I_1 I_2 \ldots I_n \quad \text{with} \quad I_1 > \cdots > I_n > I \, .$$

Now the specification of $\chi_D^{(-)^n}(I\mathcal{G})$ given in (2.6) gives $\|I_1 \ldots I_n I^3\| < D$, and (2.16) follows. In particular this holds when $\mathcal{G} = \emptyset$, in which case $n = 0$.

It may happen that

$$\begin{array}{ll} \text{either} & \|I^2\mathcal{K}_1\| < Y_1, \quad \|\mathcal{K}_2\| < Y_2 \\ \text{or} & \|I^2\mathcal{K}_2\| < Y_2, \quad \|\mathcal{K}_1\| < Y_1, \end{array} \quad (2.17)$$

where in either case the second inequality restates part of (2.15). If (2.17) is false, then $\|I^2\mathcal{K}_1\| \geq Y_1$ and $\|I^2\mathcal{K}_2\| \geq Y_2$, so (2.16) shows

$$\|I\mathcal{K}_1\| < Y_1, \quad \|I\mathcal{K}_2\| < Y_2 \, . \quad (2.18)$$

In any of the cases in (2.17) and (2.18) this gives a decomposition of $I^2\mathcal{G}$ of the type required in part (ii) of the lemma.

Next, the discretisation of the "fundamental" identity from Lemma 3.1.2 is as follows.

Lemma 3. *Suppose A is squarefree. Let $\chi_D^\pm$ be χ_D^- or χ_D^+, as given by (2.6). Write*

$$S^\pm(A) = \sum_{d|A} \mu(d)\chi_D^\pm(d) \, .$$

Then

$$S^\pm(A) = \Sigma_1^\pm(A) + S_2^\pm(A) + S_3^\pm(A) \, ,$$

where

$$\Sigma_1^\pm(A) = \sum_{\mathcal{E}\in\mathcal{C}} \mu(\mathcal{E})\chi_D^\pm(\mathcal{E}) \sum_{\substack{d|A \\ d\in\mathcal{E}}} 1 \quad (2.19)$$

$$S_2^\pm(A) = \sum_{\substack{I<\mathcal{G} \\ I\mathcal{G}\in\mathcal{C}}} \sum \mu(\mathcal{G})\chi_D^\pm(I\mathcal{G}) \sum_{\substack{p_1,p_2\in I \\ p_2<p_1}} \sum_{\substack{g\in\mathcal{G} \\ p_1p_2g|A}} \sum_{\substack{f|A/p_1p_2g \\ f|P(p_2)}} \mu(f)$$

$$S_3^\pm(A) = \sum_{\mathcal{E}\in\mathcal{C}} \mu(\mathcal{E})\bar{\chi}_D^\pm(\mathcal{E}) \sum_{\substack{g\in\mathcal{E} \\ g|A}} \sum_{\substack{f|A/g \\ f|P(q(g))}} \mu(f) \, , \quad (2.20)$$

with $\mathcal{E}$ as in (2.3).

Begin with the "fundamental" identity from Lemma 3.1.2. The most concise form of this was the equation (3.1.2.14),

$$\chi(d) = 1 - \sum_{\substack{g|d \\ q(g)>Q(d/g)}} \bar{\chi}(g) \, ,$$

6.1 The Bilinear Nature of Rosser's Construction

in which $q(n)$ and $Q(n)$ are the least and greatest prime factors of n, but for our purposes it will be more convenient in the equivalent form

$$\sum_{d|A} \mu(d) - \sum_{d|A} \mu(d)\chi(d) = \sum_{g|A} \mu(g)\bar{\chi}(g) \sum_{\substack{f|A/g \\ f|P(q(g))}} \mu(f), \qquad (2.21)$$

the case $\psi(d) = 1$ of (3.1.2.16).

In (2.21), take

$$\chi(d) = \begin{cases} \chi_D^{\pm}(\mathcal{E}) & \text{if } d \in \mathcal{E} \text{ and } \mu(\mathcal{E}) \neq 0 \\ 0 & \text{if } d \in \mathcal{E} \text{ and } \mu(\mathcal{E}) = 0, \end{cases} \qquad (2.22)$$

with $\chi_D^{\pm}$ as in the enunciation of the lemma. Thus $\chi(d) = 0$ if d is not well-separated by $\mathcal{C}$, in which case we specified $\mu(\mathcal{E}) = 0$ in (2.5). On the other hand when $\chi(d) \neq 0$ the definition (2.5) shows $\mu(d) = \mu(\mathcal{E})$. Thus in (2.21) we now have

$$\sum_{d|A} \mu(d)\chi(d) = \sum_{\mathcal{E} \in \mathcal{C}} \mu(\mathcal{E})\chi_D(\mathcal{E}) \sum_{\substack{d|A \\ d \in \mathcal{E}}} = \Sigma_1^{\pm}(A),$$

as in (2.19). Thus it is required to show that sum on the right of (2.21) can be expressed as $S_2(A) + S_3(A)$.

Distinguish three cases for the summand. Firstly, if $g/q(g)$ is not well-separated by $\mathcal{C}$ then neither is g, so that $\bar{\chi}(g) = 0$ from (2.8). These g make no contribution in (2.21).

Secondly, if $g/q(g)$ and g are both well-separated by $\mathcal{C}$ then $g \in \mathcal{E}$ for some $\mathcal{E} = I_1 I_2 \ldots I_r$. Then $\mu(g) = \mu(I) \neq 0$, and $g/q(g) \in I_1 I_2 \ldots I_{r-1}$. Now (2.22) shows

$$\chi(g) = \chi_D^{\pm}(I_1 I_2 \ldots I_r), \qquad \chi(g/q(g)) = \chi_D^{\pm}(I_1 I_2 \ldots I_{r-1}), \qquad (2.23)$$

so that (2.8) and (2.9) give $\bar{\chi}(g) = \bar{\chi}_D^{\pm}(\mathcal{E})$. Thus these g contribute to (2.21) an amount $S_3^{\pm}(A)$ as given in (2.20).

Lastly, it may happen that $g/q(g)$ is well-separated by $\mathcal{C}$ but g is not. This occurs when $\chi(g/q(g))$ is still expressible as in (2.23) but $\chi(g) = 0$, so the two smallest prime factors of g both lie in the interval I_{r-1}. Call these primes p_1 and p_2, with $p_1 > p_2$, and write $I_{r-1} = I$, $I_1 I_2 \ldots I_{r-2} = \mathcal{G}$, so that $I < \mathcal{G}$ following (2.3) and (2.2). Then

$$\mu(g) = \mu(\mathcal{G}), \qquad \bar{\chi}(g) = \chi(g/q(g)) = \chi_D^{\pm}(I\mathcal{G}).$$

Now set $g = p_1 p_2 g'$ in (2.21) to see that these g contribute an amount $S_2^{\pm}(A)$, as required in Lemma 2.

The innermost sums over f arising in Lemma 3 take the values 0 or 1, by the characteristic property of the Möbius function. Consequently the following simpler inequality follows from the identity of Lemma 3.

Lemma 4. *Let $S^\pm$, Σ_1 and S_1 be as in Lemma 3, and denote*

$$\Sigma_2^{(-)^n}(A) = \sum_{\substack{I<\mathcal{G},I\mathcal{G}\in\mathcal{C} \\ \nu(\mathcal{G})\equiv n,\,\mathrm{mod}\,2}} \bar{\chi}_D^{(-)^n}(I\mathcal{G}) \sum_{\substack{p_1,p_2\in I \\ p_2<p_1}} \sum_{\substack{g\in\mathcal{G} \\ p_1p_2g|A}} 1, \qquad (2.24)$$

Then

$$S^-(A) \geq \Sigma_1^-(A) - \Sigma_2^-(A) \qquad (2.25)$$
$$S^+(A) \leq \Sigma_1^+(A) + \Sigma_2^+(A).$$

Recall that Lemma 2 gave $S^\pm(A) = \Sigma_1^\pm(A) + S_2^\pm(A) + S_3^\pm(A)$. To establish (2.25), note from (2.9) that $\bar{\chi}_D^-(\mathcal{E}) \neq 0$ only when $r = \nu(\mathcal{E})$ is even, so that $\mu(\mathcal{E}) = 1$. This gives that $S_3^-(A) \geq 0$ in Lemma 3, and $S_3^+(A) \leq 0$ follows in a similar way. It now remains only to show

$$S_2^-(A) \geq -\Sigma_2^-(A), \qquad S_2^+(A) \leq \Sigma_2^+(A).$$

In Lemma 2, the contribution to $S_2^-(A)$ from those terms with $\mu(\mathcal{G}) = -1$ is negative, so that to show $S_2^+(A) \leq \Sigma_2^+$ we need retain only the others, for which $\nu(\mathcal{G}) \equiv 0$, mod 2. These sum to an amount not exceeding $\Sigma_2^+(A)$, as defined in (2.24). Similarly $S_2^-(A) \geq -\Sigma_2^-(A)$, so the proof of Lemma 4 is complete.

6.1.3 Specification of Details

This section specifies some of the details of the discretisation to be used, and reduces the proof of Theorem 1 to those of Lemmas 5 and 6. It will be along the following general lines. In particular, it will use the functions $\lambda^\pm$ specified in (3.6)–(3.19) below.

As in Sect. 4.4, we use a composition of two sieves along the lines outlined in Sect. 1.3.7. The first sieve uses a comparatively small level of distribution $D_1 = D^\varepsilon$ and a small product $P_1 = P(w)$ of sifting primes, employing the "Fundamental Lemma" from Sect. 3.3.4. The second and principal sieve uses the ideas of this chapter, with a comparatively large level of distribution $D_2 = D^{1-\varepsilon}$ and a sifting product $P_2 = P(z)/P(w)$ free of the small primes dividing $P(w)$.

In our standard terminology, these two sieves are specified by functions $\lambda_i^\pm$, where $i = 1$ or 2, in the way described in Theorem 1.3.1. The composite sieve is then given by the construction of Lemma 1.3.5. In this chapter the notation is changed to $\lambda_1 = \alpha$, $\lambda_2 = \lambda$, so that the notations λ_1, λ_2 are released for other purposes.

6.1 The Bilinear Nature of Rosser's Construction

According to Lemma 1.3.7, the entries $V^{\pm}(D,P)$ and $R^{\pm}(D,P)$ in Theorem 1 will then be given, for the upper bound, by

$$V^{+}(D,P) = \sum_{h|P_1} \sum_{d|P_2} \frac{\alpha^{+}(h)\lambda^{+}(d)\rho(h)\rho(d)}{hd}$$
$$R^{+}(D,P) = \sum_{h|P_1} \sum_{d|P_2} \alpha^{+}(h)\lambda^{+}(d)r_A(hd) ,$$
(3.1)

and for the lower bound by superficially more complicated expressions

$$V^{-}(D,P) = \sum_{h|P_1} \sum_{d|P_2} \frac{\lambda(h,d)\rho(h)\rho(d)}{hd}$$
$$R^{-}(D,P) = \sum_{h|P_1} \sum_{d|P_2} \lambda(h,d)r_A(hd) ,$$
(3.2)

in which

$$\lambda(h,d) = \alpha^{-}(h)\lambda^{+}(d) + \alpha^{+}(h)\lambda^{-}(d) - \alpha^{+}(h)\lambda^{+}(d) . \quad (3.3)$$

The main terms $XV^{\pm}(D.P)$ are estimated as in Lemma 4.4.1. For this, the input required consists of the corresponding estimates for the separate main terms

$$V^{\pm}(D_1, P_1) = \sum_{h|P_1} \frac{\alpha^{\pm}(h)\rho(h)}{h} , \quad V^{\pm}(D_2, P_2) = \sum_{h|P_2} \frac{\lambda^{\pm}(h)\rho(h)}{h} . \quad (3.4)$$

Of these, the first is estimated using the "fundamental" Corollary 1.3.1.1. The other is studied in Sect. 6.1.5.

In (3.1) and (3.2), the functions $\lambda^{\pm}(d)$ will be as implicitly defined by the relations

$$\Sigma_1^{-}(A) - \Sigma_2^{-}(A) = \sum_{d|A} \lambda^{-}(d) , \quad \Sigma_1^{+}(A) + \Sigma_2^{+}(A) = \sum_{d|A} \lambda^{+}(d) , \quad (3.5)$$

where Σ_1 and Σ_2 are as in (2.19) and (2.24). Thus, with C as in (2.4),

$$\lambda^{\pm}(d) = \lambda_1^{\pm}(d) + \lambda_2^{\pm}(d) , \quad (3.6)$$

where

$$\lambda_1^{\pm}(d) = \mu(\mathcal{E})\chi_{D_2}^{\pm}(\mathcal{E}) \quad \text{if} \quad d \in \mathcal{E},\ \mathcal{E} \in \mathcal{C} \quad (3.7)$$

$$\lambda_2^{(-)^n}(p_1p_2g) = (-1)^n \chi_{D_2}^{(-)^n}(I\mathcal{G}) \quad \text{if} \quad \begin{cases} p_i \in I,\ g \in \mathcal{G},\ I\mathcal{G} \in \mathcal{C} \\ I < \mathcal{G},\ \nu(\mathcal{G}) \equiv n,\ \mathrm{mod}\ 2 . \end{cases} \quad (3.8)$$

For other d, not specified in this way, set $\lambda_i^{\pm}(d) = 0$.

Because of (3.3), the remainder terms in (3.1) and (3.2) can be written as

$$R^+(D,P) = R^+_+(D,P),$$
$$R^-(D,P) = R^+_-(D,P) + R^-_+(D,P) - R^+_+(D,P).$$
(3.9)

Here, all four expressions $R^\pm_\pm$ are of the form

$$R(D,P) = \sum_{h|P_1}\sum_{d|P_2} \alpha(h)\lambda(d) r_A(hd),$$
(3.10)

where α is one of $\alpha^\pm$ and λ is of one of $\lambda^\pm$. The equation $R^+ = R^+_+$ is included in (3.9) for convenience.

The equation (3.6) induces a decomposition of the expression in (3.10) as

$$R(D,P) = R_1(D,P) + R_2(D,P),$$
(3.11)

where

$$R_i(D,P) = \sum_{h|P_1}\sum_{d|P_2} \alpha(h)\lambda_i(d) r_A(hd).$$
(3.12)

We will need to decompose the expressions $V^\pm(D_2, P_2)$ in (3.4) in an analogous way, as

$$V^-(D_2,P_2) = V_1^-(D_2,P_2) + V_2^-(D_2,P_2)$$
$$V^+(D_2,P_2) = V_1^+(D_2,P_2) + V_2^+(D_2,P_2),$$
(3.13)

where

$$V_i^\pm(D_2,P_2) = \sum_{h|P_2} \frac{\lambda_i^\pm(h)\rho(h)}{h}.$$

The only properties of $\alpha(h)$ that we will need are immediate from its construction in Sect. 3.3:

$$|\alpha(h)| \leq 1, \quad \alpha(h) = 0 \text{ if } h > D^\varepsilon.$$

To complete the description (3.6) of $\lambda^\pm$, the dissection $\mathcal{C}$ of the interval $[w, z)$ into constituent sub-intervals $[c_{j+1}, c_j)$ must be specified. An important parameter will be its *fineness*, measured by a number $\delta = \delta(\mathcal{C})$, for which

$$\max_{1\leq j \leq J} \frac{\log c_j}{\log c_{j+1}} \leq 1 + \delta.$$
(3.14)

This will be arranged in the most direct way by specifying

$$c_j = z^{1/(1+\delta)^j} \text{ for } 0 \leq j \leq J, \quad c_J = w.$$
(3.15)

in which $c_J = w$ was already specified in (2.1). This imposes a mild constraint on the choice of w, in that J, the number of intervals in $\mathcal{C}$, is an integer.

6.1 The Bilinear Nature of Rosser's Construction 237

Proof of Theorem 1. This will be completed using the following lemmas. Lemma 5 deals with the remainder terms $R^\pm$, and Lemma 6 describes the main terms $V^\pm$ in Theorem 1. These lemmas quantify the observation that the changes from the corresponding main terms in Chapters 4 and 5 tend to zero as $\delta \to 0$ in (3.14).

Lemma 5. *Let N_1, N_2 be any two numbers for which*

$$D = N_1 N_2, \quad N_1 \geq z, \quad N_2 \geq 1.$$

Suppose that $z < D^{1-2\varepsilon}$. Then the expression R_1 given by (3.12) can be estimated as

$$|R_1(D, P(z))| \ll 4^J \sup |B(N_1, N_2, z)|, \qquad (3.16)$$

where J is as in (3.15) and

$$B(N_1, N_2, z) = \sum_{\substack{n_1 < N_1 \\ n_1 n_2 | P(z)}} \sum_{n_2 < N_2} a(n_1) b(n_2) r_A(n_1 n_2), \qquad (3.17)$$

and the supremum is over arbitrary sets of real numbers $a(n_1)$, $b(n_2)$ such that $|a(n_1)| \leq 1$, $|b(n_2)| \leq 1$.
Similarly,

$$|R_2(D, P(z))| \ll 4^J \sup |B(N_1, N_2, z)|. \qquad (3.18)$$

Write $z = D^{1/s}$ as usual. Then in (3.15) take

$$J = \left\lceil \frac{\log(\delta^{-1/5}/s)}{\log(1+\delta)} \right\rceil, \qquad \delta = \frac{\log \log \log D}{\log \log D}, \qquad (3.19)$$

when D is sufficiently large. Lemma 6 depends upon this choice of J, which as inspection of the proof will show, is a natural one in the context of the argument employed. The specification of δ is slightly more arbitrary, but leads to Theorem 1 in the way described below.

Lemma 6. *Let $V^\pm(D, P)$ be as given in (3.1) and (3.2). Suppose ρ has sifting density 1, and that the parameter L in the sifting density hypothesis (1.3.5.3) satisfies*

$$L < \delta \log D, \qquad (3.20)$$

where δ is as in (3.14). Write $z = D^{1/s}$, as usual, where $s \geq 1$, and let J be as in (3.19). Then

$$V^-(D, P(z)) \geq V(P(z))(f(s) - E(D, \delta, L))$$
$$V^+(D, P(z)) \leq V(P(z))(F(s) + E(D, \delta, L)),$$

where F and f are as in Theorem 1, and

$$E(D, \delta, L) \ll L^{10} \delta^{1/5} \log \delta^{-1}.$$

The condition (3.20) is not an important one in our context, because it follows from a much stronger assumption already required in Theorem 1 for other reasons.

Once Lemmas 5 and 6 are established the proof of Theorem 1 is completed following a slightly arbitrary choice of δ. Decide first that it is acceptable if the factor 4^J appearing in Lemma 6 does not exceed a certain power of $\log D$, the exponent $1/3$ being convenient. When $s \geq 1$, as in Theorem 1, the choice (3.19) of J required for Lemma 5 satisfies

$$J \leq \frac{\log(\delta^{-1/5})}{\log(1+\delta)} \leq \frac{\log \delta^{-1}}{5} \cdot \frac{1+\delta}{\delta} \leq \tfrac{1}{5}(\delta^{-1}+1)\log \delta^{-1}.$$

Accordingly make $\delta^{-1} = \log \log D / \log \log \log D$ as in (3.19), so that

$$J < \left(\tfrac{1}{5} + \varepsilon\right) \log \log D,$$

if $\varepsilon > 0$ and D is large enough. Then $4^J \ll (\log D)^{1/3}$, so that the estimates for $R^\pm(D, P)$ required in Theorem 1 will follow from Lemma 6 and the equations (3.9), (3.11) and (3.12). Furthermore, the conclusion of Lemma 5 assumes the form

$$E(D, \delta, L) \ll \frac{L^{10}}{(\log \log D)^{1/5}},$$

so that the quantities $V^\pm$ are estimated as stated in part (i) of Theorem 1.

To complete the proof of Theorem 1 we have only to note that the functions $\lambda^\pm$ satisfy the conditions (1.3.1.2), so that they are in fact upper and lower sifting functions and Theorem 1.3.1 can be applied. This fact follows from Lemma 4 and the equations (3.5).

6.1.4 The Leading Contributions to the Main Term

The principal step in the proof of Lemma 6 is the estimation in Lemma 7 of

$$V^\pm(D_2, P_2) = \sum_{d \mid P_2} \frac{\lambda^\pm(d)\rho(d)}{d} \qquad (4.1)$$

as given in (3.4). Lemma 7 shows that the more standard expression (4.2) will be a good approximation to (4.1), provided that the measure δ of the fineness of our discretisation is good enough.

Our estimates for (4.2) were obtained in Chapter 4.

Lemma 7. *Set*

$$V_0^\pm(D, P_2) = \sum_{d \mid P_2} \mu(d)\chi_D^\pm(d)\frac{\rho(d)}{d}, \qquad (4.2)$$

where the prime factors of P_2 satisfy

$$p \mid P_2 \implies w \leq p < z. \qquad (4.3)$$

6.1 The Bilinear Nature of Rosser's Construction

As in Lemma 5, suppose that ρ has sifting density 1 and that $L < \delta \log D$, where δ satisfies (3.14). Then the sum $V^\pm(D, P_2)$ defined by (4.1) satisfies

$$V^\pm(D, P_2) = V_0^\pm(D, P_2) + O\left(\delta \frac{\log D}{\log w}\left(\frac{V(P(w))}{V(P(z))}\right)^2\right).$$

Begin by decomposing the sums (4.1) as in (3.13):

$$V^\pm(D_2, P_2) = V_1^\pm(D_2, P_2) + V_2^\pm(D_2, P_2),\qquad(4.4)$$

where

$$V_i^\pm(D_2, P_2) = \sum_{h|P_2} \frac{\lambda_i^\pm(h)\rho(h)}{h},$$

with λ_i as in (3.7) and (3.8). Thus

$$V_1^{(-)^n}(D_2, P_2) = \sum_{\mathcal{E}\in C} \mu(\mathcal{E})\chi_{D_2}^{(-)^n}(\mathcal{E}) \sum_{\substack{d|P_2 \\ d\in\mathcal{E}}} \frac{\rho(d)}{d},\qquad(4.5)$$

$$V_2^{(-)^n}(D_2, P_2) = \sum_{\substack{I<\mathcal{G}; I\mathcal{G}\in C \\ \nu(\mathcal{G})\equiv n,\,\bmod 2}} \sum \chi_{D_2}^{(-)^n}(I\mathcal{G}) \sum_{\substack{p_1,p_2\in I;\, g\in\mathcal{G} \\ p_2<p_1;\, p_1p_2g=d}} \sum \frac{\rho(d)}{d}.\qquad(4.6)$$

Here, (4.5) will provide the leading term in our estimate, while the expressions $V_2^\pm(D_2, P_2)$ are in the nature of error terms that do not have to be estimated so precisely.

In the following treatment, as elsewhere in this book, we use the expression $\chi_D^\pm(n)$, where n is an integer, to indicate a value of the function defined in Sect. 3.3.1. This should not be confused with the current notation $\chi_D^\pm(\mathcal{E})$ where $\mathcal{E}$ is a direct product of intervals. In fact the definitions in (2.6) yield the implication

$$\chi_D^\pm(\mathcal{E}) = 1,\quad d\in\mathcal{E},\quad \mu(\mathcal{E})\neq 0 \implies \chi_D^\pm(d) = 1,\qquad(4.7)$$

the notation $d \in \mathcal{E}$ having the usual significance (2.10).

Because of (4.7) the expression (4.5) is unchanged if the inner summand is replaced by $\vartheta(d) = \chi_D^\pm(d)\rho(d)/d$. Apply Lemma 3 with this ϑ, taking $A = P_2$. This shows

$$V_1^\pm(D, P_2) = V_0^\pm(D, P_2) - S_2^\pm(P_2) - S_3^\pm(P_2),\qquad(4.8)$$

where V_0 is as in (4.2) and

$$\Sigma_1^\pm(P_2) = \sum_{\mathcal{E}\in C} \mu(\mathcal{E})\bar\chi_D^\pm(\mathcal{E}) \sum_{d|P_2} \sum_{g\in\mathcal{E}} \sum_{\substack{f|P_2(q(g)) \\ gf=d}} \mu(f)\chi_D^\pm(d)\frac{\rho(d)}{d},\qquad(4.9)$$

$$\Sigma_2^\pm(P_2) = \sum_{I<\mathcal{G},I\mathcal{G}\in C} \sum \mu(\mathcal{G})\chi_D^\pm(I\mathcal{G}) \sum_{\substack{p_1,p_2\in I\ g\in\mathcal{G} \\ p_2<p_1;\, p_1p_2g=d}} \sum \chi_D^\pm(d)\frac{\rho(d)}{d}.\qquad(4.10)$$

Following (4.4), (4.6) and (4.8), the proof of the lemma will be completed by showing

$$\left|\Sigma_1^\pm(P_2)\right| + \left|\Sigma_2^\pm(P_2)\right| + \left|V_2^\pm(D, P_2)\right| \ll \delta \frac{\log D}{\log w} \left(\frac{V(P(w))}{V(P(z))}\right)^2.$$

In (4.9) we can use

$$\chi_D^\pm(d) = \chi_D^\pm(fg) \le \chi_D^\pm(g), \qquad |\mu(f)| \le 1,$$

so that we obtain

$$|\Sigma_1(P_2)| \le \sum_{\mathcal{E} \in \mathcal{C}^r} \bar{\chi}_D^\pm(\mathcal{E}) \sum_{d|P_2} \sum_{f|P_2(q(g))} \frac{\rho(f)}{f} \sum_{g \in \mathcal{E}} \chi_D^\pm(g) \frac{\rho(g)}{g}. \qquad (4.11)$$

Write $\mathcal{E} = I_1 I_2 \ldots I_r$ as in (2.3). In the last sum over g we can then set $g = p_1 p_2 \ldots p_r$ with $p_j \in I_j$, in accordance with the notation (2.10). Thus (2.3) shows $i_j > p_j \ge h_j$, where actually $[h_j, i_j] \in \mathcal{C}$, so that $p_j > i_j^{1/(1+\delta)}$ by (3.14), the characteristic property of δ. Furthermore the definition (2.9) of $\bar{\chi}$ gives $i_1 \ldots i_{r-1} i_r^3 \ge D$ when $\bar{\chi}_D^\pm(\mathcal{E}) = 1$ (the parity of r being fixed by the choice of $\pm$ as $+$ or $-$). Thus when also $\chi_D^\pm(g) = 1$ (for which see Sect. 3.3.1) we obtain

$$D > p_i \ldots p_{r-1} p_r^3 \ge \left(i_1 \ldots i_{r-1} i_r^3\right)^{1/(1+\delta)} \ge D^{1/(1+\delta)}.$$

Write $g = g_1 p$, where $p = p_r$ is the smallest prime factor of g. Then

$$\frac{D^{1/(1+\delta)}}{g_1} \le p^3 < \frac{D}{g_1},$$

so that (4.11) gives

$$|\Sigma_1(P_2)| \le \sum_{f|P_2} \frac{\rho(f)}{f} \sum_{g_1|P_2} \frac{\rho(g_1)}{g_1} \sum_{u_1 \le p < u_2} \frac{\rho(p)}{p}, \qquad (4.12)$$

where, since all primes p dividing P_2 satisfy $w \le p < z$,

$$u_1 = \max\left\{w, \left(D^{1/(1+\delta)}/g_1\right)^{1/3}\right\}, \qquad u_2 = \min\left\{z, \left(D/g_1\right)^{1/3}\right\}.$$

From Lemma 1.3.3, the sum over p satisfies the estimate

$$\sum_{u_1 \le p < u_2} \frac{\rho(p)}{p} \le \log \frac{\log u_2}{\log u_1} + \frac{L}{\log u_1}.$$

in which the term $L/\log u_1$ can be estimated as $\le L/\log w$. If $g_1 > D/w^3$ then $u_2 < w \le u_1$, so that the sum is empty. When $g_1 \le D/w^3$ (so that $g_1 \le D^{1-3\delta} \le D^{1/(1+\delta)}$) we find

$$\frac{\log u_2}{\log u_1} \le \frac{\log D - \log g_1}{\frac{\log D}{1+\delta} - \log g_1} \le \frac{\log w^3}{\log w^3 - \frac{\delta \log D}{1+\delta}} < \frac{1}{1 - \frac{\delta \log D}{3 \log w}} < 1 + \frac{\delta \log D}{2 \log w},$$

whence
$$\sum_{u_1 \le p < u_2} \frac{\rho(p)}{p} \le \frac{\delta \log D}{2 \log w} + \frac{L}{\log w} \ll \delta \frac{\log D}{\log w},$$

since $L < \delta \log D$ was specified in Lemma 7. Then (4.12) gives

$$|\Sigma_1(P_2)| \le \delta \frac{\log D}{\log w} \left(\sum_{f | P_2} \frac{\rho(f)}{f} \right)^2. \qquad (4.13)$$

The sums (4.6) and (4.10) are both majorised by

$$\left(\max_{I \in \mathcal{C}} \sum_{p \in I} \frac{\rho(p)}{p} \right) \sum_{w \le p_2 < z} \frac{\rho(p_2)}{p_2} \sum_{\substack{g | P_2 \\ q(g) > p_2}} \frac{\rho(g)}{g}$$

$$\le \left(\max_{I \in \mathcal{C}} \sum_{p \in I} \frac{\rho(p)}{p} \right) \sum_{w \le p_2 < z} \frac{\rho(p_2)}{p_2} \sum_{d | P_2} \frac{\rho(d)}{d},$$

in which the maximum is over all intervals I from $\mathcal{C}$. For any such interval $I = [c_{j+1}, c_j)$ Lemma 1.3.3 shows

$$\sum_{p \in I} \frac{\rho(p)}{p} \le \log \frac{\log c_j}{\log c_{j+1}} + \frac{L}{\log w} \le \log(1 + \delta) + \frac{L}{\log w} \ll \delta \frac{\log D}{\log w},$$

because of (3.14) and (3.20), which Lemma 7 assumes. Consequently the sums (4.6) and (4.10) are both majorised by the expression on the right of (4.13):

$$|V_2^{(-)^n}(D, P_2)| + |\Sigma_2(P_2)| \ll \delta \frac{\log D}{\log w} \left(\sum_{f | P_2} \frac{\rho(f)}{f} \right)^2. \qquad (4.14)$$

Since (4.3) leads to

$$\sum_{f | P_2} \frac{\rho(f)}{f} = \prod_{p | P_2} \left(1 + \frac{\rho(p)}{p} \right) \le \prod_{p | P_2} \left(1 - \frac{\rho(p)}{p} \right)^{-1} = \frac{V(P(w))}{V(P(z))},$$

the estimates (4.14) and (4.13) complete the proof of (3.14), and hence of Lemma 7.

6.1.5 Composition of Sieves

The remaining step in the proofs of Lemmas 6 and 7 is the composition of the main terms $V^\pm(D_2, P_2)$, as estimated in Lemma 7, with the estimates for $V^\pm(D_1, P_1)$ obtained in the "fundamental" Corollary 3.3.1.1. Here

$$D_1 = D^\varepsilon, \quad D_2 = D^{1-\varepsilon}, \quad w = D^{1/v}, \quad P_1 = P(w), \quad P_2 = P(z)/P_1, \quad (5.1)$$

as in Sect. 4.4.1.

Proof of Lemma 6. Take v and $\varepsilon > 0$ to be related by $v = \varepsilon^{-1}\log\varepsilon^{-1}$, as in the proof of Theorem 4.4.1. Recall that Corollary 3.3.1.1 gives

$$V^{\pm}(D_1, P_1) = V(P_1)(1 + \zeta_1) \quad \text{with} \quad \zeta_1 \ll \varepsilon L^{10}, \tag{5.2}$$

as was previously obtained from (4.4.1.5).

The estimate for $V^{\pm}(D_2, P_2)$ is obtained via Lemma 1 from Corollary 4.4.2.1, which gives

$$\begin{aligned} V_0^+(D_2, P_2) &\le V(P_2)\Big(F(s_2) + O(\eta(D_2, s_2))\Big) \\ V_0^-(D_2, P_2) &\ge V(P_2)\Big(f(s_2) + O(\eta(D_2, s_2))\Big), \end{aligned} \tag{5.3}$$

in which $\eta(D, s) \ll (\log\log\log D)^3/\log\log D$, and $s_2 = \log D_2/\log z$ as in (4.4.1.8). Here $V(P_2) = V(P(z))/V(P(w))$ because of (5.1).

The bound $(\log\log\log D)^3/\log\log D$ could be replaced by the entry $e^{\sqrt{L}}/(\log D)^{1-\Delta}$ from Theorem 4.4.2, but this would not improve the following argument in which it is already dominated by a larger term $\delta^{1/5}$.

Lemma 7 and (5.3) give

$$\frac{V(P(z))}{V(P(w))}(f(s_2) + \zeta_2) \le V^{\pm}(D_2, P_2) \le \frac{V(P(z))}{V(P(w))}(F(s_2) + \zeta_2),$$

where the sifting density property (1.3.5.3) leads to

$$\begin{aligned} \zeta_2 &\ll \eta(D_2, s_2) + \delta\frac{\log D}{\log w}\left(\frac{V(P(w))}{V(P(z))}\right)^3 \\ &\ll \eta(D_2, s_2) + \delta\left(\frac{\log D}{\log w}\right)^4\left(1 + \frac{L}{\log w}\right) \end{aligned} \tag{5.4}$$

for each $z \le D$. Here the entry $L/\log w$ can be dropped because of (3.20), provided that $\delta < 1/v$, as will be the case below.

Now use Lemma 4.4.1, to obtain

$$V(P(z))(f(s_2) - E) \le V(D, P(z)) \le V(P(z))(F(s_2) + E),$$

where $E \ll \zeta_1 + \zeta_2 + \zeta_1\zeta_2$, with ζ_i as in (5.2) and (5.4). Since $w = D^{1/v}$ this shows

$$E \ll (1 + \varepsilon L^{10})\left(\frac{(\log\log\log D)^3}{\log\log D} + \delta v^4\right) \tag{5.5}$$

where we absorb the effect of replacing $D_2 = D^{1-\varepsilon}$ by D into the factor $1 + \varepsilon L^{10}$.

Provided $v > e$ (so that $\varepsilon < 1/e$) the specification $v = \varepsilon^{-1}\log\varepsilon^{-1}$ gives $\log v > \log\varepsilon^{-1}$, whence $\varepsilon < (\log v)/v$. In (5.5), a reasonable choice of v will be

close to $\delta^{-1/5}$, constrained by the requirement in (3.15) that J is an integer in the equation
$$D^{1/v} = w = z^{1/(1+\delta)^J}.$$
Write $z = D^{1/s}$ as usual, so that $v = s(1+\delta)^J$. These are the considerations prompting the choice (3.19), where we declared
$$J = \left\lceil \frac{\log(\delta^{-1/5}/s)}{\log(1+\delta)} \right\rceil.$$
Thus $s(1+\delta)^J \le \delta^{-1/5} < s(1+\delta)^{J+1}$. This makes $\delta^{-1/5}/(1+\delta) < v \le \delta^{-1/5}$, whence we can use $\delta v^4 < \delta^{1/5}$ and $\varepsilon < \delta^{1/5} \log \delta^{-1}$ in (5.5).

In (3.19) we specified $\delta = \log\log\log D / \log\log D$ for reasons connected with the estimation of the remainder term in Lemma 6. With this choice the estimate (5.5) leads to
$$E \ll \left(1 + L^{10}\delta^{1/5}\log\delta^{-1}\right)\delta^{1/5} \ll L^{10}\delta^{1/5},$$
the estimate quoted in Lemma 6.

Lastly, we may replace s_2 by s in (5.3), since as in (4.4.1.10) this introduces an error $O(\varepsilon)$ that is already accounted for by the entry εL^{10} in (5.5). This establishes Lemma 6.

6.1.6 The Remainder Term

The treatment of the remainder terms $R^{\pm}$ in Theorem 1 is concluded by the following proof of Lemma 5.

It is sufficient to establish
$$|R_1(D, P(z))| \le 2 \sum_{\mathcal{E}_1 \in \mathcal{C}} \sum_{\mathcal{E}_\varepsilon \in \mathcal{C}} \sup |B(N_1, N_2, z)|, \qquad (6.1)$$
$$|R_2(D, P(z))| \le \sum_{\mathcal{E}_1 \in \mathcal{C} \cup \mathcal{C}^*} \sum_{\mathcal{E}_2 \in \mathcal{C} \cup \mathcal{C}^*} \sup |B(N_1, N_2, z)|. \qquad (6.2)$$

Then (3.16) and (3.18) follow as required, since $|\mathcal{C}^*| \le |\mathcal{C}| = 2^J$ as was recorded in (2.13).

First consider the expression (3.12) for R_1, where our concern is when $\lambda_1(d) \ne 0$, so that $d \le D_2 = D^{1-\varepsilon}$ via (3.7), (2.11) and (2.7). Take Y_1, Y_2, to be specified later, such that $Y_1 Y_2 = D_2$ and $Y_1 > z$. Then Corollary 1.1 gives that $\mathcal{E}$ can be factorised as
$$\mathcal{E} = \mathcal{E}_1 \mathcal{E}_2, \quad \text{with} \quad \|\mathcal{E}_1\| < Y_1, \ \|\mathcal{E}_2\| < Y_2.$$
Since $d \in \mathcal{E}$, this induces a factorisation of d as $d = d_1 d_2$, where $d_i \in \mathcal{E}_i$, so that $d_1 < Y_1$, $d_2 < Y_2$, again by (2.11). This procedure decomposes R_1 into
$$R_1(D, P) = \sum_{\mathcal{E}_1 \in \mathcal{C}} \sum_{\mathcal{E}_2 \in \mathcal{C}} \mu(\mathcal{E}_1 \mathcal{E}_2) \chi(\mathcal{E}_1 \mathcal{E}_2) E(\mathcal{E}_1, \mathcal{E}_2),$$

244   6. The Remainder Term in the Linear Sieve

where
$$E(\mathcal{E}_1, \mathcal{E}_2) = \sum_{\substack{d_1 \in \mathcal{E}_1 \\ d_1 < Y_1}} \sum_{\substack{d_2 \in \mathcal{E}_2 \\ d_2 < Y_2}} \sum_{\substack{h | P_1 \\ h < D^\varepsilon}} \alpha(h) r_A(h d_1 d_2) . \tag{6.3}$$

Here $\chi(\mathcal{E}_1 \mathcal{E}_2) \leq 1$, so this gives
$$|R_1(D, P)| \leq \sum_{\mathcal{E}_1 \in \mathcal{C}} \sum_{\mathcal{E}_2 \in \mathcal{C}} |E(\mathcal{E}_1, \mathcal{E}_2)| . \tag{6.4}$$

When $N_1 N_2 = D$ are as in Lemma 1, specify Y_1, Y_2, n_1, n_2 as follows. If $N_2 \geq D^\varepsilon$ take
$$Y_1 = N_1, \quad Y_2 = N_2/D^\varepsilon, \quad n_1 = d_1, \quad n_2 = h d_2 . \tag{6.5}$$

If $N_2 < D^\varepsilon$ then $N_1 > D^{1-\varepsilon} > D^{1-2\varepsilon} > D^\varepsilon$, and we can take
$$Y_1 = N_1/D^\varepsilon, \quad Y_2 = N_2, \quad m_1 = h d_1, \quad n_2 = d_2 . \tag{6.6}$$

Then in either case $Y_1 Y_2 = N_1 N_2 / D^\varepsilon = D^{1-\varepsilon}$ and $n_1 < N_1$, $n_2 < N_2$. Also $Y_1 > z$, in the first case because $Y_1 = N_1 > z$ and in the second case because $Y_1 > D^{1-2\varepsilon} > z$. These are the conditions that were required of Y_1 and Y_2.

The sum in (6.3) is expressible as a sum $E = E_1 + E_2$ of two terms of the shape
$$E_j(\mathcal{E}_1, \mathcal{E}_2) = \sum_{\substack{d_1 \in \mathcal{E}_1 \\ d_1 < N_1}} \sum_{\substack{d_2 \in \mathcal{E}_2 \\ d_2 < N_2}} \sum_{\substack{h | P_1 \\ h < D^\varepsilon}} \alpha(h) r_A(h d_1 d_2) . \tag{6.7}$$

Since $|\alpha(h)| \leq 1$, both of these can (in each of the cases (6.5) and (6.6)) be expressed in the required form (3.17). When (6.5) applies and $j = 1$ in (6.7), for example, take $a(n_1) = 1$, $b(n_2) = \alpha(h)$ to obtain
$$E_j(\mathcal{E}_1, \mathcal{E}_2) = \sum_{\substack{n_1 < N_1 \\ n_1 n_2 | P(z)}} \sum_{n_2 < N_2} a(n_1) b(n_2) r_A(n_1 n_2) .$$

Because of (6.4) the inequality (6.1) now follows as required.

The treatment of the expression R_2 given in (3.12) is along similar lines, but there are some new details to consider. From (3.12) and (3.8), our concern is now with $\lambda_2(d)$, where $d = p_1 p_2 g$ and
$$\lambda_2^{(-)^n}(p_1 p_2 g) = \chi_{D_2}^{(-)^n}(IG) ,$$
$p_1 \in I$, $p_2 \in I$, $g \in \mathcal{G}$, $I < \mathcal{G}$, $\nu(\mathcal{G}) \equiv n, \mod 2$.

Thus Lemma 2(ii) applies, and $I^2 \mathcal{G}$ can be decomposed as
$$I^2 \mathcal{G} = \mathcal{E}_1 \mathcal{E}_2, \quad \text{where} \quad \|\mathcal{E}_1\| < Y_1, \quad \|\mathcal{E}_2\| < Y_2 ,$$

now with $\mathcal{E}_i \in \mathcal{C} \cup \mathcal{C}^*$. Here $p_1 \in I$, $p_2 \in I$ and $g \in \mathcal{G}$, so $p_1 p_2 g$ can be expressed as $p_1 p_2 g = d_1 d_2$, where $d_1 < Y_1$, $d_2 < Y_2$ and $d_i \in \mathcal{E}_i$ as before.

Proceeding as already done with R_1 shows that $R_2(D,P)$ is bounded by an expression similar to the right side of (6.4), in which the inner sum is expressible as in (6.7), but now with the entries $\psi(\cdot)$ deleted. This leads to (6.2), completing the proof of Lemma 5.

6.2 Sifting Short Intervals

In this section we discuss the application of the techniques of this chapter in the simplest context, when the set $\mathcal{A}$ to be sifted is the set of integers in an interval $(Y - X, Y]$. Our first objective is to derive an upper bound for the quantity $S(\mathcal{A}, P(z))$ introduced in this context in Sect. 1.2.2. The following result is not the best of its type that is available, but has been chosen so as to establish something non-trivial but keep technicalities to a minimum. It depends on the result enunciated as Theorem 6.1.2.

Theorem 1. *Let $\mathcal{A}$ denote the set of integers in an interval $(Y - X, Y]$, where $2X < Y < X^3$. Take*

$$D = \min\left\{\frac{\sqrt{XY}}{X^\delta}, \frac{X^{7/4-\delta}}{Y^{1/4}}\right\},$$

for a suitable $\delta = \delta(\varepsilon) > 0$. Then the expression $S(\mathcal{A}, P)$ given in Definition 1.2.3 satisfies

$$S(\mathcal{A}, P(D^{1/3})) \leq \frac{2X}{\log D}\left(1 + O\left(\frac{1}{(\log\log D)^{1/5}}\right)\right) + O(X^{1-\varepsilon}),$$

in which the O-constants may depend upon $\varepsilon > 0$.

As in Sect. 1.2.2, this theorem provides an upper bound for the number of primes in the interval $(Y - X, Y]$. Note that if we write $Y = X^\gamma$, so $1 < \gamma < 3$, then the leading term in the estimate can be written as

$$\frac{2X}{\log D} = \frac{2X}{c\log X}, \quad \text{where} \quad c = \min\left\{\frac{1+\gamma}{2}, \frac{7-\gamma}{4}\right\} - \delta.$$

This improves on the upper bound given for this situation in Theorem 2.1.4, for which the results of Sect. 4.4 could also have been used.

The relatively simple analysis used to establish Theorem 1 becomes ineffective at the point $\gamma = 3$, a feature which could be removed by more detailed discussion of the exponential sums appearing in Sect. 6.2.3. On the other hand it is a hallmark of these methods that their results tend towards triviality as $\gamma \to \infty$, so that if Y is completely arbitrary then no improvement on Theorem 2.1.4 follows in this way.

Similarly, Theorem 1 does not offer a significant improvement upon Theorem 2.1.4 when γ is close to 1. In this instance not only does the treatment of exponential sums supplied here again become ineffective, but the problem would be better treated by different methods involving the use of the Riemann ζ-function and Dirichlet polynomials, which are applicable, in principle at least, when $1 < \gamma < 2$.

With R^+ as in Theorem 6.1.2, we will show $R^+(D, P(D^{1/s})) \ll X^{1-\varepsilon}$, when $\mathcal{A}$ and D are as stated in Theorem 1. For simplicity, Theorem 1 has been enunciated at the special value $s = 3$. We shall be expressing the remainder terms $r_\mathcal{A}(d)$ by Fourier analysis, in the style indicated in the introduction to this chapter. We saw that when this idea is implemented in the most straightforward way the difficulty arises that the Fourier series for the fractional part function $\{t\}$, or for its close relative $\psi(t)$, is not absolutely convergent, the hth Fourier coefficient c_h being only $\ll 1/h$ where we would prefer to find $c_h \ll 1/h^k$ for some $k > 1$. In this latter situation the Fourier series would also converge uniformly, so its sum function would necessarily be everywhere continuous. Thus the difficulty stems from the fact that in the expression

$$\sum_{\substack{Y-X<n\leq Y \\ n\equiv 0,\,\mathrm{mod}\,d}} 1 = \frac{X}{d} + r_\mathcal{A}(d)$$

the remainder $r_\mathcal{A}(d)$ necessarily inherits discontinuities from the sum on the left as Y (or $Y - X$) passes through integer values.

One way of avoiding this difficulty is to count the integers n not with the previous weight (1 if $Y - X < n \leq Y$, and zero otherwise), but with a suitable continuous weight $\alpha(n)$. We will work with an everywhere continuous function α of compact support that bounds the characteristic function of the interval $(Y - X, Y]$ from above. The cause of the difficulty over discontinuities has now been removed. It will be convenient to require that the function α is in the class $C_0^k(-\infty, \infty)$, so that besides vanishing outside a certain interval it has an everywhere continuous derivative of order k. We will choose k as large as we require.

6.2.1 The Smoothed Formulation

We shall deal with functions of the following type. Let $\Phi(u)$ be any function in the class $C^k(-\infty, \infty)$ such that $\Phi(u) = 0$ when $u \notin [0, 1]$. Make the change of variable given by $t = A + (B - A)u$, $\phi(t) = \Phi(u)$. Then

$$\left|\phi^{(k)}(t)\right| \leq \frac{1}{|B - A|^k} \sup_u \left|\Phi^{(k)}(u)\right| \ll \frac{1}{|B - A|^k} \quad \text{for all } t,\qquad(1.1)$$

$$\phi(t) = 0 \text{ if } t \notin (A, B)\,.$$

Here the constant implied by the $\ll$ symbol depends on k and Φ.

Construct the smoothing function α as follows. Here (also in Lemma 4 later) the rôle of the number $\frac{1}{5}$ is that of a conveniently small positive constant. Take

$$A = Y_1 = Y - \tfrac{6}{5}X\,, \quad B = Y_2 = Y + \tfrac{1}{5}X\,,$$

6.2 Sifting Short Intervals

and make $\alpha(t) = \phi(t)$. By choice of the function Φ we can arrange in addition that α bounds the characteristic function of $[Y - X, Y]$ from above, so that

$$\begin{aligned} \alpha(y) &\geq 1 & \text{if } & y \in [Y - X, Y] \\ \alpha^{(k)}(y) &\ll 1/X^k & \text{if } & x \in [Y_1, Y_2] \\ \alpha(y) &= 0 & \text{if } & y \notin [Y_1, Y_2] \,. \end{aligned} \qquad (1.2)$$

For future reference observe

$$Y \ll Y_1 \leq Y_2 \ll Y \qquad (1.3)$$

when the condition $2X \leq Y$ of Theorem 1 is satisfied.

Following the slightly generalised formulation of a sieve introduced in Sect. 1.3.3, observe

$$S(\alpha, P) = \sum_{(n, P) = 1} \alpha(n) \geq \sum_{\substack{Y - X \leq n < Y \\ (n, p) = 1}} 1 = S(\mathcal{A}, P) \,,$$

in which $S(\mathcal{A}, P)$ has the significance used in Theorem 1. Then the bound required will follow from a corresponding bound for $S(\alpha, P)$.

In Sect. 1.3.3 we denoted

$$|A_d| = \sum_{n \equiv 0, \bmod d} \alpha(n) \,. \qquad (1.4)$$

Lemma 1 provides the information required in Definition 1.3.4., with $\rho(p) = 1$ for all p, so that ρ has sifting density 1 and Theorem 6.1.1 will apply.

Lemma 1. *Let α be as in (1.2). Then the expression (1.4) satisfies*

$$|A_d| = \frac{X}{d} + r_A(d) \,,$$

with

$$X = \int_{Y_1}^{Y_2} \alpha(y) \, dy \,, \qquad r_A(d) = \frac{1}{d} \sum_{\substack{h = -\infty \\ h \neq 0}}^{\infty} \int_{Y_1}^{Y_2} \alpha(y) e\left(\frac{hy}{d}\right) dy \,. \qquad (1.5)$$

Lemma 1 will follow easily from the Poisson sum formula, to be given in Lemma 3.

In the situation described in Lemma 1 we may apply Theorem 6.1.2, with $\rho(p) = 1$, so that $L = O(1)$ in (1.3.5.3). Thus we obtain

$$S\big(\alpha, P(D^{1/s})\big) \leq X\, F(s) V\big(P(D^{1/s})\big) \big(1 + \eta(D, s)\big) + R^+\big(D, P(D^{1/s})\big) \,,$$

with $\eta(D,s) \ll 1/(\log\log D)^{1/5}$, and $F(s) = 2e^\gamma/s$ when $1 \leq s \leq 3$. Theorem 1 (with D as stated) will follow once we establish

$$R^+(D, P(D^{1/s})) \ll X^{1-\varepsilon}. \tag{1.6}$$

The expression for R^+ in Theorem 6.1.2 is

$$R^\pm(D, P(D^{1/s})) = \log^{1/3} D \sum_{m<M} \sum_{n<N} a_m b_n r_A(mn),$$

where $MN = D$ and $|a_m| \leq 1$, $|b_n| \leq 1$. The sums over m and n can be dissected as

$$\sum_{0 \leq u \ll \log M} \sum_{0 \leq v \ll \log N} \sum_{M_u \leq m < 2M_u} \sum_{N_v \leq n < 2N_v},$$

where $M_u = M/2^u$, $N_v = N/2^v$. This expresses R^+ as a sum of $\ll \log MN$ terms,

$$R^\pm(D, P(D^{1/s})) = \log^{1/3} D \sum_{0 \leq u \ll \log M} \sum_{0 \leq v \ll \log N} R(M_u, N_v), \tag{1.7}$$

where, with r_A as in Lemma 1,

$$R(M_u, N_v) = \sum_{M_u \leq m < 2M_u} \sum_{N_v \leq n < 2N_v} a_m b_n r_A(mn). \tag{1.8}$$

These expressions will be estimated as in Lemma 2, in which the suffices u, v have been suppressed.

Lemma 2. *Let $R(M, N)$ be as in (1.8), so that $|a_m| \leq 1$, $|b_n| \leq 1$, as in Theorem 6.1.1, and r_A is as in Lemma 1. Suppose $2X \leq Y \leq X^3$, as in Theorem 1. Let M and N satisfy*

$$MN^2 \leq \frac{X^{5/2-\delta}}{Y^{1/2}}, \quad M^{1+\delta} N^2 \leq Y, \quad M \leq X^{1-\delta},$$

where $\varepsilon > 0$ and $\delta = \delta(\varepsilon) > 0$ is to be chosen suitably. Then

$$R(M, N) \ll X^{1-\varepsilon}.$$

Lemma 2, once established, leads to Theorem 1 as follows. In the first place it gives $R(M_u, N_v) \ll X^{1-\varepsilon}$ in (1.7), since $M_u \leq M$, $N_v \leq N$. With the aim of making $D = MN$ as large as possible choose

$$M = X^{1-\delta}, \quad N = \min\left\{\frac{X^{3/4}}{Y^{1/4}}, \sqrt{\frac{Y}{X}}\right\},$$

consistently with the requirements of Lemma 2. This leads to the expression for $D = MN$ stated in Theorem 1. On the other hand the inequality (1.6), which is sufficient for Theorem 1, follows from Lemma 2, the number of terms in (1.7) now being $\ll \log^2 X$.

6.2 Sifting Short Intervals

Lemma 1 rests on the following version of the Poisson summation formula, to the general effect that the sum of the values at integers of a function equals the sum of its Fourier transform. Lemma 3 uses the usual conventions

$$e(t) = e^{2\pi i t}, \qquad \sum_{h=-\infty}^{\infty} = \lim_{H \to \infty} \sum_{h=-H}^{H},$$

and will be used again in the treatment of Lemma 4 given in Sect. 6.2.3.

Lemma 3. *Suppose that $f(x)$ is everywhere continuous and $f'(x)$ is piecewise continuous on an interval $a < x < b$, and that $f(x) = 0$ when $x \notin [a, b]$. Then*

$$\sum_{a \leq n \leq b} f(n) = \sum_{h=-\infty}^{\infty} \int_a^b f(x) e(hx) \, dx \ . \qquad (1.9)$$

When n is an integer the given conditions are more than sufficient to ensure that the Fourier series for $f(x)$ over the interval $(n, n+1)$, viz.

$$\sum_{h=-\infty}^{\infty} c_h e(-hx), \quad \text{where} \quad c_h = \int_n^{n+1} f(x) e(hx) \, dx \ ,$$

converges to $f(x)$ when $n < x < n+1$. On the other hand when $x = n$ or $x = n+1$ the sum of the series is

$$\tfrac{1}{2}\bigl(f(n) + f(n+1)\bigr) = \sum_{h=-\infty}^{\infty} c_h = \sum_{h=-\infty}^{\infty} \int_n^{n+1} f(x) e(hx) \, dx \ .$$

Sum this relation over $N_1 \leq n \leq N_2$, where N_1 and N_2 are integers with $N_1 < a$, $b < N_2$. This establishes Lemma 3.

Essentially the same argument establishes a more usual version of the Poisson formula

$$\sideset{}{'}\sum_{N_1 \leq n \leq N_2} f(n) = \sum_{h=-\infty}^{\infty} \int_{N_1}^{N_2} f(x) e(hx) \, dx \ ,$$

in which N_1, N_2 are integers and the symbol $\sum'$ indicates that the entries $f(N_1)$ and $f(N_2)$ are to be counted with weight $\tfrac{1}{2}$.

Proof of Lemma 1. This follows directly from Lemma 3. Rewrite (1.4) as

$$|A_d| = \sum_m \alpha(md) \ .$$

Then Lemma 3 gives

$$|A_d| = \sum_{h-\infty}^{\infty} \int_a^b \alpha(ud)e(hu)\,du = \frac{1}{d}\sum_{h=-\infty}^{\infty} \int_{Y_1}^{Y_2} \alpha(y)e\left(\frac{hy}{d}\right)dy,$$

where $Y_1 = ad$, $Y_2 = bd$. Separating off the term with $h = 0$ gives

$$|A_d| = \frac{X}{d} + r_A(d),$$

with X and r_A as stated in Lemma 1.

6.2.2 The Remainder Sums

To complete the proof of Theorem 1 it remains to establish the estimate enunciated in Lemma 2 for the sums (1.8). This will be accomplished via a representation of $R(M,N)$ in terms of trigonometrical sums. In this section Lemma 2 is deduced from Lemma 4, which is established later.

Lemma 4. *Let ϕ satisfy (1.1) with $A = \frac{4}{5}M$, $B = \frac{11}{5}M$, and $k = k(\varepsilon)$ sufficiently large. Suppose $|t| \geq M^{1+\varepsilon}$, where $M \geq 1$.*

$$\sum_{4M/5 \leq m \leq 11M/5} \phi(m)e(t/m) \ll \sqrt{\frac{|t|}{M}},$$

the implied constant depending upon $\varepsilon > 0$.

The reader familiar with the subject area will see that the essential content of Lemma 4 is that $\left(\frac{1}{2}, \frac{1}{2}\right)$ is an exponent pair. In Sect. 6.2.3 we will give a self-contained treatment of Lemma 4 that uses only the Poisson summation formula (Lemma 3) and multiple integrations by parts.

Proof of Lemma 2. To deduce Lemma 2 from Lemma 4, start from the expression (1.5) for r_A, which gives

$$r_A(mn) = \frac{1}{mn}\sum_{\substack{h=-\infty \\ h \neq 0}}^{\infty} \widehat{\alpha}\left(\frac{h}{mn}\right), \quad \text{where} \quad \widehat{\alpha}(u) = \int_{Y_1}^{Y_2} \alpha(y)e(yu)\,dy,$$

$\widehat{\alpha}$ being the Fourier transform of α. Then (1.8) gives

$$R(M,N) = \sum_{M \leq m < 2M} \sum_{N \leq n < 2N} \frac{a_m b_n}{mn} \sum_{\substack{h=-\infty \\ h \neq 0}}^{\infty} \widehat{\alpha}\left(\frac{h}{mn}\right). \tag{2.1}$$

We require to show $R(M,N) \ll X^{1-\varepsilon}$, under the conditions on M and N stated in Lemma 2.

6.2 Sifting Short Intervals

The contribution from large values of $|h|$ ($\geq H$, say) is easily handled. Since α satisfies the conditions (1.2), integration by parts k times gives

$$\widehat{\alpha}(u) = \int_{Y_1}^{Y_2} e(yu)\alpha^{(k)}(y) \frac{dy}{(2\pi i u)^k} \ll \frac{Y}{(Xu)^k}.$$

Specify

$$H = \frac{MN}{X}Y^\varepsilon, \qquad (2.2)$$

about as small as is practicable. Then, when $m < 2M$, $n < 2N$,

$$\sum_{|h| \geq H} \widehat{\alpha}\left(\frac{h}{mn}\right) \ll X \sum_{|h| \geq H} \left(\frac{MN}{Xh}\right)^k \ll XH\left(\frac{MN}{HX}\right)^k \ll \frac{MN}{Y^{\varepsilon(k-1)}}.$$

Hence

$$\sum_{M \leq m < 2M} \sum_{N \leq n < 2N} \sum_{|h| \geq H} \frac{a_m b_n}{mn} \widehat{\alpha}\left(\frac{h}{mn}\right) \ll \frac{MN}{Y^{\varepsilon(k-1)}} \ll \frac{MN}{Y}, \qquad (2.3)$$

if we make $k > 1 + \varepsilon^{-1}$. Since $MN \leq Y$ in Lemma 2, (2.3) and (2.1) give

$$R(M,N) = \int_{Y_1}^{Y_2} \alpha(y) S(M,N,H,y)\, dy + O(1),$$

where we denote

$$S(M,N,H,y) = \sum_{M \leq m < 2M} \frac{a_m}{m} \sum_{N \leq n < 2N} \sum_{0 < |h| < H} \frac{b_n}{n} e\left(\frac{hy}{mn}\right).$$

To establish Lemma 2 it is now sufficient (since $M \leq X$ is given) to show

$$S(M,N,H,y) \ll 1/X^\varepsilon \quad \text{when} \quad Y \ll y \ll Y, \qquad (2.4)$$

with H is as in (2.2), the condition on y being that stated in (1.3).

To deal with the smaller $|h|$ use Cauchy's inequality in the style outlined in the introduction to this chapter. Let $\phi(m)$ satisfy the conditions specified in Lemma 4, so that in particular $|a_m|^2 \leq 1 \leq \phi(m) \ll 1$ when $M \leq m < 2M$. Then

$$|S(M,N,H,y)|^2 \leq \left(\sum_m \frac{\phi(m)}{m^2}\right) \sum_m \phi(m) \left|\sum_{N \leq n < 2N} \sum_{0 < |h| < H} \frac{b_n}{n} e\left(\frac{hy}{mn}\right)\right|^2$$

$$\ll \frac{1}{M} \sum_m \phi(m) \sum_{\substack{N \leq n_1, n_2 \leq 2N \\ 0 < |h_1|, |h_2| \leq H}} \frac{b_{n_1} b_{n_2}}{n_1 n_2} e\left(\left(\frac{h_1}{n_1} - \frac{h_2}{n_2}\right)\frac{y}{m}\right). \qquad (2.5)$$

252   6. The Remainder Term in the Linear Sieve

Since H is as in (2.2) the contribution to (2.5) from the "diagonal" terms with $h_1 n_2 = h_2 n_1$ is

$$\frac{1}{M} \sum_m \phi(m) \sum_{\substack{N \leq n < 2N \\ 0 < |h| < H}} \frac{d(nh)}{N^2} \ll \frac{H}{N}(HN)^\varepsilon \ll \frac{M}{X}(MNY)^\varepsilon \ll X^\varepsilon ,$$

under the conditions of Lemma 2, in particular because $M < X^{1-\delta}$.

The off-diagonal terms with $1 \leq h_1 n_2 - h_2 n_1 \leq 2HN$ contribute at most

$$\ll \frac{1}{M} \sum_{\substack{N \leq n_1, n_2 \leq 2N \\ 0 < |h_1|, |h_2| \leq H}} \frac{b_{n_1} b_{n_2}}{n_1 n_2} \sum_m \phi(m) e\left(\frac{t}{m}\right), \qquad (2.6)$$

in which $t = (h_1 n_2 - h_2 n_1) y / n_1 n_2 \neq 0$. This t satisfies $Y/N^2 \ll |t| \ll HY/N$ when $Y \ll y \ll Y$ as in (2.4). But $Y/N^2 > M^{1+\varepsilon}$ in Lemma 2, so Lemma 4 is applicable, to give

$$\sum_m \phi(m) e\left(\frac{t}{m}\right) \ll \sqrt{\frac{t}{M}} \ll \sqrt{\frac{HY}{MN}} .$$

Thus the quantity (2.6) is

$$\ll \frac{H^2}{M} \sqrt{\frac{HY}{MN}} \ll \frac{MN^2}{X^2} \sqrt{\frac{Y}{X}} Y^{2\varepsilon} .$$

The contribution to (2.5) with $h_1 n_2 - h_2 n_1$ negative is treated similarly. Since $MN^2 < X^{5/2-\delta}/Y^{1/2}$ in Lemma 2 this establishes (2.4), and the result of the lemma follows.

6.2.3 Trigonometrical Sums

The proof of Theorem 1 will be completed by the following treatment of Lemma 4, as enunciated in Sect. 6.2.2. Lemma 4 will follow from Lemmas 5 and 6, which are stated and proved below.

Lemma 5 gives a truncated version of the Poisson summation formula in the case that concerns us. Here $f(n) = \phi(n) e(t/n)$, where $\phi(n)$ is a function in $C^k(-\infty, \infty)$ supported on $[a, b]$. Provided k is large enough this gives the good error term in Lemma 5. It is better than we need, but it seems that nothing sufficient for our requirements follows any more simply.

Lemma 5. *Suppose $t > M^{1+\varepsilon}$. Assume that ϕ is as in Lemma 4, so that it satisfies (1.1) and is supported on $\left(\frac{4}{5}M, \frac{11}{5}M\right)$. Then*

$$\sum_{M/2 \leq m \leq 5M/2} \phi(m) e(t/m) = \sum_{t/5M^2 \leq l \leq 5t/M^2} I(t, l) + O\left(\frac{1}{M}\right),$$

where

$$I(t, l) = \int_{4M/5}^{11M/5} \phi(x) e\left(\frac{t}{x} + lx\right) dx .$$

6.2 Sifting Short Intervals   253

Because of Lemma 3 we need to show only that when $f(x) = \phi(x)e(t/x)$ the terms in (1.9) with $l < t/5M^2$ or $l > 5t/M^2$ sum to an amount bounded by the O-term indicated in Lemma 5. We may suppose $A \leq x \leq B$ (so that in particular $x \gg M$), since otherwise $\phi(x) = 0$.

Set $\eta(x) = t/x + lx$. The integral in Lemma 5 becomes

$$I(t,l) = \int_{\frac{4}{5}M}^{\frac{11}{5}M} \frac{\phi(x)}{\eta'(x)} \frac{de(\eta(x))}{2\pi i} = (-)^k \int_{\frac{4}{5}M}^{\frac{11}{5}M} e(\eta(x)) \phi_k(x) \frac{dx}{(2\pi i)^k}, \qquad (3.1)$$

after k integrations by parts. Here ϕ_n denotes the sequence of functions given by

$$\phi_0 = \phi, \qquad \phi_{n+1}(x) = \frac{d}{dx} \frac{\phi_n(x)}{\eta'(x)}. \qquad (3.2)$$

This used the fact that $\phi_n(a) = \phi_n(b) = 0$ when $0 \leq n \leq k$, a consequence of (1.1).

Observe that when $\frac{4}{5}M \leq x \leq \frac{11}{5}M$ we find

if $l > 5t/M^2$ then $\eta'(x) = l - t/x^2 > l - \frac{25}{16}t/M^2 \gg l$,

if $l < \frac{1}{5}t/M^2$ then $-\eta'(x) = t/x^2 - l > \frac{25}{121}t/M^2 - l \gg t/M^2$.

Thus in either case

$$|\eta'(x)| \gg |l| + t/M^2, \qquad \eta^{(n)}(x) \ll \frac{t}{M^{n+1}}, \quad \text{if } n \geq 2, \qquad (3.3)$$

the second inequality, in which the implied constant depends on n, following since $x \gg M$.

We can show that the integrand in (3.2) satisfies

$$\phi_n(x) \ll \frac{1}{(|l| + t/M^2)^n} \cdot \frac{1}{M^n}. \qquad (3.4)$$

To see this, one may reverse the substitution leading to (1.1), so that when A and B are as in Lemma 4

$$\phi(x) = \Phi(u), \quad \text{where} \quad x = A + (B-A)u = A + \tfrac{7}{5}Mu,$$

and Φ is independent of M. Then set $\eta'(x) = t\, z(u)/M^2$, so that (3.3) gives

$$z(u) \gg 1 + \frac{lM^2}{t}. \qquad z^{(n)}(u) \ll 1 \quad \text{when } n \geq 1. \qquad (3.5)$$

Denote

$$\Phi_n(u) = \left(\frac{7t}{5M}\right)^n \phi_n(x),$$

so that $\Phi_0 = \phi_0$, and (3.2) gives

$$\Phi_{n+1}(u) = \frac{d}{du} \frac{\Phi_n(u)}{z(u)}.$$

Then an easy induction shows $\Phi_n(u) = P_n(\Phi, z)/(z(u))^n$, where P_n is a certain polynomial in Φ and the derivatives of Φ and z of non-zero order. Because of (3.5) all these derivatives are bounded and we obtain

$$\phi_n(x) \ll \left(\frac{M}{t}\right)^n \Big/ \left(1 + \frac{lM^2}{t}\right)^n ,$$

from which (3.4) follows.

Now (3.1) gives

$$I(t,l) \ll \frac{M}{(|l| + t/M^2)^k} \cdot \frac{1}{M^k} . \qquad (3.6)$$

Use this with $k = 2$ when $l \neq 0$ to see that the sum of these terms is $\ll 1/M$.

The term with $l = 0$ is somewhat special, because if t/M^2 is small then l lies close to the point t/x^2 at which η' vanishes. Choose $k > 2/\varepsilon$. Then the estimate (3.6) reduces, because $t > M^{1+\varepsilon}$, to

$$M(M/t)^k \ll M^{1-k\varepsilon} \ll 1/M ,$$

the implied constants now depending upon ε. This proves Lemma 5.

The integrals in Lemma 5, in which l is not bounded away from the interval in which the derivative of $t/x + lx$ could vanish, are examined in a more explicit way (but again employing integration by parts) as follows.

Lemma 6. *Suppose the conditions of Lemma 5 are satisfied, so that in particular $A = \frac{4}{5}M$, $B = \frac{11}{5}M$ and $\phi \in C_0^2(-\infty, \infty)$ is supported on (A, B). Then the integral in Lemma 5 is expressible as $I(t,l) = J(t,l)e(2\sqrt{lt})$, for a certain $J(t,l)$ for which*

$$J(t,l) \ll \sqrt{M^3/t} \quad \text{when} \quad \tfrac{1}{5}t/M^2 \le l \le 5t/M^2 .$$

Begin by observing

$$\frac{t}{x} + lx = 2\sqrt{lt} + \left(\sqrt{t/x} - \sqrt{lx}\right)^2 ,$$

the squared factor vanishing at the critical point where $l = t/x^2$. This shows that the quantity $J(t,l)$ in Lemma 6 is actually

$$J(t,l) = \int_{4M/5}^{11M/5} \phi(x) e\left(\left(\sqrt{t/x} - \sqrt{lx}\right)^2\right) dx .$$

Change the variable of integration into $u = u(x) = \sqrt{t/x} - \sqrt{lx}$, and denote the solution of this equation by $x = x(u)$. Then

$$J(t,l) = \int_a^b \phi(x(u)) x'(u) \Omega'(u) du ,$$

6.2 Sifting Short Intervals

with $\phi(x(a)) = \phi(x(b)) = 0$ and

$$\Omega(u) = \int_0^u e(v^2)\, dv\,.$$

Then $\Omega(u)$ is bounded, indeed convergent as $u \to \infty$, as can be shown using the well-worn technique of integration by parts.

Since $\phi(x(A)) = \phi(x(B)) = 0$ we find

$$J(t,l) = -\int_a^b \Omega(u)\Big(\phi'(x(u))(x'(u))^2 + \phi(x(u))x''(u)\Big)\, du\,.$$

To estimate the order of magnitude of the integrand observe

$$u(x) \ll \frac{t^{1/2}}{M^{1/2}}\,, \qquad -u'(x) = \tfrac{1}{2}\frac{t^{1/2}}{x^{3/2}} + \tfrac{1}{2}\frac{l^{1/2}}{x^{1/2}} \gg \frac{t^{1/2}}{M^{3/2}}\,, \qquad (3.7)$$

since $l \gg t/M^2$ in Lemma 6. Hence $x'(u) \ll M^{3/2}/t^{1/2}$. Also, since $l \ll t/M^2$,

$$u''(x) = \tfrac{1}{2}\cdot\tfrac{3}{2}\frac{t^{1/2}}{x^{5/2}} + \tfrac{1}{2}\cdot\tfrac{1}{2}\frac{l^{1/2}}{x^{3/2}} \ll \frac{t^{1/2}}{M^{5/2}}\,,$$

whence

$$x''(u) = \frac{d}{dx}\left(\frac{1}{du/dx}\right)\frac{dx}{du} = -\frac{u''(x)}{(u'(x))^3} \ll \frac{t^{1/2}/M^{5/2}}{(t^{1/2}/M^{3/2})^3} = \frac{M^2}{t}\,. \qquad (3.8)$$

Then (1.1), (3.7) and (3.8) give

$$J(t,l) \ll \frac{t^{1/2}}{M^{1/2}}\left(\frac{1}{M}\cdot\frac{M^3}{t} + \frac{M^2}{t}\right) \ll \frac{M^{3/2}}{t^{1/2}}\,,$$

as required for Lemma 6.

Proof of Lemma 4. Use only the observation $|I(t,l)| \le |J(t,l)|$, taking no advantage of the oscillating nature of $e(2\sqrt{lt})$ in Lemma 6. The sum over l in Lemma 5 has $\le 5t/M^2$ terms (this remaining true when $t \le M^2/5$). Use of Lemmas 5 and 6 in this way gives

$$\sum_{a \le m \le b} \phi(m) e\left(\frac{t}{m}\right) \ll \frac{t}{M^2}\sqrt{\frac{M^3}{t}} + \frac{1}{M}\,.$$

This gives the case $t > 0$ of Lemma 4 because $1/M \le \sqrt{t/M}$ when $t \ge M \ge 1$, the case $t < 0$ then following by complex conjugation.

6.3 Notes on Chapter 6

Sect. 6.1. The material in this section is based on [Iwaniec (1980b)] and the subsequent exposition by Y. Motohashi (1981), and also draws on [Halberstam and Richert (1985b)]. This paper extended the bilinear form of the error term to the sieve with weights described in Sect. 5.3. Halberstam and Richert showed that in this sieve the remainder term can again be written in the form given in Theorem 1.

We have not given details of this extension of Theorem 1 here, partly because the structure of the argument that we have chosen (for other reasons) is not ideally suited for this purpose. It is characteristic of an appeal to the lemma of Sect. 1.3.7, as was made in Sect. 6.1.5, that an upper as well as a lower bound sieve result is required, and an upper bound version of the result was not supplied in Sect. 5.3. This could be done, but it is *a priori* clear that it should not be essential, and indeed plays no part in the argument in the argument in [Halberstam and Richert (1985b)]. This is organised along the alternative (and more usual) lines indicated in the note to Sect. 4.4.1.

A bilinear remainder term of a rather special type appeared in the earlier paper by J.-R. Chen (1975). Chen's work was based on a application of Buchstab's identity to the main term, with the result that one of the Iwaniec's variables m and n was, in Chen's paper, constrained to be a prime. There is a informative discussion of Chen's method in Iwaniec's lecture to the 1978 International Congress of Mathematicians [Iwaniec (1978b)].

Sect. 6.1.1. For the extension of Theorem 1 to Theorem 2, see [Iwaniec (1980a)] or [Motohashi (1981)]. The additional assumption relating to $\rho(p^2)$ has been stated in the weaker form enunciated by Motohashi. In Iwaniec's paper, the corresponding assumption was stronger, but still so weak that its verification in any situation of interest is unlikely to be a problem.

A study that predates the one described here is that of a bilinear form for the remainder in Selberg's upper bound method. This is derived starting from (2.1.1.6), but inherits the constraint that the numbers M and N must be equal (to $\sqrt{D}$, in the notation of Sect. 2.1). This idea appeared in [Motohashi (1975)]. An updated account, including a general theorem analogous to our Theorem 1, is in [Motohashi (1999)]. This paper notes that the improvement to the Brun-Titchmarsh theorem (*cf.* Sect. 2.3.1) obtained by the method of [Motohashi (1975)] could now be upgraded to

$$\pi(x;k,l) \leq (2+o(1))\frac{x}{\log(xk^{-3/8})} \quad \text{when} \quad k \leq x^{9/20+\varepsilon}$$

as $x \to \infty$. For further results applicable for larger k the reader may consult [Friedlander and Iwaniec (1997)].

In the meantime, the papers [Salerno (1991)] and [Salerno and Vitolo (1994)] also discussed the question of bilinear remainders in Selberg's method, and in some of the lower bound methods derived from it that we discuss in Chap. 7.

Sect. 6.2. The smoothing device of Sect. 6.2.1 and the treatment of the trigonometrical sums in Sect. 6.2.3 are drawn from a paper by Iwaniec and Laborde (1981). In this paper, actually dealing with the different problem discussed below, due advantage was taken of the oscillatory nature of the factor $e(2\sqrt{lt})$ that we treated in a trivial way in the text. This step invoked exponent pairs other than the prototype pair $(\frac{1}{2}, \frac{1}{2})$ implicit in our Lemma 4.

The smoothing introduced in Sect. 6.1.1 operates by changing the sifting question to be considered, but it is also possible to leave the sifting question unmodified and introduce a smoothing operation at a later stage. It is also possible to be more specific about the details, rather than to refer to the fairly arbitrary Φ of Sect. 6.2.1. Chen (1975) used an idea from Vinogradov (1952). Heath-Brown (1978) used a treatment referring to an explicit smoothing function $\exp(-t^2/H^2)$.

The problem of the upper bound for the number of primes in a short interval that was discussed in this section does not appear to have received a great deal of attention in the literature. It would, however, be a routine exercise to determine what could be proved in this direction using all the weapons described in the papers referred to below. These are concerned with the more interesting question of determining for which exponents θ the interval $[Y - Y^\theta, Y]$ can, for sufficiently large Y, be guaranteed to contain an almost-prime of specified order, and in particular when it contains a P_2 number. If one uses only a trivial treatment of the remainder term, in which each entry $r_A(d)$ is estimated via its absolute value, Then the best result available would be $\theta > 1/(2 - \delta_2)$, where $\delta_2 = 0.04456\ldots$ is the number from the author's paper [Greaves (1986)] referred to in the introduction to Chap. 5. If the conjectured result on the sieve with weights were known, whereby we could take $\delta_2 = 0$, then the range of available θ derived in this way would improve to $\theta > \frac{1}{2}$.

It had therefore been of considerable significance when it was shown in [Chen (1975)] that the result for P_2 numbers holds when $\theta = \frac{1}{2}$. This paper opened a line of enquiry that has been pursued almost up to the present day, in which further improvements are presumably possible.

Laborde (1978) used his weighted sieve, published in Laborde (1979) and outlined in Sect. 5.2, in conjunction with a treatment of the remainder terms following Chen (1975), to reach $\theta = 0.4867$.

Chen (1979) introduced a number of improvements on the methods of Chen (1975) leading to $\theta = 0.477$.

Halberstam, Heath-Brown and Richert (1981) used the weighted sieve slightly inferior to that of Sect. 5.3 that was alluded to in the notes on that section, together with Iwaniec's bilinear form described in Theorem 1 and a treatment of trigonometrical sums that drew on earlier work in [Heath-Brown (1978)]. In this way they showed that any $\theta > 0.455$ was permissible.

258   6. The Remainder Term in the Linear Sieve

Iwaniec and Laborde (1981) used the slightly weaker weighted sieve of Laborde (1979) discussed in Sect. 5.2. Apart from using a different treatment of exponential sums, along the lines introduced in the text, this paper contained a further ingenious device. The use of the bilinear error term allows the traditional level of distribution X ($= Y^\theta$ in the present context) to be replaced by a larger level of distribution MN, as in Lemma 2. It does not, however, permit the prime p that supports the weight function $w(p)$ in the weighted sieve being used to exceed the traditional level. In Iwaniec and Laborde's approach this fact is rather clear because they used the expression of the Buchstab-Laborde weights given in Lemma 5.2.4 to reduce the estimations to a number of applications of the sieve without weights. Consequently the restriction $M \leq X^{1-\delta}$ appearing in our Lemma 2 translates into $M \leq (X/p)^{1-\delta}$. Since $M \geq 1$ appears already in Theorem 6.1.1 this is impossible when $p > X$. Iwaniec and Laborde introduce a new idea, depending on Selberg's upper bound with $\kappa = 2$, for dealing with the weight $w(p)$ when $p > X$. In this way they reached a value $\theta = 0.45$.

The work of Halberstam, Heath-Brown and Richert (1981) similarly rested on the bilinear form of the remainder term in the sieve without weights, because the estimations involved in the weighted sieve they used could also be reduced to estimations from the sieve without weights in the style of Lemma 5.2.4. On the other hand for the sieve in Sect. 5.3 a corresponding reduction appears not to be possible, so that this sieve could not instantly be used in conjunction with a bilinear remainder term in a similar way. This situation was remedied in the paper of Halberstam and Richert (1985b), where a bilinear remainder term for this particular weighted sieve was supplied. In conjunction with the analysis of trigonometrical sums used in Halberstam, Heath-Brown and Richert (1981), and the device of Iwaniec and Laborde, this allowed them to reduce θ to 0.4476.

Further progress has depended on improved estimates for trigonometrical sums. Fouvry (1988), using Laborde's weights, reduced θ to 0.4436, which his student Jie Wu (1992) improved to 0.44 by appealing to the results of Halberstam and Richert (1985b). More recently Hong Ze Li (1994) obtained a value 0.4386, and H.-Q. Liu (1996) has reduced this to 0.436.

7. Lower Bound Sieves when $\kappa > 1$

The methods of this chapter stem from the observation, derived in part from numerical evidence, that, when the sifting density κ exceeds 1 and for the smaller values of s, Selberg's upper bound for $S(\mathcal{A}, P(D^{1/s}))$ is better than that derived by Rosser's method. We therefore seek to apply Selberg's ideas also to the lower bound sifting problem. There is more than one way in which this can be attempted, of which we will introduce two. In neither of these cases do we attempt a complete account of all the work to be found in the existing literature, partly for reasons of space and partly because it seems not to be clear that the best approach to the question has yet been found.

In one approach a key idea is iterative use of Buchstab's identity (3.1.3.3):

$$S(\mathcal{A}, P(z)) = S(\mathcal{A}, P(w)) - \sum_{\substack{w \leq p < z \\ p | P}} S(\mathcal{A}_p, P(p)) \quad \text{if} \quad w < z.$$

An upper bound for the summand over p on the right would be needed to infer a lower bound for the sum on the left, and vice versa. In principle, this remark leads to an iterative process whereby improved upper and lower bound sieve results can successively be derived. We will describe the first iteration of such a process, in which Selberg's upper bound is used to bound the summand over p. This first step already gives quite good results.

If we had used Rosser's upper bound in place of Selberg's we would have simply have recovered Rosser's sieve (so that Rosser's method is invariant under Buchstab iteration). Thus the effectiveness of the proposed procedure depends on Selberg's upper bound having been, over at least part of the range where it has been used, better than Rosser's.

A technical question that immediately arises is that an upper bound is required for $S(\mathcal{A}, P(D^{1/s}))$ for values of s larger than 2, a question that was not considered in Chapter 2 except in the limiting case $s \to \infty$. In Sect. 7.1 we therefore investigate the behaviour of Selberg's upper estimate for all values of s. As elsewhere in this volume, our approach will be designed so that it refers only to a one-sided sifting density hypothesis on the underlying function ρ.

Detailed determination of the results of the first (and subsequent) iterations of the use of Buchstab's identity in the proposed way requires fairly considerable numerical computations, the results of which are described in the notes. A case that is susceptible to a more mathematical analysis is that where the sifting density κ is allowed to become large, We will describe an analysis which shows that the sifting limit for the method is asymptotic to $2C\kappa$ as $\kappa \to \infty$, where $2C$ is a certain constant whose value is close to $2 \cdot 44 \ldots$. Note that the constant $2C$ is rather smaller than the constant $3 \cdot 591 \ldots$ that arose in the corresponding analysis of Rosser's method in Sect. 4.2.5. Incidentally, this remark substantiates (for large κ) the initial suggestion that Selberg's upper bound is, over part of its range, superior to Rosser's.

A different way of making use of Selberg's ideas is described in Sect. 7.3. In this method one starts with an expression

$$\sum_{a \in \mathcal{A}} \left(1 - \sum_{p|(a,P(z))} 1\right) \left(\sum_{d|(a,P(z))} \lambda(d)\right)^2,$$

which is, as required for our purpose, positive only when $(a, P(z)) = 1$. In this approach the imprecisions of the extremely simple combinatorial construction on the left are dampened down, for composite $(a, P(z))$, by the ensuing squared factor. The available numerical evidence indicates that this method is, as it stands, not very effective for small κ (say for $\kappa = 2$).

For large κ the situation is very different. We will exhibit a choice of the numbers $\lambda(d)$ that lead to a sifting limit asymptotic to 2κ as $\kappa \to \infty$. Thus for sufficiently large κ this approach is superior to that of Sect. 7.2.

The approaches described in Sects. 7.2 and 7.3 have both been the subject of development beyond the stage that we describe. An indication of these developments is given in the notes.

7.1 An Extension of Selberg's Upper Bound

Sect. 2.2 provided an asymptotic estimation of the sum $G_z(x)$ in two extreme cases: one in Theorem 2.2.1, when $s = \log x / \log z \to \infty$, and the other in Theorem 2.2.2, where $s = 1$ (so that $x = z$), and $z \to \infty$. This section extends Theorem 2.2.2 to the general case, where (without loss of generality) $x \geq z$, so that $s \geq 1$. Both of the special cases from Chap. 2 will be used in the treatment of Theorem 1 below.

As in Chap. 2, the expression $g(n)$ is non-negative, multiplicative and defined over squarefree numbers n. Initially, the function g was presented as $g(p) = \rho(p)/(p - \rho(p))$, but this will not be of first-rate importance in the following discussion. We will, as usual, be making an assumption to the effect that ρ has sifting density κ, but this will (as in Theorem 2.2.2) be expressed in terms of g rather than of ρ.

7.1.1 The Integral Equation and the Function $\sigma(s)$

The integral equation referred to was already established in Chap. 2, though it will now be rewritten in the form (1.6). This suits our current purpose, because we will be using the relationship with the equation (1.9) satisfied by a continuous function σ, the essential properties of which will be described.

As in Sect. 2.2.2 write

$$\sum_{p<v} g(p) \log p = \kappa \log v + \eta(v) \quad \text{when} \quad v \geq 2, \tag{1.1}$$

$$G_z(x) = \sum_{\substack{n<x \\ n|P(z)}} g(n). \tag{1.2}$$

When $z \geq x$, Lemma 2.2.2 gave the integral equation

$$G_z(x) \log x = \int_1^x G_z(t) \frac{dt}{t} + \kappa \int_{x/z}^x G_z(t) \frac{dt}{t} + \delta_z(x), \tag{1.3}$$

with $\delta_z(x)$ of the shape

$$\delta_z(x) = \sum_{n<x} g_z(n) E_{z,n}\left(\frac{x}{n}\right) \quad \text{with} \quad E_{z,n}(t) \leq \eta(\min\{t,z\}). \tag{1.4}$$

In (1.3), make the change of variable given by

$$u = \frac{\log t}{\log z}, \quad s = \frac{\log x}{\log z}, \quad G_z(t) = F_z(u), \quad \delta_z(t) = \Delta_z(u). \tag{1.5}$$

Then $\log z \, du = dt/t$, so that (1.3) assumes the form

$$s F_z(s) = \int_0^s F_z(u) \, du + \kappa \int_{s-1}^s F_z(u) \, du + \frac{1}{\log z} \Delta_z(s) \quad \text{if} \quad s \geq 1. \tag{1.6}$$

The situation when $s \to \infty$ is almost trivial, because (2.2.1.4) and the remark immediately preceding it imply

$$F_z(s) = F_z(\infty) = 1/V(P(z)) \quad \text{if} \quad s > \log P(z)/\log z. \tag{1.7}$$

On the other hand when $u \leq 1$ in (1.5), so that $t \leq z$, observe

$$F_z(u) = G_z(t) = G_t(t) = F_t(1), \tag{1.8}$$

because when $n < t \leq z$ the conditions $n|P(z)$ and $n|P(t)$, arising from (1.2), say the same, both being vacuous. Thus the estimations when $u < 1$ reduce to those at $u = 1$, but with a change from z to t. Due account of this change will be taken in Lemma 4.

7. Lower Bound Sieves when $\kappa > 1$

The Function σ and its Adjoint. The expression $\sigma(s)$, which will provide the leading term in Theorem 1, is as follows. When $s > 0$ it will satisfy an equation obtained by ignoring the term $\Delta_z(s)/\log z$ in (1.6):

$$s\sigma(s) = \int_0^s \sigma(u)\,du + \kappa \int_{s-1}^s \sigma(u)\,du . \tag{1.9}$$

This implies

$$s\sigma'(s) = \kappa\bigl(\sigma(s) - \sigma(s-1)\bigr) , \tag{1.10}$$

so that

$$\frac{d}{ds}\left(\frac{\sigma(s)}{s^\kappa}\right) + \frac{\kappa\sigma(s-1)}{s^{\kappa+1}} = 0 \quad \text{if} \quad s \neq 0 . \tag{1.11}$$

This has an everywhere continuous solution for which $\sigma(s) = 0$ if $s < 0$ and

$$\sigma(s) = Cs^\kappa \quad \text{when} \quad 0 \leq s \leq 1 ; \quad C = e^{-\gamma\kappa}/\Gamma(\kappa+1) . \tag{1.12}$$

For larger s the function $\sigma(s)$ is defined on $[0, N]$ by induction on N, using successive integrations of (1.11). Then (1.10) follows, and an integration leads to (1.9). We will see in Lemma 1 that $\sigma(s)$ increases with s, and that the constant C has been chosen in such a way that $\sigma(s) \to 1$ as $s \to \infty$.

The equation (1.10) is the case $b = -a = \kappa$ of the situation studied in Sect. 4.2.2, where we used an adjoint equation

$$\frac{d}{ds}\bigl(sr(s)\bigr) + \kappa\bigl(r(s) - r(s+1)\bigr) = 0 . \tag{1.13}$$

This equation has a solution in terms of Laplace transforms as given by (4.2.3.6):

$$r(s) = \int_0^\infty e^{-sx} \exp\left(\kappa \int_0^x \frac{1 - e^{-t}}{t}dt\right) dx \quad \text{if} \quad s > 0 . \tag{1.14}$$

In the present situation (4.2.3.8) and Lemma 4.2.4(i) say

$$sr(s) \to 1 \text{ as } s \to \infty , \quad s^{\kappa+1}r(s) \to 1/C \text{ as } s \to 0+ , \tag{1.15}$$

where C is the constant in (1.12). Note also from (1.14) that $r(s)$ decreases as s increases.

The connection between σ and r involves an inner product which in the present context, where r satisfies (1.13), is defined by

$$\langle Q, r\rangle(s) = sr(s)Q(s) - \kappa \int_{s-1}^s r(x+1)Q(x)\,dx . \tag{1.16}$$

Then Lemma 4.2.1 shows that $\langle \sigma, r\rangle(s)$ is independent of s, in fact that

$$\langle \sigma, r\rangle(s) = 1 \quad \text{when} \quad s \geq 1 . \tag{1.17}$$

7.1 An Extension of Selberg's Upper Bound

Here the value of the constant is as determined by Lemma 4.2.1(ii), using (1.15) and the values in (1.12) of $\sigma(s)$ when $0 < s < 1$. Actually the lemma was phrased rather with the requirements of Sect. 4.2 in mind, with the consequence that it is now required to take $\beta = 0$ in part (i), but $\beta = 1$ in part (ii).

Observe also that

$$s\,r(s) - \kappa \int_{s-1}^{s} r(x+1)\,dx = 1. \tag{1.18}$$

This could be obtained in the same way as (1.17), using the fact that (1.10) also has a solution $\sigma(s) = 1$ for all s, but a more direct procedure is straightforward integration of (1.13). In this case the constant 1 is most easily determined by the behaviour of $r(s)$ as $s \to \infty$, as given in (1.15).

Lemma 1 gives an important property of $\sigma(s)$.

Lemma 1. *The function $\sigma = \sigma_\kappa$, defined by (1.11) and (1.12), increases when $s \geq 0$ and satisfies*

$$\sigma(s) = 1 + O\!\left(e^{-s\log s}\right) \quad \text{when} \quad s > 0, \tag{1.19}$$

where the implied constant may depend upon κ.

Because of (1.10) and (1.12) the derivative σ is continuous in $s > 0$. Let s_1 be the infimum of those $s > 1$ for which $\sigma'(s) < 0$, if any exist. Then $\sigma'(s_1) = 0$, so that (1.10) shows $\sigma(s_1) = \sigma(s_1 - 1)$. Then $\sigma'(s) = 0$ for some $s < s_1$, a contradiction. Thus $\sigma'(s) \geq 0$ when $s > 1$, which with (1.12) shows that $\sigma(s)$ increases as stated.

For (1.19), proceed as in the corresponding Lemma 4.2.6 about the functions F and f in Rosser's sieve. Set $U(s) = 1 - \sigma(s)$, so that $U(s) \geq 0$. Subtraction of (1.18) from (1.17) leads to

$$s\,r(s)U(s) = \kappa \int_{s-1}^{s} r(x+1)U(x)\,dx \leq \kappa r(s) \int_{s-1}^{s} U(x)\,dx, \tag{1.20}$$

since $r(s)$ decreases. Lemma 4.2.7 now gives $U(s) \ll e^{-s\log s}$, and Lemma 1 follows.

Before proceeding, note that because of (1.18) the inner product (1.16) can be rewritten as

$$\langle Q, r \rangle(s) = Q(s) + \kappa \int_{s-1}^{s} \bigl(Q(s) - Q(x)\bigr) r(x+1)\,dx, \tag{1.21}$$

which will be more suitable later.

It is clear that if the expression (1.21) is negative then $Q(s)$ cannot increase through positive values. Lemma 2 asserts rather more.

264   7. Lower Bound Sieves when $\kappa > 1$

Lemma 2. *Assume that $Q(s)$ is continuous in $s > 0$, apart from possible jump discontinuities at which it may decrease. Suppose that $Q(s) < 0$ when $0 < s \le 1$ and that $\langle Q, r \rangle(s) < 0$ when $s \ge 1$. Then $Q(s) \le 0$ for all $s > 0$.*

If $Q(s) > 0$ for any $s > 0$, let s_1 be the infimum of these s. Then s_1 is not a discontinuity of Q, so that $Q(s_1) = 0$ and $Q(x) \le 0$ when $x < s_1$. Furthermore $s_1 \ge 1$. Since $r(x) > 0$ when $x > 0$ the expression (1.16), or alternatively (1.21), would then give $\langle Q, r \rangle(s_1) \ge 0$, a contradiction. Lemma 2 follows.

The following lemma establishes a technical property of σ which will be used in Sect. 7.2.2.

Lemma 3. *Set*
$$H(t) = \frac{1}{\sigma\left(\frac{1}{2}(t-1)\right)} - 1.$$
Suppose $\alpha \ge 0$. Then $t^\alpha H(t)$ is decreasing when $t \ge 2\alpha + 2\kappa + 1$.

It is sufficient to show that the expression $K(s) = s^\alpha (1/\sigma(s) - 1)$ decreases when $s \ge \alpha + \kappa$, because
$$t^\alpha H(t) = 2^\alpha \left(\frac{t}{t-1}\right)^\alpha \cdot \left(\left(\tfrac{1}{2}(t-1)\right)^\alpha \left(\frac{1}{\sigma\left(\frac{1}{2}(t-1)\right)} - 1\right) \right),$$
in which the factor $t/(t-1)$ is also decreasing.

In fact
$$K'(s) = \alpha s^{\alpha-1}\left(\frac{1}{\sigma(s)} - 1\right) - \frac{s^\alpha \sigma'(s)}{\sigma^2(s)},$$
so that use of (1.10) gives
$$\sigma^2(s) K'(s) = \alpha s^{\alpha-1} \sigma(s)(1 - \sigma(s)) - \kappa s^{\alpha-1}(\sigma(s) - \sigma(s-1)).$$
Thus it will suffice to show $\alpha \sigma(s)(1 - \sigma(s)) \le \kappa(\sigma(s) - \sigma(s-1))$. In the notation used in (1.20), this asks
$$\alpha(1 - U(s))U(s) \le \kappa(U(s-1) - U(s)),$$
which will follow provided $\alpha U(s) \le \kappa(U(s-1) - U(s))$. In fact $U(s)$ decreases, because of Lemma 1, so that (1.20) gives $sU(s) \le \kappa U(s-1)$. This gives
$$(\alpha + \kappa) U(s) \le \kappa U(s-1),$$
when $s \ge \alpha + \kappa$, so the proof of Lemma 3 is complete.

7.1.2 The estimation of $G_z(x)$

In this section we adopt the hypothesis

$$\sum_{u \leq p < v} g(p) \log p \leq \kappa \log \frac{v}{u} + A \quad \text{when} \quad 2 \leq u < v, \tag{2.1}$$

where $A > 1$. As shown in Lemma 1.3.4, this implies our usual form (1.3.5.3) of the sifting density hypothesis. The special case $u = 2$ states that in (1.1) we can use $\eta(v) \leq A$ for each $v \geq 2$. Consequently, in the expression (1.4) for the error term $\delta_z(s)$ in the integral equation (1.3) we may use

$$E_{z,n}(t) \leq A. \tag{2.2}$$

The full force of (2.1), where $u > 2$ may be close to v, is used in Lemma 4, which will be an essential ingredient in the proof of Theorem 1.

The statement and proof of Theorem 1 will be conducted in terms of the expression $F_z(u) = G_z(z^u)$ given in (1.5). The central idea is to apply considerations involving the inner product with the function r, already used with the continuous approximation $\sigma(s)$, to the object $F_z(s)/F_z(\infty)$ actually under investigation. This process will appear in the proof of Lemma 5, which has been left to the end of this section.

Theorem 1 refers to the expression

$$\psi_B(s) = \int_{B < t < s} \log \frac{t}{B} \, dt = \max\left\{0, s \log \frac{s}{B} - s + B\right\} \tag{2.3}$$

appearing in Theorem 2.2.1, where B was any positive number with the property

$$B \geq B(z) = \frac{1}{\log z} \sum_{p < z} \frac{\rho(p) \log p}{p}, \tag{2.4}$$

for example, $B = B(z)$.

Theorem 1. *Assume the sifting density hypothesis about the non-negative function g in the form (2.1), and suppose $\kappa \geq 1$. Let ξ_z be defined by*

$$\frac{F_z(s)}{F_z(\infty)} = \sigma(s) - \xi_z(s), \tag{2.5}$$

where σ is as in Lemma 1. Then, if $z \geq 2$ and $s \geq 1$,

$$\xi_z(s) \leq O\left(\frac{A^\kappa e^{-\psi_B(s)}}{\log z}\right), \tag{2.6}$$

when B satisfies (2.4). The implied constant may depend upon κ.

In Theorem 1 and elsewhere the notation $f \leq O(g)$, used when $g > 0$, means $f \leq cg$ for some constant $c > 0$.

Theorem 1 and the trivial bound $\psi_B(s) \geq 0$ imply Corollary 1.1 below, but this will be proved directly in a simpler way. Actually in the application in Sect. 7.2 it is only Corollary 1.1 that will be used, the useful fact about the s-dependence in (2.6) being that it does not increase with increasing s. The admissible value of B is then not relevant.

Corollary 1.1. *In Theorem 1*,
$$\xi_z(s) \leq O\left(\frac{A^\kappa}{\log z}\right). \tag{2.7}$$

Corollary 1.2. *In Theorem 1*,
$$\xi_z(s) \leq O\left\{\frac{A^\kappa}{\log z} \exp\left(-s\log s + s + s\log \kappa + O\left(\frac{As}{\log z}\right)\right)\right\}.$$

To deduce Corollary 1.2, note that $\rho(p)/p < \rho(p)/(p-\rho(p)) = g(p)$. Since Theorem 1 assumes (2.1), it follows that (2.4) holds when
$$B = \kappa + \frac{A - \kappa \log 2}{\log z} = \kappa + O\left(\frac{A}{\log z}\right).$$
Corollary 1.2 now follows, giving an asymptotic expression for $\xi_z(s)$ (and hence for $F_z(s)$) when s is large, provided z is large enough in terms of the constant A.

A necessary preliminary that we dispose of directly is the behaviour of $F_z(u)$ when $0 < u < 1$. This rests on the remark (1.8) that for these u $F_z(u)$ reduces to $G_t(t)$, with $t = z^u$.

Lemma 4. *Assume the hypotheses of Theorem 1. Then, when $0 < u \leq 1$,*
$$\xi_z(u) \leq \frac{cA^\kappa}{\log z}, \tag{2.8}$$
for a certain c depending only on κ.

In fact when $0 < u \leq 1$ Theorem 2.2.2 shows
$$F_z(u) = \sum_{n < z^u} g(n) \geq \frac{C}{V(P(z^u))}\left(1 + O\left(\frac{A}{\log z^u}\right)\right),$$
where C is the constant in (1.12). But because of Lemma 1.3.4 the hypothesis (2.1) implies
$$\frac{V(P(z^u))}{V(P(z))} \leq \left(\frac{\log z}{\log z^u}\right)^\kappa \left(1 + O\left(\frac{A}{\log z^u}\right)\right) \quad \text{if} \quad \log z^u > A.$$

7.1 An Extension of Selberg's Upper Bound

We noted $1/V(P(z)) = F_z(\infty)$, and $\sigma(u) = Cu^\kappa$ when $0 < u \leq 1$, in (1.7) and (1.12) respectively. Thus

$$\frac{F_z(u)}{F_z(\infty)} \geq Cu^\kappa \left(1 + O\left(\frac{A}{u \log z}\right)\right)$$

$$= \sigma(u) + O\left(\frac{Au^{\kappa-1}}{\log z}\right) \quad \text{if} \quad 1 \geq u > A/\log z \,. \tag{2.9}$$

When $u \leq A/\log z$ (and $u \leq 1$ as before) the trivial estimate $F_z(u) \geq 0$ gives

$$\frac{F_z(u)}{F_z(\infty)} \geq \sigma(u) + O(u^\kappa) = \sigma(u) + O\left(\frac{A^\kappa}{\log^\kappa z}\right)\,.$$

With (2.9) and (2.5) this gives the result in Lemma 4, because $\kappa \geq 1$.

The case when $\kappa < 1$ is less important because one would be unlikely to be using Selberg's method in this context. The argument then leads to a version of Lemma 3 in which the O-term is weakened to $O(A^\kappa/\log^\kappa z)$.

A key ingredient in the treatment of Theorem 1 is Lemma 5, in which the inner product $\langle \cdot, \cdot \rangle$ and the function r are as in (1.16) and (1.14), respectively.

Lemma 5. *Assume the hypothesis of Theorem 1, and suppose $s \geq 1$. Then ξ_z, as given in (2.5), satisfies*

$$\langle \xi_z, r \rangle(s) \leq \frac{A\, r(s)}{\log z} e^{-\psi_B(s)}\,,$$

where ψ_B is as in (2.3) with B as in (2.4).

In particular $\psi_B(s) \geq 0$, so that Lemma 5 implies

$$\langle \xi_z, r \rangle(s) \leq \frac{cA}{\log z}\,, \tag{2.10}$$

provided $c \geq r(1)$.

Proof of Corollary 1.1. This will follow fairly easily from Lemma 5. Let $c \geq 1$ be a constant valid in both (2.8) and (2.10). Define

$$Q_z(s) = U_z(s) - 1\,, \quad \text{where} \quad U_z(s) = \frac{\xi_z(s)}{cA^\kappa/\log z}\,. \tag{2.11}$$

Then $Q_z(s) < 0$ when $0 < s \leq 1$, because of Lemma 4. On the other hand (2.10) and an inspection of the form (1.21) of the inner product shows $\langle Q_z, r \rangle(s) < 0$ when $s \geq 1$. Furthermore, (2.5) shows that $Q_z(s)$ and $\xi_z(s)$ differ only by a continuous function from a negative multiple of the sum $F_z(s) = G_z(x)$ given by (1.5) and (1.2). But $G_z(x)$ is, as in (1.2), a sum of non-negative terms. Thus $Q_z(s)$ decreases at its discontinuities as assumed in Lemma 2, which now gives that $Q_z(s) \leq 0$ for all $s > 0$. Hence $\xi_z(s) \ll A^\kappa/\log z$, which is the inequality (2.7) required in Corollary 1.1.

For the complete treatment of Theorem 1 replace Lemma 2 by Lemma 6, in which the required s-dependence appears.

Lemma 6. *Assume that a function U_z is continuous in $s \geq 0$, apart from possible jump discontinuities at which it may decrease. Suppose $U_z(s) \leq 1$ when $0 \leq s \leq 1$, and*

$$\langle U_z, r \rangle(s) < r(s)e^{-\psi_B(s)} \quad \text{when} \quad s \geq 1, \tag{2.12}$$

where $\psi_B(s)$ is as in (2.3) and r satisfies (1.13). Then

$$U_z(s) \ll e^{-\psi_B(s)} \quad \text{when} \quad s \geq 1,$$

where B and ψ_B are as in Lemma 5, and the implied constant depends only on κ.

Proof of Theorem 1. To deduce Theorem 1 from Lemmas 5 and 6, observe that the expression $U_z(s)$ given in (2.11) satisfies the conditions of Lemma 6, the condition (2.12) following from Lemma 5. Theorem 1 follows.

Proof of Lemma 6. It remains to establish Lemmas 5 and 6. In proving Lemma 6 we may suppose $s > c_0 B$, for a suitable constant c_0 depending on κ. For when $s \leq c_0 B$ we find $\psi_B(s) \leq c_1 B$, for a corresponding c_1 depending on c_0, so that it suffices to show $U_z(s) = O(1)$. But if $c > 1$ and $c > r(1)$ then $Q(s) = U_z(s) - c$ will again satisfy the requirements of Lemma 2, so that in fact $Q(s) < 0$ and $U_z(s) < c$ when $s > 0$, as required.

When the inner product is expressed as in (1.16) the hypothesis (2.12) of Lemma 6 gives

$$s\, r(s) U_z(s) < \kappa \int_{s-1}^{s} U_z(x) r(x+1)\, dx + r(s) e^{-\psi_B(s)}.$$

Here $r(x+1) \leq r(s)$. Since ψ_B is as in (2.3) this gives

$$\frac{U_z(s)}{e^{-B}} < \frac{\kappa}{s} \int_{s-1}^{s} \frac{U_z(x)}{e^{-B}}\, dx + \frac{e^{-\phi(s)}}{s} \quad \text{when} \quad s > B,$$

with $\phi(x) = \psi_B(x) - B = x \log x - Nx$, where $N = 1 + \log B$. Adopt the notation $f(x) = U_z(x)e^B e^{\phi(x)}$, as in Lemma 4.2.8, which now gives

$$\frac{U_z(s)}{e^{-B}} < e^{-\phi(s)} \left(\frac{1}{s} + \frac{1}{2} \sup_{s-1 \leq x \leq s} f(x) \right),$$

7.1 An Extension of Selberg's Upper Bound

provided $s > c_2 e^N$, for a suitable absolute constant c_2. Thus

$$f(s) < \tfrac{1}{2} \sup_{s-1 \leq x \leq s} f(x) + \tfrac{1}{2},$$

when $s > c_2 eB$, as we have shown may suppose. Now Lemma 4.2.9 gives that $f(s)$ is bounded for these s, so that $U_z(s) \ll e^{-\phi(s)-B} = e^{-\psi_B(s)}$. This completes the proof of Lemma 6.

Proof of Lemma 5. Finally we deal with the central part of the argument. The procedure is to use the expression $r(s)$, specified in (1.14), as an integrating factor for (1.6).

When $s > 1$, the integral equation (1.6) gives

$$\frac{1}{\log z} \int_1^s r(u)\, d\Delta_z(u) = \int_1^s r(u) \left(u\, dF_z(u) - \kappa\, F_z(u)\, du + \kappa\, F_z(u-1)\, du \right).$$

Since r satisfies (1.13) the first entry on the right is

$$\int_1^s u\, r(u)\, dF_z(u) = s\, r(s) F_z(s) - r(1) F_z(1) + \kappa \int_1^s F_z(u) \bigl(r(u) - r(u+1) \bigr)\, du.$$

Thus, as an analogue of (1.17), we obtain

$$\frac{1}{\log z} \int_1^s r(u)\, d\Delta_z(u)$$
$$= s\, r(s) F_z(s) - r(1) F_z(1) - \kappa \int_1^s F_z(u) r(u+1) - F_z(u-1) r(u)\, du$$
$$= \langle F_z, r \rangle(s) + C_z, \tag{2.13}$$

where the inner product notation is as in (1.16), and where C_z is independent of s. Here (1.21) and (1.7) show $\langle F_z, r \rangle(s) \to F_z(\infty)$ as $s \to \infty$. Hence

$$C_z = \frac{1}{\log z} \int_1^\infty r(u)\, d\Delta_z(u) - F_z(\infty), \tag{2.14}$$

so that

$$\langle F_z, r \rangle(s) = -\frac{1}{\log z} \int_s^\infty r(u)\, d\Delta_z(u) + F_z(\infty). \tag{2.15}$$

The inner product $\langle Q, r \rangle(s)$ in (1.16) is linear in Q, and $\langle \sigma, r \rangle(s) = 1$ when $s > 1$, as in (1.17). Hence ξ_z, as specified in (2.5), satisfies

$$\langle \xi_z, r \rangle(s) = \frac{1}{F_z(\infty) \log z} \int_s^\infty r(u)\, d\Delta_z(u). \tag{2.16}$$

The Stieltjes integral in (2.16) can be examined in the style already employed in Sect. 2.2.2. Set $y = z^v$, and $x = z^s$, $\delta_z(t) = \Delta_z(u)$, $t = z^u$, as in (1.5). Then (1.4) gives

$$\int_s^v r(u)\,d\Delta_z(u) = \int_x^y r\left(\frac{\log t}{\log z}\right) d\delta_z(t)$$

$$= \int_x^y r\left(\frac{\log t}{\log z}\right) d \sum_{x \le n < t} g_z(n) E_{z,n}\left(\frac{t}{n}\right).$$

Lemma 5 assumes (2.1), so we may use its weaker consequence (2.2). Then partial summation in the form given by Lemma 1.3.2 gives

$$\int_s^v r(u)\,d\Delta_z(u) \le A \sum_{x \le n < y} r\left(\frac{\log n}{\log z}\right) g_z(n) \le A\,r(s) \sum_{x \le n < y} g_z(n),$$

where we used the fact that $r(s)$ decreases as s increases.

The last sum over n can be estimated using Lemma 2.2.1. In the current notation, this gives

$$\sum_{n \ge x} g_z(n) \le F_z(\infty) e^{-\psi_B(s)},$$

with $\psi_B(s)$ as in (2.9), the property of $\rho(p)$ required for Lemma 2.2.1 being just (2.4). Thus we derive

$$\int_s^v r(u)\,d\Delta_z(u) \le A\,F_z(\infty) r(s) e^{-\psi_B(s)},$$

so that (2.16) now gives the inequality asserted in Lemma 5.

7.2 A Lower Bound Sieve via Buchstab's Identity

Buchstab's identity, as recorded in Sect. 3.1.3, states

$$S(a, P(z)) = 1 - \sum_{p < z} S(a/p, P(p)),$$

in the usual notation defined in (1.2.1.2). The principle followed in this section is that we can obtain a lower bound sieve by applying an upper bound sieve to the summands on the right. In a similar way one can derive an upper sieve from a lower one. These processes can be iterated, and such an iteration was used by Buchstab, who started from Brun's bounds and obtained a sequence of improved sieves.

The details of Buchstab's improved sieves are now of less interest, because Rosser's sieve described in Chap. 4 gives direct access to the limiting case of these iterations. The situation changes, and (if $\kappa > 1$) better results can be obtained, if we start Buchstab's iterative process from Selberg's bounds. The first iteration of such a process is described in some detail in Sect. 7.2.2.

In general the sieve constructed in this way leads to some computational work when numerical values of the sifting limit are required (the results of some such calculations are recorded in the notes to this chapter) but the limiting case when the sifting density κ tends to infinity is susceptible to a more mathematical analysis. We will show in Sect. 7.2.3 that the sifting limit β obtained in this way satisfies $\beta \sim 2C\kappa$ as $\kappa \to \infty$, where $2C = 2 \cdot 44 \ldots$ is a certain constant, better than the constant $3 \cdot 591 \ldots$ obtained in Chapters 3 and 4 from Rosser's ideas. This reflects the fact that, for some values of the parameters involved, Selberg's bounds provide a better starting point for Buchstab's iterative process than is implied by Rosser's sieve.

On the other hand it is not a universal truth that Selberg's method is always better than Rosser's, or Brun's. Indeed, the upper bound in the Fundamental Lemma derived from Brun's ideas in Corollary 3.3.1.2 is better than the corresponding bound obtained from Selberg's method in Theorem 2.2.1. The object of further iterations of Buchstab's procedure would be to use Selberg's bounds only when it is advantageous to do so. There is a report on further work in this direction in the notes.

7.2.1 Buchstab's Iterations

If an upper sifting function λ^+ has been constructed, so that it satisfies the basic inequality (1.3.1.2), then Buchstab's identity (as recorded above) will give

$$S(a, P(z)) \geq 1 - \sum_{p < z} \sum_{d_1 | (a/p, P(p))} \lambda^+_{D/p}(d_1) = \sum_{d | (a, P(z))} \lambda^-_D(d), \qquad (1.1)$$

where the suffix D/p indicates that λ^+ will have to be of level D/p if the resulting expression $\lambda^-_D(d)$ is to give a lower sifting function of level D.

In (1.1), the explicit expression for $\lambda^-_D(d)$ is not crucial for our purposes, but is given by

$$\lambda^-_D(1) = 1, \qquad \lambda^-_D(d) = -\lambda^+_{D/p}(d/Q(d)) \quad \text{if} \quad d > 1,$$

where (as in Chapter 5) $Q(d)$ denotes the greatest prime factor of d. The resulting main term, arising as described in Theorem 1.3.1, is given by

$$V^-(D, P(z)) = \sum_{d | P(z)} \frac{\lambda^-_D(d) \rho(d)}{d} = 1 - \sum_{p | P(z)} \frac{\rho(p)}{p} \sum_{d_1 | P(p)} \frac{\lambda^+_{D/p}(d_1) \rho(d_1)}{d_1}$$

$$= 1 - \sum_{p | P(z)} \frac{\rho(p)}{p} V^+\left(\frac{D}{p}, P(p)\right). \qquad (1.2)$$

7. Lower Bound Sieves when $\kappa > 1$

In the prototype case when $\lambda^+(d) = \mu(d)$ this identity reduces to one already noted is Sect. 4.1,

$$V(P(z)) = 1 - \sum_{p|P(z)} \frac{\rho(p)}{p} V(P(p)),$$

where $V(P)$ is, as usual, as in (1.3.1.8). Then a subtraction yields

$$V^-(D, P(z)) = V(P(z)) - \sum_{p|P(z)} \frac{\rho(p)}{p} \left(V^+\left(\frac{D}{p}, P(p)\right) - V(P(p)) \right). \quad (1.3)$$

Suppose now that the main term in the upper sieve is the subject of an estimate

$$V^+(D, P(D^{1/s})) \leq V(P(D^{1/s})) \left(F_1^+(s) + O(\eta(D, s)) \right),$$

where $\eta(D, s)$ is an appropriately small error term whose contribution to the following discussion we propose to ignore meanwhile. Then (1.3) yields

$$V^-(D, P(D^{1/s})) \geq V(P(z)) \left(1 - \sum_{p|P(z)} \frac{\rho(p)}{p} \frac{V(P(p))}{V(P(z))} (F_1^+(s_p) - 1) \right), \quad (1.4)$$

where $s_p = \log(D/p)/\log p$. The advantage gained from using (1.3) rather than (1.2) is that (1.4) is suitable for application of partial summation in the form given in Lemma 4.1.2, under the usual sifting density hypothesis. This leads, apart from contributions from further suppressed error terms, to

$$V^-(D, P(D^{1/s})) \geq V(P(z)) G_s^-(s),$$

where

$$G_2^-(s) = 1 - \int_s^\infty \left(F_1^+(t-1) - 1 \right) \frac{dt^\kappa}{s^\kappa}, \quad G_2^+(s) = 1 - \int_s^\infty \left(F_1^-(t-1) - 1 \right) \frac{dt^\kappa}{s^\kappa},$$

in which G_2^+ describes the upper bound that would follow from applying this process to previously known lower bound given by F_1^-. In practice one would use the functions $G_2^\mp$ only for those s for which they gave an improvement over $F_1^\pm$, so $F_2^\mp$ can be replaced by

$$F_2^-(s) = \max\{F_1^-(s), G_2^-(s)\}, \qquad F_2^+(s) = \min\{F_1^+(s), G_2^+(s)\}.$$

We may call the pair F_2^+, F_2^- the *Buchstab transform* of F_1^+, F_1^-.

Starting with, say, Brun's bounds, the functions F_2^+, F_2^- could now improved upon by a further application of the Buchstab transform, and the process repeated iteratively. One may, however, check that the pair of functions F, f introduced in Sect. 4.2.1 are invariant under a Buchstab transform, so that Rosser's sieve described in Chapter 4 gives direct access to the limiting case of these iterations.

The situation changes if the iterative procedure is started with Selberg's upper bound. In the next section we describe the first iteration of this process.

7.2.2 The Buchstab Transform of the λ^2 Method

In this section we examine the lower bound sieve obtained by taking the upper sieve λ^+ in (1.1) to be that supplied by Selberg's λ^2 method, as described in Sect. 7.1.1. We assume the sifting density hypothesis in the form (7.1.2.1), so that free use may be made of its consequence (1.3.5.3). We may also use

$$\sum_{p<z} \frac{\rho(p)\log p}{p} < B\log z \quad \text{when} \quad z \geq 2, \tag{2.1}$$

with $B = \kappa + O(A/\log z)$ as in (7.1.2.4).

In Selberg's method the expression V^+ appearing in (1.3) is

$$V^+(D, P(z)) = 1/G_z(\sqrt{D}). \tag{2.2}$$

Theorem 1. *Let $V^-(D, P(z))$ be constructed as in (1.3), with V^+ being provided by Selberg's method as in (2.2). Then*

$$V^-(D, P(z)) \geq V(P(z))\left(f(s) + O\left(\frac{A\log\log D}{\log D}\right)\right),$$

where

$$f(s) = 1 - \int_s^\infty \left(\frac{1}{\sigma(\frac{1}{2}(t-1))} - 1\right)\frac{dt^\kappa}{s^\kappa},$$

with σ as in Sect. 7.1.1.

In the treatment of Theorem 1 we may suppose

$$\frac{\log D}{\log\log D} > A,$$

since the theorem is trivial otherwise.

We will require to estimate the sum over p appearing in (1.3). For this purpose denote

$$\Sigma(w, z) = \sum_{w \leq p < z} \frac{\rho(p)}{p}\left(V^+\left(\frac{D}{p}, P(p)\right) - V(P(p))\right). \tag{2.3}$$

Lemma 1 shows that the contribution from the smaller values of p can be estimated satisfactorily using the methods of Chap. 2.

Lemma 1. *Write $w = D^{1/v}$, and suppose $v > 2\kappa + 5$, $\kappa \geq 1$, and $v > 2B+1$. Then the expression (2.3) satisfies*

$$\Sigma(2, w) \ll V(P(w))B\exp\left(-\psi_B(\tfrac{1}{2}(v-1))\right),$$

where B satisfies (2.1) and ψ_B is as in Theorem 2.2.1.

7. Lower Bound Sieves when $\kappa > 1$

When (2.1) holds, Theorem 2.2.1 gives

$$\frac{1}{G_p(\sqrt{D/p})} \geq \frac{V(P(p))}{1 - \exp(-\psi_B(\frac{1}{2}\tau))} \quad \text{if } \tfrac{1}{2}\tau > B,$$

when $p = (D/p)^{1/\tau}$. With (2.2) this means that in (1.3) we may use

$$V^+\left(\frac{D}{p}, P(p)\right) - V(P(p)) \ll V(P(p)) \exp\left(-\psi_B\left(\tfrac{1}{2}\frac{\log D/p}{\log p}\right)\right),$$

so that (2.3) satisfies

$$\Sigma(2,w) \ll V(P(w)) \sum_{p \mid P(w)} \frac{\rho(p) \log p}{p \log w} \left(\frac{\log w}{\log p}\right)^{\kappa+1} \exp\left(-\psi_B\left(\tfrac{1}{2}\frac{\log D/p}{\log p}\right)\right)$$

$$\ll \frac{V(P(w))}{\log w} \sum_{p \mid P(w)} \frac{\rho(p) \log p}{p} \frac{\phi_w(p)}{v^{\kappa+1}}, \tag{2.4}$$

where $w = D^{1/v}$ as stated and

$$\phi_w(p) = \left(\frac{\log D}{\log p}\right)^{\kappa+1} \exp\left(-\psi_B\left(\tfrac{1}{2}\frac{\log D/p}{\log p}\right)\right).$$

We will use the fact that $\phi_w(p)$ increases with p in the situation of Lemma 1, so that when $p < D^{1/v}$

$$\phi_w(p) \leq v^{\kappa+1} \exp\left(-\psi_B\left(\tfrac{1}{2}(v-1)\right)\right). \tag{2.5}$$

To see this, write $p = D^{1/u}$, so that $u > v$, and observe

$$u^{\kappa+1} \exp\left(-\psi_B\left(\tfrac{1}{2}(u-1)\right)\right) = \exp\left(-\tfrac{1}{2}\int_{B \leq t \leq (u-1)/2} \log t \, dt + (\kappa+1)\log u\right)$$

provided $u > 2B + 1$. The derivative of the exponent is negative if u is large enough in terms of κ and B. Actually $u > v > 2\kappa + 5$, $v > 2B + 1$ and $\kappa \geq 1$ in Lemma 1, so that

$$\tfrac{1}{2}\log\frac{u-1}{2} > \tfrac{1}{2}\log(\kappa+2) > \tfrac{1}{2} > \frac{\kappa+1}{u},$$

which is sufficient for our purpose. Now (2.5) follows.

Use of (2.5) in (2.4) gives

$$\Sigma(2,w) \ll \frac{V(P(w))}{\log w} \sum_{p \mid P(w)} \frac{\rho(p) \log p}{p} \exp\left(-\psi_B\left(\tfrac{1}{2}(v-1)\right)\right),$$

from which Lemma 1 follows after using (2.1).

7.2 A Lower Bound Sieve via Buchstab's Identity

When p is not particularly small when compared with D it is necessary to use the estimate from Theorem 7.1.1 when handling the expression (2.3). Part (i) of the theorem gives

$$G_z(x) \geq V(P(z))\sigma(u)\left(1 + O\left(\frac{A^\kappa}{\log z}\right)\right) \quad \text{when} \quad z = x^{1/u}.$$

Take $x = \sqrt{D}$ and write $p = D^{1/u}$. Then (2.2) and Corollary 7.1.1.1 give, when $u \geq 1$ as may be supposed,

$$V^+(D, P(p)) \leq \frac{V(P(p))}{\sigma(\frac{1}{2}s)}\left(1 + O\left(\frac{A^\kappa}{\log p}\right)\right). \tag{2.6}$$

The O-term can be written as shown because $\sigma(\frac{1}{2}u) \geq \sigma(\frac{1}{2}) > 0$ for these s. As in Theorem 1, the O-constant depends only on κ.

In Lemma 2, the smallest primes $p < w$ (in respect of which Lemma 1 will be used) have been removed, so that the O-term in (2.6) will be a satisfactory one.

Lemma 2. *Assume that ρ has sifting density κ, as in Theorem 7.1.1. Denote*

$$H(t) = \frac{1}{\sigma(\frac{1}{2}(t-1))} - 1.$$

When $\log w \geq A^\kappa$ the expression (2.3) satisfies

$$\Sigma(w, z) \leq V(P(z)) \int_s^\infty H(t) \frac{dt^\kappa}{s^\kappa}\left(1 + O\left(\frac{A^\kappa}{\log w}\right)\right).$$

Use of the estimate (2.6) for each p gives

$$\Sigma(w, z) = V(P(z))\Sigma^*(w, z)\left(1 + O\left(\frac{A^\kappa}{\log w}\right)\right), \tag{2.7}$$

where

$$\Sigma^*(w, z) = \sum_{w \leq p < z} \frac{\rho(p)}{p} \frac{V(P(p))}{V(P(z))} H\left(\frac{\log D}{\log p}\right),$$

with H as stated.

The sum $\Sigma^*(w, z)$ can be estimated using partial summation in the form given by Lemma 4.1.2. Let $z_1 = \min\{z, D^{4\kappa}\}$, and write $z_1 = D^{1/s_1}$. If $z \geq D^{1/(4\kappa)}$ it is necessary to dissect $\Sigma^*(w, z)$ as

$$\Sigma^*(w, z) = \frac{V(P(z_1))}{V(P(z))} \Sigma^*(w, z_1) + \Sigma^*(z_1, z). \tag{2.8}$$

Use Lemma 4.1.2(i) in respect of the primes p exceeding $D^{1/(4\kappa)}$. The contribution of these p is

$$\Sigma^*(z_1, z) \leq \int_{s<t<s_1} H(t) \frac{dt^\kappa}{s^\kappa} + O\left(\frac{A}{\log w}\right). \tag{2.9}$$

Here the error term is as stated because Lemma 1.3.4 shows that we may use $K - 1 \ll A/\log D$ in Lemma 4.1.2 when $\log D > A$, the implied constant depending on κ, and we may use $D \geq w$ since otherwise the sum in question is empty.

Lemma 7.1.3 gives that $t^\kappa H(t)$ decreases when $t > s_1$ and $H(t)$ is as in Lemma 2. Then Lemma 4.1.2(ii) shows that the contribution of primes p with $w < p < z_1$ to $\Sigma^*(w, z)$ is

$$\frac{V(P(z_1))}{V(P(z))} \Sigma^*(w, z_1) \leq \left(\frac{\log z}{\log z_1}\right)^\kappa \int_{s_1<t} H(t) \frac{dt^\kappa}{s_1^\kappa} + O\left(\frac{A}{\log w}\right), \tag{2.10}$$

the range of integration having been extended to ∞ since we plan that w will be relatively small, and the quotient of logarithms being the bounded factor s_1/s.

Use of (2.9) and (2.10) in (2.8) gives

$$\Sigma^*(w, z) \leq \int_{s<t} H(t) \frac{dt^\kappa}{s^\kappa} + O\left(\frac{A}{\log w}\right),$$

from which Lemma 2 follows using (2.7).

Proof of Theorem 1. Write $w = D^{1/v}$ as in Lemma 1, and specify

$$v = 1 + 2eB + 4\log\log D .$$

With this choice of v, Lemma 1 will show that the contribution to (1.3) from the primes p with $p < w$ is absorbed by the error term in Theorem 1. Note that when $u > eB$

$$\psi_B(u) \geq u \log \frac{u}{eB} \geq u ,$$

so that when $\log \log D > 0$ we obtain

$$\exp\left(-\psi_B\left(\tfrac{1}{2}(v-1)\right)\right) \leq \exp\left(-\psi_B(eB + 2\log\log D)\right) \leq \frac{1}{\log^2 D} .$$

Then Lemma 1 shows

$$\Sigma(1, w) \ll A \frac{V(P(w))}{\log^2 D} \ll A \frac{V(P(z))}{\log^2 D} \left(\frac{\log z}{\log w}\right)^\kappa \ll A^{\kappa+1} \frac{V(P(z))}{\log D} ,$$

the factor A estimating the contribution from the term $A \log w$ occurring in Lemma 1, and the last step following because

$$\frac{\log z}{\log w} \leq \frac{\log D}{\log w} \leq v \ll A + \log\log D .$$

Theorem 1 now follows, because the remaining contribution to the sum over primes p in (1.3) is as estimated by Lemma 2, which provides the integral over t occurring in Theorem 1.

7.2.3 The Sifting Limit as $\kappa \to \infty$

Theorem 1 raises the question of determining the associated sifting limit, the value $\beta = \beta(\kappa)$ at which $f(\beta) = 0$. In general this is a somewhat recondite matter (the results of some numerical computations are described in the notes on this chapter), but in the limiting case $\kappa \to \infty$ a relatively clean asymptotic analysis is possible. Theorem 2 implies that this sifting limit is asymptotic to $2C\kappa$, where $2C$ is a certain constant close to $2\cdot 44\ldots$.

Theorem 2. *Let $f(s) = f_\kappa(s)$ be the expression arising in Theorem 1. Set*

$$C = \exp\left(-1 + \int_0^{\log 2} \frac{e^u - 1}{u} du - \log\log 2\right),$$

and let A be a positive constant (independent of κ).
  (i) *If $A > 0$ and $s < 2C\kappa - A\log\kappa$, then $f_\kappa(s) \to -\infty$ as $\kappa \to \infty$.*
  (ii) *If $A > C$ and $s > 2C\kappa + A\log\kappa$, then $f_\kappa(s) \to 1$ as $\kappa \to \infty$.*

The requirement $A > C$ in part (ii) could be relaxed, or removed, at the cost of more extensive calculations than appear below. The key ingredient in the proof of Theorem 2 is the following asymptotic expression for the function $\sigma = \sigma_\kappa$ introduced in Sect. 7.1.1.

Lemma 3. *Suppose that κ is sufficiently large and that $c_0 \kappa \le x \le c_1 \kappa$, where c_0 and c_1 are absolute constants with $c_1 > C > c_0 > 1$. Define v by*

$$\frac{x}{\kappa} = \frac{e^v - 1}{v}, \tag{3.1}$$

so that $v > 0$ because $x/\kappa > 1$. Denote

$$g(-v) = -(e^v - 1) + \int_0^v \frac{e^u - 1}{u} du.$$

Then the function σ appearing in Theorem 1 satisfies

$$\sigma(x) = 1 - \frac{e^{\kappa g(-v) + O(1)}}{\sqrt{x}}.$$

In Lemma 3, observe that $g(-v) < 0$, for example from its power series expansion.

Lemma 3 will be derived by a suitable use of the saddle-point method. Once this has been done, Theorem 2 can be deduced relatively quickly, as follows.

Proof of Theorem 2. It is sufficient to derive Theorem 2 for the case when $c_1 \kappa \geq \frac{1}{2}(s-1) \geq c_0 \kappa$, where c_0 and c_1 are as in Lemma 3. Then Theorem 2 follows for other s since $f(s)$ rather obviously increases towards 1 as s increases. We suppose, throughout the proof, that κ is sufficiently large.

The expression for $f(s)$ appearing in Theorem 1 implies

$$1 - f(s) = \frac{\kappa}{s^\kappa} \int_s^\infty \left(\frac{1}{\sigma(\frac{1}{2}(t-1))} - 1 \right) t^{\kappa-1} \, dt \, .$$

Substitute $\frac{1}{2}(t-1) = x$, and specify $X = 3\kappa$. Then

$$1 - f(s) = I_s(X) + E_s(X) \, , \tag{3.2}$$

where $I_s(X)$ denotes the contribution from the interval $\frac{1}{2}(s-1) < x < X$,

$$I_s(X) = \frac{\kappa 2^\kappa}{s^\kappa} \int_{(s-1)/2}^X \left(\frac{1}{\sigma(x)} - 1 \right) x^{\kappa+1} \left(1 + \frac{1}{2x} \right)^{\kappa-1} \frac{dx}{x^2} \, , \tag{3.3}$$

and $E_s(X)$ denotes the remaining contribution from (X, ∞). Since $X > 2\kappa + 1$ (provided $\kappa > 1$), Lemma 7.1.3 implies that the factor $x^{\kappa+1}(\sigma^{-1}(x) - 1)$ decreases in (X, ∞), as does the factor $2 + 1/x$. Majorising these factors by their values at X and integrating $1/x^2$ from X to ∞ gives

$$0 \leq E_s(X) \leq \frac{\kappa 2^\kappa}{s^\kappa} \left(\frac{1}{\sigma(X)} - 1 \right) X^\kappa \left(1 + \frac{1}{2X} \right)^{\kappa-1} . \tag{3.4}$$

Under the conditions of Lemma 3 we obtain

$$\frac{1 - \sigma(x)}{\sigma(x)} = \frac{1}{\sqrt{x}} e^{\kappa g(-v) + O(1)} \, ,$$

because $e^{\kappa g(-v)+O(1)} + e^{2\kappa g(-v)+O(1)} = e^{\kappa g(-v)+O(1)}$ when $g(-v) < 0$. In (3.3) the factor $(1 + 1/2x)^\kappa$ is bounded because $x > \frac{1}{2}(s-1) > \kappa$. Thus

$$I_s(X) = \frac{\kappa 2^\kappa}{s^\kappa} \int_{(s-1)/2}^X e^{\kappa g(-v) + \kappa \log x + O(1)} \frac{dx}{x^{3/2}} \, ,$$

where $v = v(x)$ is defined by (3.1). Expressing the entry $\log x$ in terms of v by means of (3.1) gives

$$I_s(X) = \int_{(s-1)/2}^X \exp\left(-\kappa \left(\phi(v(x)) + \log \frac{s}{2\kappa} \right) + \log \kappa + O(1) \right) \frac{dx}{x^{3/2}} \, , \tag{3.5}$$

where

$$\phi(v) = e^v - 1 - \int_0^v \frac{e^u - 1}{u} \, du - \log \frac{e^v - 1}{v} \quad \text{if} \quad v > 0 \, . \tag{3.6}$$

7.2 A Lower Bound Sieve via Buchstab's Identity

A similar procedure applied to (3.4) shows

$$0 \leq E_s(X) \leq \exp\left(-\kappa\left(\phi(v(X)) + \log \frac{s}{2\kappa}\right) + \log \kappa + O(1)\right). \tag{3.7}$$

We will see that the behaviour of the integral (3.5) is dominated by those x close to the minimum point of the expression $\phi(v(x))$. Observe from (3.6) that

$$\phi'(v) = e^v - \frac{e^v - 1}{v} + \frac{1}{v} - \frac{e^v}{e^v - 1} = (e^v - 2)\left(\frac{e^v}{e^v - 1} - \frac{1}{v}\right), \tag{3.8}$$

in which the second factor is positive in $v > 0$, so that $\phi(v)$ has a unique minimum at $v = \log 2$. The minimum value of $\phi(v)$ is thus

$$\phi(\log 2) = 1 - \int_0^{\log 2} \frac{e^u - 1}{u} du + \log \log 2 < 1 - \log\left(\frac{2}{\log 2}\right) < 0,$$

because $e^u > 1 + u$ and the minimum of $y/\log y$ for $y > 1$ is attained at $y = e$. A more accurate computation leads to $\phi(\log 2) = -0.2009\ldots$, obtainable, for example, by using more terms of the power series for e^u.

Note also from (3.8) that $\phi''(v)$ is bounded in $v > v_0$, for each $v_0 > 0$, whence

$$\phi(v) \leq \phi(\log 2) + \kappa^{-1} \quad \text{if} \quad |v - \log 2| < c_2/\sqrt{\kappa},$$

for a suitable constant c_2. In terms of x this gives via (3.1)

$$\phi(v(x)) \leq \phi(\log 2) + \kappa^{-1} \quad \text{if} \quad |x - \kappa/\log 2| < c_3 \sqrt{\kappa}. \tag{3.9}$$

To establish part (i) of the theorem it suffices, because of (3.2), to show that $I_s(X) \to \infty$ as $\kappa \to \infty$, when $A > 0$ and $s < 2C\kappa - A\log \kappa$, so that $s/2\kappa < C(1 - A(\log \kappa)/(2C\kappa))$. The constant appearing in Theorem 2 is just $C = e^{-\phi(\log 2)}$. For those x appearing in (3.9) and for sufficiently large κ the exponent in (3.5) satisfies

$$-\kappa\left(\phi(v(x)) + \log \frac{s}{2\kappa}\right) + \log \kappa + O(1) \geq \left(1 + \frac{A}{2C}\right)\log \kappa + O(1).$$

Taking only the contribution from such an interval shows

$$I_s(X) \gg \sqrt{\kappa} \cdot \kappa^{1+A/2C} \cdot \frac{1}{\kappa^{3/2}} \to \infty \quad \text{as } \kappa \to \infty,$$

so that Theorem 2(i) follows.

To prove part (ii) of the theorem, use only the fact that $\phi(v) \geq \phi(\log 2)$ for all $v > 0$. When $A > C$ and $s > 2C\kappa + A\log \kappa$ a similar procedure shows that the exponents in (3.5) and (3.7) satisfy

$$-\kappa\left(\phi(v) + \log \frac{s}{2\kappa}\right) + \log \kappa + O(1) \leq \left(1 - \frac{A}{2C}\right)\log \kappa + O(1).$$

Since $I_s(X)$ is over an interval of length $\ll \kappa$ this gives

$$I_s(X) + E_s(X) \le \kappa \cdot \kappa^{1-A/2C} \cdot \frac{1}{\kappa^{3/2}} \to 0 \quad \text{as} \quad \kappa \to \infty.$$

so that (3.2) shows that $f(s) \to 1$ as $\kappa \to \infty$.

Proof of Lemma 3. This will follow from Lemmas 4 and 5, if we choose any Y satisfying the condition required in Lemma 5.

Lemma 4. *Suppose that $0 \le Y \le \pi$ and that x satisfies the conditions of Lemma 3. Define*

$$g(z) = g_x(z) = \frac{xz}{\kappa} - \int_0^z \frac{1 - e^{-w}}{w} \, dw, \tag{3.10}$$

Then, for a certain constant $c > 0$, the expression $\sigma(x)$ in Lemma 3 satisfies

$$\sigma(x) = 1 - \frac{1}{2\pi} \int_{-Y}^{Y} \frac{e^{\kappa g(-v+iy)}}{v - iy} \, dy + O\left(e^{\kappa g(-v)} e^{-\kappa c Y^4}\right), \tag{3.11}$$

where $v = v(x)$ is as in Lemma 3.

Lemma 5. *Suppose x and $v = v(x)$ are as in Lemma 3. Then*

$$\frac{1}{2\pi} \int_{-Y}^{Y} \frac{e^{\kappa g(-v+iy)}}{v - iy} \, dy = \frac{e^{\kappa g(-v) + O(1)}}{\sqrt{x}}$$

whenever $\kappa^{-1/3} \ll Y \le Y_0$, for a certain positive constant Y_0.

In Lemmas 4 and 5, the condition $c_0 \kappa \le x \le c_1 \kappa$ appearing in Lemma 3 implies that the number v given by (3.1) satisfies

$$0 < v_0 \le v \le v_1, \tag{3.12}$$

where v_0 and v_1 are constants depending only on c_0 and c_1. We have taken advantage of this fact, in that a possible v-dependence of the constants occurring in Lemma 4 need not be written down. Note that (3.1) implies that the definition (3.10) of $g(z) = g_x(z)$ is consistent with the expression for $g(-v)$ already given in Lemma 3. Observe also that, given the moderate level of accuracy attained in Lemma 5, the factor $\sqrt{x}$ could equally well have been written $\sqrt{\kappa}$, since (3.1) now gives that x/κ is bounded above and below by positive constants.

7.2 A Lower Bound Sieve via Buchstab's Identity

Proof of Lemma 4. Begin by expressing $\sigma(x)$ in terms of its Laplace transform

$$\hat{\sigma}(z) = \int_0^\infty \sigma(x) e^{-xz}\,dx\ .$$

Then

$$z\,\hat{\sigma}(z) = \int_0^\infty \sigma'(x) e^{-xz}\,dx\ ,$$

since $\sigma(0) = 0$. Apart from a scale factor, $\hat{\sigma}(z)$ is determined by the difference-differential equation (7.1.1.10). We obtain

$$-\frac{d}{dz} z\,\hat{\sigma}(z) = \int_0^\infty x\,\sigma'(x) e^{-xz}\,dx = \kappa \int_0^\infty (\sigma(x) - \sigma(x-1))e^{-xz}\,dx$$
$$= \kappa\,\hat{\sigma}(z)(1 - e^{-z})\ ,$$

in which we used $\sigma(u) = 0$ when $-1 < u < 0$. This appropriate integral of this equation is

$$z\,\hat{\sigma}(z) = \exp\left(\kappa \int_0^z \frac{e^{-w} - 1}{w}\,dw\right),$$

because the boundary conditions $\sigma(0) = 0$, $\sigma(x) \to 1$ as $x \to \infty$ show

$$\lim_{z \to 0+} z\,\hat{\sigma}(z) = \lim_{z \to 0+} \int_0^\infty \sigma'(x) e^{-xz}\,dx = \int_0^\infty \sigma'(x)\,dx = 1\ .$$

The expression $\sigma(x)$ is now given by the inversion integral for Laplace transforms,

$$\sigma(x) = \frac{1}{2\pi i} \int_{1-i\infty}^{1+i\infty} e^{xz} \hat{\sigma}(z)\,dz = \frac{1}{2\pi i} \int_{1-i\infty}^{1+i\infty} e^{\kappa g(z)} \frac{dz}{z}\ , \tag{3.13}$$

where $g(z) = g_x(z)$ is as in (3.10). We examine the integral (3.13) by an application of the saddle-point method. In accordance with the principles of the method, we seek to approximate to $g(z)$ near its stationary point. It follows from (3.10) that when $s > \kappa$ this arises at $z = -v$, where $v > 0$ is as stated in Lemma 3. The reader familiar with the saddle-point method will expect to find that the integral (3.13) is approximated by a multiple of $e^{g(-v)}$. This is the content of Lemmas 4 and 5.

Lemma 4 will follow using the following technical estimate, the proof of which we postpone.

Lemma 6. *Suppose $v_0 < v < v_1$, where $v_0 > 0$ and v_1 are absolute constants. Then the function g defined in (3.10) satisfies*

$$\left|e^{g(-v+iy)}\right| \ll \frac{e^{g(-v)}}{|y|} \quad \text{if } |y| > \pi\ , \tag{3.14}$$

$$\int_{|y|>Y} \frac{e^{\kappa g(-v+iy)}}{v - iy}\,dy \ll e^{\kappa g(-v)} e^{-c\kappa Y^4} \quad \text{if } Y \leq \pi\ , \tag{3.15}$$

where $c > 0$ is a certain absolute constant.

282   7. Lower Bound Sieves when $\kappa > 1$

Note from (3.10) that $\operatorname{Re} g(t + iy) - \operatorname{Re} g(-v + iy) \ll 1/|y|$ when $-v \leq t \leq 1$, uniformly in y. Thus Lemma 6 gives $\left|e^{\kappa g(t+iy)}\right| \ll 1/|y|^\kappa$ for these t, not uniformly in κ. Lemma 4 can now be obtained by shifting the path of integration in (3.13) to the line $\operatorname{Re} z = -v$. The pole at $z = 1$ contributes a residue 1 to the expression (3.11) in Lemma 4, in which the error term is as stated because of (3.15).

Proof of Lemma 5. Estimate the integral by approximating the expression (3.10) for $g(z)$ by the first few terms of its Taylor series about the point $z = -v$. For this purpose observe

$$g'(z) = \frac{x}{\kappa} - \frac{1-e^{-z}}{z}, \qquad g''(z) = \frac{1-e^{-z}}{z^2} - \frac{e^{-z}}{z}. \qquad (3.16)$$

Then

$$g''(-v) = \frac{1}{v^2}\left(ve^v - e^v + 1\right) = \frac{e^v}{v^2}\int_0^v 1 - e^{-t}\, dt\ .$$

Because of (3.12) this lies between positive multiples of e^v/v, and in particular

$$1 \ll g''(-v) \ll 1\ . \qquad (3.17)$$

When $Y_1 < |y| < Y$ we may use $g(-v+iy) = g(-v) - \tfrac{1}{2}y^2 g''(-v+\theta iy)$, where $0 < \theta < 1$, since (3.1) gives $g'(-v) = 0$. Observe that there is a Y_0 (independent of κ) such that $\left|g''(-v+\theta iy) - g''(-v)\right| < \tfrac{1}{2}g''(-v)$ if $|y| < Y_0$, since (3.16) shows that $g''(z)$ is independent of κ and continuous except at $z = 0$. Then

$$\left|\int_{Y_1 < |y| < Y} \frac{e^{\kappa g(-v+iy)}}{-v+iy}\,dy\right| \leq \frac{e^{\kappa g(-v)}}{v}\int_{Y_1 < |y| < Y} e^{-\tfrac{1}{4}\kappa y^2 g''(v)}\,dy$$

$$\ll \frac{e^{\kappa g(-v) - \tfrac{1}{4}\kappa Y_1^2 g''(-v)}}{Y_1 \kappa g''(-v)} \quad \text{if } Y_1 < Y < Y_0,\qquad (3.18)$$

because $\operatorname{Re} g''(-v + \theta iy) \geq \tfrac{1}{2}g''(-v)$. At the last step we used

$$2U\int_U^\infty e^{-u^2}\,du \leq \int_U^\infty e^{-u^2}\cdot 2u\,du \leq e^{-U^2}\ .$$

Set $Y_1 = \kappa^{-1/3}$. When $|y| < Y_1$ use more terms of the Taylor series. Further differentiations show

$$g^{(3)}(z) = O(1)\ ,\quad g^{(4)}(z) = O(1) \quad \text{when} \quad \operatorname{Re} z = -v\ ,$$

so that $\kappa y^n g^{(n)}(-v) \ll 1$ when $n = 3$ or 4. Expansion of $e^{\kappa y^n g^{(n)}(-v)/n!}$ gives that the integrand in Lemma 4 satisfies

$$\frac{e^{\kappa g(-v+iy)}}{v - iy} = \frac{e^{\kappa g(-v)}}{v}\frac{e^{-\tfrac{1}{2}\kappa y^2 g''(-v)}}{1 - iy/v}\left(1 - \tfrac{1}{6}\kappa i y^3 g^{(3)}(-v) + O(\kappa y^4 + \kappa^2 y^6)\right)\ .$$

7.2 A Lower Bound Sieve via Buchstab's Identity

Since the integrals of the terms of odd degree vanish this gives

$$\int_{-Y_1}^{Y_1} \frac{e^{\kappa g(-v+iy)}}{v-iy} dy = \frac{e^{\kappa g(-v)}}{v} \int_{-Y_1}^{Y_1} e^{-\frac{1}{2}\kappa y^2 g''(-v)} \left(1 - \frac{y^2}{v^2} + O(\kappa y^4 + \kappa^2 y^6)\right) dy.$$

Since $1 \ll \kappa y^3$ in the range $|y| > Y_1$, the range of integration can be extended to $(-\infty, \infty)$ while keeping the same O-term, to give

$$\int_{-Y_1}^{Y_1} \frac{e^{\kappa g(-v+iy)}}{v-iy} dy = \frac{\sqrt{2\pi} e^{\kappa g(-v)}}{\sqrt{\kappa g''(-v)}} \left\{1 + O\left(\frac{1}{\kappa}\right)\right\}.$$

After use of (3.18) and the rather approximate estimate (3.17) for $g''(-v)$ this gives

$$\frac{1}{2\pi} \int_{-Y}^{Y} \frac{e^{\kappa g(-v+iy)}}{-v+iy} dy = \frac{e^{\kappa g(-v)+O(1)}}{\sqrt{\kappa}}.$$

But x/κ is bounded above and below by positive constants under the conditions of Lemma 5, so the conclusion of the lemma follows.

Proof of Lemma 6. Observe from (3.10) that

$$g(-v+iy) = g(-v) - \int_{-v}^{-v+iy} \frac{1-e^{-w}}{w} dw + \frac{ixy}{\kappa}.$$

Set $w = -v + iu$. This leads to

$$\frac{e^{g(-v+iy)} v(e^v - 1)}{(v-iy)(e^v-1)} = e^{g(-v)} \exp\left(-\int_0^y \frac{e^v - e^{v-iu}}{-v+iu} i\, du + \frac{ixy}{\kappa}\right), \quad (3.19)$$

because

$$\int_0^y \frac{i\, du}{-v+iu} = \log \frac{v-iy}{v}.$$

Substitute $u = t$ or $u = -t$ according as $y > 0$ or $y < 0$ in (3.10). This gives

$$\text{Re}\left(-\int_0^y \frac{e^v - e^{v-iu}}{-v+iu} i\, du\right) = -e^v \int_0^{|y|} \frac{t(1-\cos t) + v\sin t}{v^2+t^2} dt$$

$$\leq -e^v \int_0^{|y|} \frac{t(1-\cos t)}{v^2+t^2} dt. \quad (3.20)$$

The last step follows because $\phi(t) = 1/(v^2+t^2)$ is positive decreasing, and uses considerations of the graph of $\phi(t)\sin t$, or alternatively an integration by parts,

$$\int_0^{|y|} \phi(t)\sin t\, dt = \phi(|y|)(1-\cos|y|) - \int_0^{|y|}(1-\cos t)\phi'(t)\, dt \geq 0.$$

284   7. Lower Bound Sieves when $\kappa > 1$

When $0 \le |y| \le \pi$ use

$$\int_0^{|y|} \frac{t(1-\cos t)}{v^2+t^2}\,dt = 2\int_0^{|y|} \frac{t\sin^2 \tfrac{1}{2}t}{v^2+t^2}\,dt$$

$$\ge \frac{2}{\pi^2}\int_0^{|y|} \frac{t^3\,dt}{v^2+t^2} = \frac{1}{\pi^2}\int_0^{y^2} \frac{u\,du}{v^2+u}\,.$$

If $y^2 \le v^2$ use $v^2 + u < 2v^2$ when $0 < u < y^2$ to obtain

$$\int_0^{y^2} \frac{u\,du}{v^2+u} \ge \int_0^{y^2} \frac{u\,du}{2v^2} = \frac{y^4}{4v^2}\,.$$

If $y^2 > v^2$ then $v^2 + u < 2u$ when $v^2 < u < y^2$, whence

$$\int_0^{y^2} \frac{u\,du}{v^2+u} > \frac{v^4}{4v^2} + \int_{v^2}^{y^2} \frac{u\,du}{2u} = \frac{y^2-v^2}{2} + \frac{v^2}{4} = \frac{y^2}{2} - \frac{v^2}{4} > \frac{y^2}{4}\,.$$

This shows

$$\int_0^{|y|} \frac{t(1-\cos t)}{v^2+t^2}\,dt > \frac{y^2}{4\pi^2}\min\left\{1, \frac{y^2}{v^2}\right\} \quad \text{if} \quad 0 \le |y| \le \pi\,.$$

An integration by parts shows that $\int_\pi^\infty t\cos t/(v^2+t^2)\,dt$ converges, to a value bounded uniformly in v. Hence

$$\int_0^{|y|} \frac{t(1-\cos t)}{v^2+t^2}\,dt \ge \log|y| + O(1) \quad \text{if} \quad |y| > \pi\,.$$

With (3.19) and (3.20) these inequalities give

$$\frac{|e^{g(-v+iy)}|v^{(e^v-1)}}{|v-iy|^{(e^v-1)}} \le \begin{cases} e^{g(-v)}e^{-c_4 y^4} & \text{if } |y| \le \min\{v,\pi\} \\ e^{g(-v)}e^{c_5}/|y|^{e^v} & \text{if } |y| > \pi, \end{cases}$$

for certain constants c_4 and c_5. Furthermore, the expression (3.20) decreases as $|y|$ increases, so that

$$\frac{|e^{g(-v+iy)}|v^{(e^v-1)}}{|v-iy|^{(e^v-1)}} \le e^{g(-v)}e^{-c_4 Y^4} \quad \text{if} \quad |y| \ge Y \text{ and } 0 \le Y \le \min\{v,\pi\}\,.$$

Since $|v-iy| \ge \max\{v,|y|\}$ we now obtain

$$|e^{g(-v+iy)}| \le \begin{cases} e^{g(-v)}e^{-c_4 Y^4} & \text{if } |y| \ge Y \text{ and } 0 \le Y \le \min\{v,\pi\} \\ e^{g(-v)}e^{c_6}/|y| & \text{if } |y| > \pi\,. \end{cases} \quad (3.21)$$

The second inequality in (3.21) gives (3.14) as required in Lemma 6.

When $|y| > Y$, $0 \le Y \le \pi$ and κ is sufficiently large this implies

$$\frac{|e^{\kappa g(-v+iy)}|}{e^{\kappa g(-v)}} \ll e^{-c_4 \kappa Y^4/2}\min\left\{1, \frac{1}{y^2}\right\}\,. \quad (3.22)$$

To see this, note that $\kappa - 2 > \kappa/2$ when $\kappa > 4$, use (3.21), and observe that $e^{c_6\kappa}/|y|^{\kappa-2} \leq e^{-c_4\kappa Y^4}$ when $\frac{1}{2}\log|y| \geq c_6 + c_4 Y^4$. For smaller y with $|y| > Y$ observe

$$\frac{|e^{\kappa g(-v+iy)}|}{e^{\kappa g(-v)}} \leq e^{-c_4\kappa Y^4} \leq \frac{1}{y^2} e^{-c_4\kappa Y^4 + 4c_6 + 4c_4 Y^4} \ll \frac{1}{y^2} e^{-(\kappa-4)c_4 Y^4},$$

from which (3.22) follows if κ is large enough. Integrating (3.22) from Y to ∞ gives (3.15) and completes the proof of Lemma 6.

7.3 Selberg's $\Lambda^2 \Lambda^-$ Method

The Buchstab transform described in Sect. 7.2 is not the only way in which a lower bound sieve can be obtained from Selberg's λ^2 method. The method described below gives better results in the case when the sifting density κ is sufficiently large. On the other hand the question of how the ideas in this section might be developed in the hope of improving the results of Sect. 7.2 (or developments thereof) for κ of moderate size, such as $\kappa = 2$, appears to be a difficult one.

We saw in Sect. 7.2.3 that the sifting limit β associated with the method of Sect. 7.2 is asymptotic to $2C\kappa$ as $\kappa \to \infty$, where $2C = 2.44\ldots$ is a certain constant (greater than 2). It turns out to be rather easier to show that the sifting limit β_κ in the method to be described satisfies $\beta_\kappa \leq 2\kappa + o(\kappa)$ as $\kappa \to \infty$. This will follow from the theorem in Sect. 7.3.2.

The central idea is to count the numbers a in a suitable sequence $\mathcal{A}$ with a weight

$$m(a) = \left(1 - \sum_{\substack{p|a \\ p|P(z)}} 1\right)\left(\sum_{\substack{d|a \\ d|P(z)}} \lambda(d)\right)^2,$$

subject to the usual constraint $\lambda(1) = 1$. Then

$$m(a) \leq S(a, P(z)) = \begin{cases} 1 & \text{if } (a, P(z)) = 1 \\ 0 & \text{if } (a, P(z)) > 1. \end{cases}$$

More generally, one might consider a weight

$$m(a) = \left(\sum_{\substack{d|a \\ p|P(z)}} \lambda_1(d)\right)\left(\sum_{\substack{d|a \\ d|P(z)}} \lambda(d)\right)^2,$$

where λ_1 is any lower sifting function, for example one of Brun's type as discussed in Chap. 3. This gives a new lower sifting function λ^- for which

$$m(a) = \sum_{d|a} \lambda^-(d), \qquad \lambda^-(d) = \sum_{[d_1,d_2,d_3]=d} \lambda_1(d_1)\lambda(d_2)\lambda(d_3).$$

Here $[\cdot,\cdot]$ denotes a least common multiple, as in the discussion of Selberg's λ^2 upper bound in Sect. 2.1.

On invoking Theorem 1.3.1 in the usual way we need to estimate

$$V^-(\lambda) = \sum_{d|P} \frac{\lambda^-(d)\rho(d)}{d}.$$

7.3.1 An Identity for the Main Term

Lemma 1 gives an identity for the sum V^-, parallel to that derived in Lemma 2.1.1 for the corresponding V^+ that arose in Selberg's upper bound λ^2 method. At this stage the numbers $\lambda(d)$ and $\lambda_1(d_1)$ may be quite arbitrary, subject only to a condition that they are supported on finite intervals, so that the sums in Lemma 1 are finite ones.

Lemma 1. *Denote*

$$V^-(\lambda) = \sum_{d_1}\sum_{d_2}\sum_{d_2} \lambda_1(d_1)\lambda(d_2)\lambda(d_3) \frac{\rho([d_1,d_2,d_3])}{[d_1,d_2,d_3]}.$$

Let ρ^* and $g = \rho/\rho^*$ be as in Lemma 2.1.1, and define $x(h)$, $y(h)$ so that

$$\mu(h)g(h)y(h) = x(h) = \sum_{d\equiv 0,\,\text{mod}\,h} \frac{\lambda(d)\rho(d)}{d}. \tag{1.1}$$

Then

$$V^-(\lambda) = \sum_{d_1} \frac{\lambda(d_1)\rho(d_1)}{d_1} \sum_{(h,d_1)=1} g(h)\left(\sum_{k|d_1}\mu(k)y(hk)\right)^2.$$

In particular, if

$$\lambda_1(1) = 1, \qquad \lambda_1(p) = -1 \quad \text{when} \quad p|P(z), \tag{1.2}$$

and $\lambda(d) = 0$ otherwise, then

$$V^-(\lambda) = \sum_h g(h)y^2(h) - \sum_{p|P(z)} \frac{\rho(p)}{p} \sum_{(h,p)=1} g(h)\bigl(y(h) - y(ph)\bigr)^2. \tag{1.3}$$

If $g(h) = 0$ then $\rho(d) = 0$ whenever $d|h$, so that $y(h)$ is not specified by (1.1) (although it is clear what the natural choice would be). These $y(h)$ make no contribution to Lemma 1. The notation $x(h)$ is as in (2.1.1.12), and $y(h)$ has been specified in such a way that the standard choice (2.1.1.18) of $x(h)$ made in the usual λ^2 upper bound method (Theorem 2.1.1) is of the type

$$y(h) = C \quad \text{if} \quad 1 \le h < H, \qquad y(h) = 0 \quad \text{otherwise}. \tag{1.4}$$

In this chapter, however, a different choice of $y(h)$ will eventually be made.

7.3 Selberg's $\Lambda^2\Lambda^-$ Method

Proof of Lemma 1. Begin by writing

$$V^-(\lambda) = \sum_{d_1} \frac{\lambda(d_1)\rho(d_1)}{d_1} \sum_{d_2}\sum_{d_3} \frac{\lambda(d_2)\lambda(d_3)\rho_{d_1}([d_2,d_3])}{[d_2,d_3]}, \quad (1.5)$$

where $\rho_{d_1}(d)$ is defined for squarefree $d = [d_2, d_3]$ so that

$$\frac{\rho(d_1)}{d_1}\frac{\rho_{d_1}(d)}{d} = \frac{\rho([d_1,d])}{[d_1,d]} = \frac{\rho([d_1,d])(d_1,d)}{d_1 d}.$$

When $\rho(d_1) \neq 0$ this implies

$$\rho_{d_1}(d) = \frac{\rho([d_1,d])(d_1,d)}{\rho(d_1)} = \prod_{\substack{p|d_1 \\ p|d}} p \prod_{\substack{p\nmid d_1 \\ p|d}} \rho(p),$$

all of which is satisfied if ρ_{d_1} is the multiplicative function for which

$$\rho_{d_1}(p) = \begin{cases} p & \text{if } p|d_1 \\ \rho(p) & \text{if } p\nmid d_1 . \end{cases} \quad (1.6)$$

Thus the "conjugate" function $\rho_{d_1}^*$ satisfies $\rho_{d_1}^*(p) = p - \rho_{d_1}(p) = 0$ if $p|d_1$, so that $\rho_{d_1}^*(h) \neq 0$ only when $(h,d_1) = 1$. Furthermore

$$\frac{\rho_{d_1}^*(h)}{\rho_{d_1}(h)} = \begin{cases} 0 & \text{if } (h,d_1) > 1 \\ \rho^*(h)/\rho(h) & \text{if } (h,d_1) = 1 \text{ and } \rho(h) \neq 0 . \end{cases} \quad (1.7)$$

Note also from (1.6) that when $\rho(d) \neq 0$

$$\frac{\rho_{d_1}(d)}{\rho(d)} = \frac{(d_1,d)}{\rho((d_1,d))} = \sum_{k|(d_1,d)} \frac{\rho^*(k)}{\rho(k)} . \quad (1.8)$$

Lemma 2.1.1 and (1.7) now show that the inner sum in (1.5) is

$$\sum_{d_2}\sum_{d_3} \frac{\lambda(d_2)\lambda(d_3)\rho_{d_1}([d_2,d_3])}{[d_2,d_3]} = \sum_{(h,d_1)=1} \frac{x_{d_1}^2(h)}{g(h)}, \quad (1.9)$$

where

$$x_{d_1}(h) = \sum_{d\equiv 0,\bmod h} \frac{\lambda(d)\rho_{d_1}(d)}{d} = \sum_{d\equiv 0,\bmod h} \frac{\lambda(d)\rho(d)}{d} \sum_{k|(d_1,d)} \frac{\rho^*(k)}{\rho(k)},$$

the last step following from (1.8). Thus, when $(h,d_1) = 1$ as in (1.9),

$$x_{d_1}(h) = \sum_{k|d_1} \frac{\rho^*(k)}{\rho(k)} \sum_{d\equiv 0,\bmod hk} \frac{\lambda(d)\rho(d)}{d} = \sum_{k|d_1} \frac{\rho^*(k)}{\rho(k)} x(hk)$$

$$= \frac{\mu(h)\rho(h)}{\rho^*(h)} \sum_{k|d_1} \mu(k) y(hk) = \mu(h)g(h) \sum_{k|d_1} \mu(k)y(hk),$$

where y is as in (1.1). With (1.9) this gives

$$\sum_{d_2}\sum_{d_3}\frac{\lambda(d_2)\lambda(d_3)\rho_{d_1}([d_2,d_3])}{[d_2,d_3]} = \sum_{\substack{(h,d_1)=1\\\rho(h)\neq 0}} g(h)\left(\sum_{k|d_1}\mu(k)y(hk)\right)^2,$$

so that (1.5) gives the identity for $V^-(\lambda)$ enunciated in Lemma 1.

7.3.2 The Improved Sifting Limit for Large κ

It will appear that a quite simple argument is sufficient to establish the following theorem, which is enunciated using the general notations of Sect. 1.1.3.

Theorem 1. *Suppose that the function ρ satisfies*

$$\sum_{p<z}\rho(p)\log p < \kappa\log z + O(1). \tag{2.1}$$

Write $z = D^{1/s}$, and assume

$$s > \beta = 2\kappa + c\sqrt{\kappa\log\kappa}, \tag{2.2}$$

where c is a constant satisfying $c > 2\sqrt{2}$. Then there exists a sieve of level D for which

$$V^-(D, P(z)) > 0,$$

provided $\kappa > \kappa_0(c)$ for a certain $\kappa_0(c)$ depending only upon c.

In particular this theorem shows, for sufficiently large sifting density κ, the existence of a sieve with sifting limit $\beta < 2C\kappa$ for each constant $C > 1$. For these κ this is significantly better than the result obtained in Theorem 7.2.2 from the Buchstab transform of Selberg's method. Observe also that the hypothesis (2.1) is much less demanding than our usual sifting density hypothesis, requiring as it does merely a suitable bound for the sum in (2.1) for the value of z actually appearing in the theorem. An assumption of this weaker type was also sufficient for the estimates in Sect. 2.2.1, which will be called upon below.

Proof of Theorem 1. The construction used for Theorem 1 employs the very simple $\lambda_1(d_1)$ given in (1.2), so that $\lambda_1(1) = 1$ and $\lambda_1(p) = -1$ for primes $p < z$. For the major input $\lambda(d)$ we will require

$$\lambda(d) \neq 0 \quad \text{only if} \quad d < \xi = \sqrt{D/z},\ d|P(z). \tag{2.3}$$

Then the sifting function

$$\lambda^-(d) = \sum_{[d_1,d_2,d_3]=d}\lambda_1(d_1)\lambda(d_2)\lambda(d_3)$$

7.3 Selberg's $\Lambda^2\Lambda^-$ Method

is of level D, as required in Theorem 1. Since it is also supported on divisors of $P(z)$ the sum written as $V^-(\lambda)$ in Lemma 1 will now be rewritten as $V^-(D, P(z))$.

We need to estimate the expression V^- appearing in (1.3). Ideally one might seek to optimise the choice of the function $\lambda(d)$, in the style successfully pursued in Chap. 2, but this would turn out to be an ambitious aim. Accordingly we settle for the following choice, which we specify in terms of the quantities $y(h)$ introduced in (1.1). Define

$$y(h) = \begin{cases} 1 & \text{if } h < \xi/z \\ \dfrac{\log \xi/h}{\log z} & \text{if } \xi/z \leq h < \xi \\ 0 & \text{if } h > \xi. \end{cases} \quad (2.4)$$

Since $z < D^{1/2\kappa}$ and $\xi = \sqrt{D/z}$ in Theorem 1, in which κ is large, this choice of $y(h)$ amounts only to a natural and not very radical smoothing of the choice (1.4) made in the upper bound method of Chap. 2. The normalisation is not such that $\lambda^-(1) = 1$, but for Theorem 1 as stated this is of no importance.

Observe from (2.4) that

$$y(h) - y(ph) = 0 \quad \text{if} \quad p < z \text{ and } h < \xi/z^2.$$

If $p < z$ and $\xi/z^2 \leq h < \xi$ then

$$0 \leq y(h) - y(ph) \leq \frac{\log p}{\log z}, \quad (2.5)$$

for $y(h) - y(ph)$, if not zero, satisfies

$$y(h) - y(ph) \leq \begin{cases} 1 - \dfrac{\log(\xi/ph)}{\log z} = \dfrac{\log p - \log(\xi/hz)}{\log z} & \text{if } h < \xi/z < ph \\ \dfrac{\log(\xi/h) - \log(\xi/ph)}{\log z} & \text{if } \xi/z \leq h < \xi. \end{cases}$$

Use the information (2.5) in the identity (1.3) for the main term V^-, obtained in Lemma 1. First observe

$$\sum_{h<\xi; h|P(z)} g(h)y^2(h) \leq \sum_{h|P(z)} g(h) = \prod_{p|P(z)} (1 + g(p)) = \frac{1}{V(P(z))},$$

this being just the inequality (2.2.1.4). In the opposite direction

$$\sum_{\substack{h<\xi \\ h|P(z)}} g(h)y^2(h) \geq \sum_{\substack{h<\xi/z^2 \\ h|P(z)}} g(h) = \frac{1}{V(P(z))} - \sum_{\substack{h\geq \xi/z^2 \\ h|P(z)}} g(h).$$

290   7. Lower Bound Sieves when $\kappa > 1$

For the second entry in (1.3) we obtain from (2.5)

$$\sum_{p|P(z)} \frac{\rho(p)}{p} \sum_{\substack{h<\xi \\ h|P(z);(h,p)=1}} g(h)\Big(y(h) - y(ph)\Big)^2$$

$$< \sum_{p|P(z)} \frac{\rho(p)}{p} \left(\frac{\log p}{\log z}\right)^2 \sum_{\substack{\xi/z^2 \leq h < \xi \\ h|P(z)}} g(h) < B(z) \sum_{\substack{h \geq \xi/z^2 \\ h|P(z)}} g(h),$$

where

$$B(z) = \frac{1}{\log z} \sum_{p|P(z)} \frac{\rho(p) \log p}{p}, \qquad (2.6)$$

a notation used in Sect. 2.2.1. Thus (1.3) leads to

$$V^-(D, P(z)) \geq \frac{1}{V(P(z))} - (1 + B(z)) \sum_{\substack{h \geq \xi/z^2 \\ h|P(z)}} g(h).$$

This procedure was somewhat approximate, but reduces all the estimations to those of the last sum.

A suitable estimate, in terms of $B(z)$, was provided in Lemma 2.2.1. Since ξ is as in (2.3) and $z = D^{1/s}$ in Theorem 1, this gives

$$V^-(D, P(z)) \geq \frac{1}{V(P(z))} \left(1 - (1 + B(z))e^{-\psi_{B(z)}(v)}\right), \qquad (2.7)$$

where

$$v = \frac{\log(\xi/z^2)}{\log z} = \frac{s-5}{2}, \qquad \psi_B(v) = \int_{B<t<v} \log \frac{t}{B} \, dt, \qquad (2.8)$$

so that $\psi_B(v) = \max\{0, v\log(v/B) - v + B\}$.

Suppose, for the moment, that $v < 2\kappa$. The hypothesis (2.1) shows that the quantity (2.6) satisfies

$$B(z) \leq \kappa + O\left(\frac{1}{\log z}\right),$$

and (2.2) implies that $v > B(z)$ as soon as κ is large enough. Note $\partial \psi_B(v)/\partial B = 1 - v/B$. If $B > \kappa$, then

$$\frac{\psi_B(v) - \psi_\kappa(v)}{B - \kappa} \geq 1 - \frac{v}{\kappa} \geq -1.$$

Thus $\psi_B(v) \geq \psi_\kappa(v) + O(1/\log z)$, while if $B \leq \kappa$ then $\psi_B(v) \geq \psi_\kappa(v)$. Thus we obtain

$$(1 + B(z))e^{-\psi_{B(z)}(v)} \leq (\kappa + 1)\exp\left(-\kappa \int_1^{v/\kappa} \log x \, dx\right)\left(1 + O\left(\frac{1}{\log z}\right)\right).$$

Write $v/\kappa = (s-5)/\kappa = 1 + h$. Then the last integral is $\frac{1}{2}h^2 + O(h^3)$, and in Theorem 1 we have

$$h > \tfrac{1}{2}c\sqrt{\frac{\log(\kappa+1)}{\kappa}} + O\Big(\frac{1}{\kappa}\Big).$$

Here $c > 2\sqrt{2}$, so that

$$\int_1^{v/\kappa} \log x\, dx > \frac{c^2}{8}\frac{\log(\kappa+1)}{\kappa}\Big(1 + O\Big(\frac{1}{\sqrt{\kappa}}\Big)\Big) > \frac{c_0 \log(\kappa+1)}{\kappa},$$

where $c_0 > 1$, provided $\kappa > \kappa_0(c)$ as in Theorem 1. Then

$$(1+B(z))e^{-\psi_{B(z)}(v)} \le (\kappa+1)\exp\bigl(-c_0\log(\kappa+1)\bigr)\Big(1 + O\Big(\frac{1}{\log z}\Big)\Big)$$

$$\le \frac{1}{(\kappa+1)^{c_0-1}}\Big(1 + O\Big(\frac{1}{\log z}\Big)\Big).$$

This conclusion persists when $v > 2\kappa$ since $\psi_B(v)$ increases with v. Theorem 1 now follows from (2.7), using only that $z \ge 2$ and taking κ sufficiently large.

7.4 Notes on Chapter 7

Sect. 7.1. The extension of the estimates of Sect. 2.2 to the so-called "incomplete" situation considered here was discussed in [Ankeny and Onishi (1964)], and is also the subject of subsequent accounts in [Halberstam and Richert (1974)] and [Selberg (1991)].

The problem of estimating the sum $G_z(x)$ is susceptible to attack by using an identity of Buchstab's type

$$\sum_{\substack{n<N \\ n|P(N^{1/\sigma})}} g(n) = \sum_{\substack{n<N \\ n|P(N^{1/s})}} g(n) - \sum_{N^{1/\sigma} \le p < N^{1/s}} g(p) \sum_{\substack{m<N/p \\ m|P(p)}} g(n),$$

when $\sigma > s$, as a basis for an induction on $[s]$, the integer part of s, but from our point of view such a procedure would suffer from two disadvantages. In the first place the error terms in such an estimate would increase with s, in the style we have already seen in Sect. 4.5. In the second place the presence of the negative sign means that to induce an estimate for the sum on the left we would need an estimate from the opposite side for the inner sum on the right, and it is not clear how this is to be done using a hypothesis that bounds g from only one side as Theorem 1 aims to do. Lastly, the identity given above does not directly address the problem in hand, of estimating the quotient $G_z(s)/G_z(\infty)$ rather than the sum $G_z(s)$ itself.

The account in [Halberstam and Richert (1974)] is, like in the text, based on a use of the integral equation (1.3), but in other respects shares the properties just ascribed to an appeal to an identity of Buchstab's type. In particular, the error term in their analogue of our Theorem 1 grows like $s^{2\kappa}$, which is a rather unsatisfactory state of affairs given that it is known from Sect. 2.2 to be nearly as small as $e^{-s\log s}$ for large s.

This undesirable growth of error terms was a symptom encountered in Sect. 4.5.1, for similar reasons.

These difficulties are familiar from the literature relating to the problem of estimating the number of integers $< x$ divisible by no prime $< y$ (for which see [Tenenbaum (1990)] for a recent account) and might well be attacked in an analogous style. The approach in the text is somewhat different, but seems to be germane to the problem being discussed. As the text should make clear, the integral equations for $G_z(s)$ and for the associated continuous function $\sigma(s)$ are very closely related, and it seems to be entirely appropriate that they should be treated in a parallel fashion as in the proof of Theorem 1.

Sect. 7.1.1. The function σ appears in [Ankeny and Onishi (1964)], not in terms of the precise notation used here. These authors established various properties of this function, including most of those given in the text. They did not, however, supply the error term quoted in Lemma 1, of which a more refined version is given in [Iwaniec (1980a)].

Sect. 7.2. A version of Theorem 1 was the principal theorem in [Ankeny and Onishi (1964)], and a further account appears in [Halberstam and Richert (1974)]. Our account follows these authors, with some adjustments to take account of our one-sided hypothesis.

Sect. 7.2.1. Buchstab's iterations were introduced (in connection with the method of V. Brun) in [Buchstab (1938)].

Sect. 7.2.2. In [Halberstam and Richert (1974)] there is a report on calculations of the sifting limit β_κ attained by the method of Theorem 1. A selection of the values they give is as follows.

Table 7.4.1. Sifting Limits for the Buchstab Transform of the λ^2 Sieve

κ	1	2	3	4	5	6	7	8	9	10
β_κ	2·06..	4·42..	6·84..	9·32..	11·80..	14·28..	16·77..	19·25..	21·74..	24·22..

As these authors remark, the numerical evidence suggests that β_κ/κ is increasing towards the value 2·44.. determined in Sect. 7.2.3. As soon as κ exceeds 1 the sifting limits attained are significantly better than those obtained using Rosser's sieve.

More recently, the result of further iterations of the Buchstab transform described in Sect. 7.2.1 has been extensively investigated. Expected values of the sifting limits were calculated in a purely numerical investigation in [Iwaniec, van de Lune and te Riele (1980)], an extract from the results of which is tabulated below. The corresponding sieve theorem was established in [Diamond, Halberstam, and Richert (1988)] subject to a hypothesis, at that stage unproved, concerning the existence of functions F_κ, f_κ playing a rôle analogous to that of the functions of Sect. 4.2. The existence of these functions was established in a series of papers by the same authors, concluded in [Diamond, Halberstam, and Richert (1996)]. These papers also depended on a considerable amount of numerical work. The sifting limits obtained in this way are summarised in the following table.

Table 7.4.2. Sifting Limits for the Buchstab Iteration Sieve

κ	1	2	3	4	5	6	7	8	9	10
β_κ	2	4·26..	6·64..	9·07..	11·53..	14·01..	16·50..	18·99..	21·49..	23·99..

The improvement over the values in Table 7.4.1 is quite noticeable for moderate values of κ, but fades away as κ increases.

In the case $\kappa = 1$ the method used reduces to that treated in a rather special context in [Jurkat and Richert (1965)], and later handled in [Halberstam and Richert (1974)] in a more general situation similar to that discussed in Chap. 4. In this instance the sifting limit attained is the same as that reached by Rosser's method.

Sect. 7.2.3. The relationship between the sifting limit for large κ and the constant $2C = 2\cdot44\ldots$ was discussed in [Ankeny and Onishi (1964)], using a rather different method based on an asymptotic estimation of the moments $\int_0^\infty s^n \sigma(s)\,ds$ as $\kappa \to \infty$. The more direct method used here is derived from [Selberg (1991)]. Not surprisingly, many of the calculations needed in these two approaches are the same.

Sect. 7.3. The method discussed in this section appears to be not very effective for κ of moderate size, though the author is not aware of any detailed calculations in this situation having been carried out. [Halberstam and Richert (1974)] discuss an analogous approach to the problem of the sieve with weights, in which one starts with an expression

$$\left(1 - \sum_{p|a} w(p)\right)\left(\sum_{d|a} \lambda(d)\right)^2.$$

For other accounts of this approach one may consult [Miech 1964], or the relevant section in [Bombieri 1974]. In a note in [Halberstam and Richert

(1974)] it is reported that, for all κ, better results were obtained using a weighted analogue of Theorem 7.2.1. One may note that for large κ this situation is not analogous to that described in the text for the unweighted case.

A more refined version of Theorem 1 appears in [Selberg 1991], of which there is a preliminary account in [Selberg 1989]. Selberg conducts a more precise asymptotic analysis, free of the slightly rough approximations of the version in the text (which is also taken from [Selberg 1991]). The calculations involved are related to those in Sect. 7.2.3. A different choice of $\lambda(d)$ is used, in which the expression $y(h)$ is much more closely related to $1 - \log h/\log z$ than is that used in Theorem 1. The result is that the inequality of Theorem 1 can be replaced by the sharper version $s > 2\kappa + \frac{19}{36}$.

One might have ambitions to choose the numbers $\lambda(d)$ in this treatment in an optimal way, is the style of Sect. 2.1. This appears to be a hopeless undertaking, but in [Selberg 1991] and also in [Selberg 1989] a discussion is given of a simpler idealised sifting problem, in which it is possible to obtain a clear guide to what a good choice of $\lambda(d)$ might look like.

The theorem of this section raises some tantalising possibilities. In the first place the sifting limit is known, from Chap. 4, to be 2κ in the cases $\kappa = 1$ and $\kappa = \frac{1}{2}$. It has now been shown to be asymptotic to a value no worse than 2κ as $\kappa \to \infty$. On the other hand a guess that it might equal 2κ for all κ is too simplistic, because numerical calculations carried out on the results from Rosser's method indicate that better than this is true when $\frac{1}{2} < \kappa < 1$.

Suppose that (as suggested in the text) one were to consider a more general expression

$$\left(\sum_{d|a} \mu(d)\chi(d)\right)\left(\sum_{d|a} \lambda(d)\right)^2,$$

possibly with χ as in Rosser's sieve. In Sect. 7.3 this was done with a very simple χ, supported only at 1 and the primes, and with a fairly elaborate function λ, thereby obtaining good results for large κ. On the other hand in Rosser's sieve an elaborate χ is used, and a completely trivial λ, supported only at 1, giving results at $\kappa = \frac{1}{2}$ or 1 which turned out to be optimal, as in Sect. 4.5. One might expect that good results for κ of intermediate size might follow by combining these approaches, perhaps with λ still relatively simple when $\kappa = 2$, say. Unfortunately the difficulties involved in executing such a programme appear to be severe.

At the time of writing the best results obtained when $\kappa = 2$, for example, are those derived from further iterations of the Buchstab transform referred to in the notes to Sect. 7.2. It seems clear, from remarks in [Selberg (1991)] which are supported by numerical work quoted in [Diamond, Halberstam, and Richert (1988)], that the sifting limit is this method has the same asymptotic value as $\kappa \to \infty$ as appears in Theorem 7.2.2. The method is thus not optimal when κ is large, having been surpassed by that of Sect. 7.3. There therefore

7.4 Notes on Chapter 7

seems to be no particular reason to suppose that its results are optimal for κ of moderate size. It would be very satisfactory if a method that gave improved results for these κ could be discovered.

The "Vector Sieve" of Brüdern and Fouvry. This method is applicable in some of the situations where the methods of Chap. 7 apply, and accordingly deserves at least a mention here. It has the attractive feature of treating a problem where the sifting density (or "dimension") is an integer K in a K-dimensional way. It is because the other methods of Chap. 7 do not have this feature that the author was reluctant to follow a recent tradition and speak of them as constituting a sieve of dimension K.

For simplicity we will consider the specimen situation in which

$$\mathcal{A} = \left\{ n(n+2) : 1 \leq n < X,\ 2 \nmid n \right\}, \qquad P = \prod_{3 \leq p < z} p,$$

so that $K = 2$. Here

$$S(\mathcal{A}, P) \leq \sum_{n \in \mathcal{A}} \left(\sum_{d_1 | (n,P)} \lambda_1^+(d_1) \right) \left(\sum_{d_2 | (n+2,P)} \lambda_2^+(d_2) \right),$$

where λ_i are upper sifting functions with $|\lambda(d)| \leq 1$ supported on the divisors of P. Then

$$S(\mathcal{A}, P) \leq \sum_{d_1 \leq D_1} \sum_{d_2 \leq D_2} \lambda(d_1)\lambda(d_2) \left(\frac{X}{d_1 d_2} + O(1) \right)$$

$$= X \left(\sum_{d_1 \leq D_1} \frac{\lambda(d_1)}{d_1} \right) \left(\sum_{d_2 \leq D_2} \frac{\lambda(d_2)}{d_2} \right) + O(D_1 D_2),$$

since d_1 and d_2 are automatically coprime in this situation. Then the main term n the resulting upper bound is

$$F\left(\frac{\log D_1}{\log z} \right) F\left(\frac{\log D_2}{\log z} \right) \prod_{p < z} \left(1 - \frac{1}{p} \right)^2.$$

To construct a lower bound sieve in this style we may compose two sieves as in Sect. 1.3.7, and use

$$S(\mathcal{A}, P) \geq \sum_{n \in \mathcal{A}} \sum_{\substack{f|(n,P) \\ g|(n+2,P)}} \left(\lambda_1^+(f)\lambda_2^-(g) + \lambda_1^-(f)\lambda_2^+(g) - \lambda_1^+(f)\lambda_2^+(g) \right).$$

Here we can make the functions $\lambda_i^+(d)$ be supported on $d < D_1$ and $\lambda_i^-(d)$ be supported on $d < D_2$. In this way the main term in the resulting estimate is

$$F(s_1)\bigl(2 f(s_2) - F(s_1)\bigr) \prod_{p < D_1} \left(1 - \frac{1}{p} \right) \prod_{p < D_2} \left(1 - \frac{1}{p} \right).$$

296   7. Lower Bound Sieves when $\kappa > 1$

If we keep to the "elementary" range $s_2 < 4$, $s_1 < 3$ then this expression is positive with
$$2\frac{\log(s_2-1)}{s_2} - \frac{1}{s_1}.$$
Take $s_2 = 3$, which is not far from the optimal choice. Then the limiting value for s_1 is $3/(2\log 2)$. This leads to a value for $s_1 + s_2$ close to 4·365. This is not as good as the corresponding value at $\kappa = 2$ in Table 2 for the Buchstab iteration sieve, but is better than the value in Table 1 for the first Buchstab transform of the λ^2 method. The method has the attraction that heavy computations have not been involved.

This vector sieve has the great advantage, however, that the sifting functions $\lambda^\pm$ can be of Rosser's type, so that the bilinear form of the remainder term from Chap. 6 becomes available. For details, see [Brüdern and Fouvry (1996)]. The problem considered there has some extra embellishments, one effect of which is that the alternative method of sifting the sequence $p + 2$, where p is prime, would not be applicable.

The only other application of the vector sieve currently in the literature is the original [Brüdern and Fouvry (1994)], where it was used on the problem of representing a large integer as a sum of 4 squares of almost-prime variables.

Quite recently, in a paper [Ford and Halberstam (2000)] discussing a development of the "almost-pure" sieve of C. Hooley [Hooley (1994)] referred to in the notes on Chap. 3, it was remarked that a dual form of the sieve discussed therein might lead to a method superior to that of Sect. 1.3.7 for composing sieves, as used in the work of Brüdern and Fouvry. It would be extremely interesting if it turns out that significant improvements in the method can be obtained in this or other ways.

References

Ankeny, N.C., Onishi, H. (1964): The general sieve. Acta Arith **10**, 31–62
Baker, R.C., Harman, G.: The difference between consecutive primes. Proc. London Math. Soc. (3) **72**, 261–280
Bombieri, E. (1974): Le grand crible dans la théorie analytique des nombres (Asterisque **18**, i+87 pp.) (2nd ed. 1987, 103 pp.) Société Mathematique de Paris
Bombieri, E. (1976): The asymptotic sieve. Mem. Acad. Naz. dei XL, **1/2**, 243–269
Bombieri, E., Friedlander, J.B., Iwaniec, H. (1986): Primes in arithmetic progressions to large moduli, II. Acta Math. **156**, 203–251
Brüdern, J., Fouvry, E. (1994): Lagrange's four squares theorem with almost prime variables. J. Reine Angew. Math. **454**, 59–96
Brüdern, J., Fouvry, E. (1996): Le crible à vecteurs. Compositio Math. **102**, 337–355.
de Bruijn, N.G. (1950): On the number of uncancelled elements in the sieve of Eratosthenes. Proc. Konink. Nederl. Akad. Wetensch. **53**, 803–812 = Indag. Math. **12**, 247–256
Brun, V. (1915): Über das Goldbachsche Gesetz und die Anzahl der Primzahlpaare. Archiv for Math. og Naturvid. B **34**, no. 8. 19 pp.
Brun, V. (1920): Le crible d'Eratosthène et le théorème de Goldbach. Videnskaps. Skr., Mat.-Naturv. Kl. Kristiania. No. 3. 36pp.
Buchstab, A.A. (1937): An asymptotic estimate of a general number-theoretic function (Russian). Mat. Sbornik (2) **44**, 1239–1246
Buchstab, A.A. (1938): New improvements in the method of the sieve of Eratosthenes (Russian). Mat. Sbornik (2) **4 (46)**, 375–387
Buchstab, A.A. (1965): A combinatorial intensification of the sieve of Eratosthenes. Uspehi Mat. Nauk **22**, 199–226 (Russian). [English transl.: Russ. Math. Surv. **22**, 205-233]
Buchstab, A.A. (1985): A new type of weighted sieve. Abstracts of papers, All-Union Conference, Tbilisi, 22–24
Chebyshev, P.L. (1853): Sur la fonction qui détermine la totalité des nombres premiers inferieures à une limite donnée. J. Math. Pures Appl. (1) **17**, 366–390.
Chen, J.-R. (1973): On the representation of a larger even integer as the sum of a prime and the product of at most two primes. Sci. Sinica **16**, 157–176
Chen, J.-R. (1975): On the distribution of almost primes in an interval. Scientia Sinica **18**, 611–627
Chen, J.-R. (1979): On the distribution of almost primes in an interval, II. Scientia Sinica **22**, 253–275
Diamond, H.G., Halberstam, H. (1997): Some applications of sieves of dimension exceeding 1. Sieve Methods, Exponential Sums, and their Applications in Number Theory (Cardiff 1995) (G.R.H. Greaves, G. Harman, M.N. Huxley, eds.) (London Math. Soc. Lecture Notes **237**), 101–107. Cambridge University Press.

Diamond, H.G., Halberstam, H., Richert, H.-E. (1988): Combinatorial sieves of dimension exceeding one. J. Number Theory **28**, 306–346.
Diamond, H.G., Halberstam, H., Richert, H.-E. (1996): Combinatorial sieves of dimension exceeding one, II. Analytic Number Theory, Vol. I (Allerton Park, 1995) (B.C. Berndt, H.G. Diamond, A.J. Hildebrand, eds.) (Progress in Math. **138**), 265–308. Birkhäuser, Boston Basel Berlin
Dickman, K. (1930): On the frequency of numbers containing prime factors of a certain relative magnitude. Ark. Mat. Astr. Fys. **22**, 1–14
Dickson, L.E. (1920): History of the theory of numbers. Chicago
Euler, L. (1775): De tabule numerorum primorum. Novi Commentarii Acad. Petropol. **19**, 132–133
Fluch, W. (1959): Verwendung der zeta-Funktion beim Sieb von Selberg. Acta Arith. **5**, 381–405.
Ford, K., Halberstam, H. (2000): The Brun–Hooley sieve. J. Number Theory **81**, 335–350
Fouvry, E. (1987): Autour du théorème de Bombieri-Vinogradov, II. Ann. Scient. Éc. Norm. Sup. (4) **20**, 617–640
Fouvry, E. (1990): Nombres presque premiers dans les petits intervalles. Analytic Number Theory (Tokyo 1988) (K. Nagasaka, E. Fouvry, eds.) (Lecture Notes in Mathematics **1434**), 65–85. Springer, Berlin Heidelberg New York
Fouvry, E., Iwaniec, H. (1997): Gaussian primes. Acta Arith **79**, 249–287
Friedlander, J., Iwaniec, H. (1978): On Bombieri's asymptotic sieve. Ann. Scuola Norm. Sup. Pisa Cl. Sci. (4) **5**, 719–756
Friedlander, J., Iwaniec, H. (1997): The Brun-Titchmarsh theorem. Analytic Number Theory (Kyoto, 1996) (Y. Motohashi, ed.) (London Math. Soc. Lecture Note Series **247**), 85–93. Cambridge Univ. Press
Friedlander, J., Iwaniec, H. (1998a): The polynomial $X^2 + Y^4$ captures its primes. Annals of Math. **148**, 965–1040
Friedlander, J., Iwaniec, H. (1998b): Asymptotic sieve for primes. Annals of Math. **148**, 1041–1065
Gallagher, P.X. (1971): A larger sieve. Acta Arith. **18**, 77–81.
Gallagher, P.X (1973): The large sieve and probabilistic Galois theory. Proc. Sympos. Pure Math. **24** (St. Louis 1972), 91–101. Amer. Math. Soc., Providence, R.I.
Greaves, G. (1971): Large prime factors of binary forms. J. Number Theory **3**, 35–59, and corrigendum, ibid. **9** (1977), 561–562
Greaves, G. (1976): On the representation of a number in the form $x^2 + y^2 + p^2 + q^2$ where p,q are odd primes. Acta Arith. **29**, 257–274
Greaves, G. (1982a): A weighted sieve of Brun's type. Acta Arith. **40**, 297–332
Greaves, G. (1982b): An algorithm for the Hausdorff moment problem. Numerische Math. **39**, 231–238
Greaves, G. (1982c): An algorithm for the solution of certain difference-differential equations of advanced type. Math. Comp. **38**, 237–247
Greaves, G. (1985): A comparison of some weighted sieves. Banach Centre Publ. **17**, 143–153. Warsaw
Greaves, G. (1986): The weighted linear sieve and Selberg's λ^2-method. Acta Arith. **47**, 71–96
Greaves, G (1989): Some remarks on the sieve method. Number Theory, Trace Formulas and Discrete Groups (Oslo 1987) (K.E. Aubert, E. Bombieri, D. Goldfeld, eds.), 289–308.
Greaves, G. (1992): Power-free values of binary forms. Quart. J. Math. Oxford (2) **43**, 45–65

Grupp, F., Richert, H.-E. (1986): The functions of the linear sieve. J. Number Theory **22**, 208–239
Halberstam, H. (1982): Lectures on the linear sieve. Topics in Analytic Number Theory, 165–220. Univ. Austin, Tex.
Halberstam, H., Heath-Brown, D.R., Richert, H.-E. (1981): Almost-primes in short intervals. Recent Progress in Number Theory **1** (Durham 1979), 69–113. Academic Press, London
Halberstam, H., Richert, H.-E. (1971): Mean value theorems for a class of arithmetic functions. Acta Arith. **18**, 243–256
Halberstam, H., Richert, H.-E. (1974): Sieve methods. Academic Press, London
Halberstam, H., Richert, H.-E. (1985a): A weighted sieve of Greaves' type, I. Banach Centre Publ. **17**, 155–182. Warsaw
Halberstam, H., Richert, H.-E. (1985b): A weighted sieve of Greaves' type, II. Banach Centre Publ. **17**, 183–215. Warsaw
Heath-Brown, D.R. (1978): Almost-primes in arithmetic progressions and short intervals. Proc. Camb. Phil. Soc. **83**, 357–375.
Hooley, C. (1994): On an almost-pure sieve. Acta Arith. **66**, 359–368
Iwaniec, H. (1971): On the error term in the linear sieve. Acta Arith. **19**, 1–30
Iwaniec, H. (1972): Primes of the type $\phi(x,y) + A$ where ϕ is a quadratic form. Acta Arith. **21**, 203–234
Iwaniec, H. (1974): Primes represented by quadratic polynomials in two variables. Acta Arith. **24**, 435–459
Iwaniec, H. (1976): The half dimensional sieve. Acta Arith. **29**, 69–95.
Iwaniec, H. (1978a): Almost-primes represented by quadratic polynomials. Invent. Math. **47**, 171–188
Iwaniec, H. (1978b): Sieve methods. Internat. Congress of Math. Proc., Helsinki, 357–364.
Iwaniec, H. (1978c): On the problem of Jacobsthal. Demonstratio Math. **11**, 225–230
Iwaniec, H. (1980a): Rosser's sieve. Acta Arith. **36**, 171–202
Iwaniec, H. (1980b): A new form of the error term in the linear sieve. Acta Arith. **37**, 307–320
Iwaniec, H., Jutila, M. (1979): Primes in short intervals. Ark. Mat. **17**, 167–176
Iwaniec, H., Laborde, M. (1981): P_2 in short intervals. Annales Institut Fourier **31**, 37–56
Iwaniec, H., van de Lune, J., te Riele, H.J.J. (1980): The limits of Buchstab's iteration sieve. Proc. Konink. Nederl. Akad. Wetensch. ser. A **83**, 409–417 = Indag. Math. **42**, 409–417
Jurkat, W.B., Richert, H.-E. (1965): An improvement of Selberg's sieve method, I. Acta Arith. **11**, 217–240
Kuhn, P. (1941): Zur Viggo Brunschen Siebmethode, I. Norske Vid. Selsk. Forh. Trondheim **14**, 145–148
Kuhn, P. (1954): Neue Abschätzungen auf Grund der Viggo Brunschen Siebmethode. 12 Skand. Mat. Kongr., Lund (1953), 160–168.
Laborde, M. (1978): Nombres presque-premiers dans les petits intervalles. Sém. de théorie des nombres 1977-1978, no. 15. CNRS, Talence
Laborde, M. (1979): Buchstab's sifting weights. Mathematika **26**, 250–257
Landau, E. (1909): Handbuch der Lehre von der Verteilung der Primzahlen, II. Teubner, Leipzig Berlin
Landau, E. (1927): Vorlesungen über Zahlentheorie, I. Hirzel, Leipzig
Landau, E. (1937): Über einige neure Fortschritte der additiven Zahlentheorie. Cambridge Tracts in Math. no. 35. Cambridge Univ. Press, London
Legendre, A.M. (1808): Théorie des Nombres, 2me ed. Paris

LeVeque, W.J. (1956): Topics in number theory, II. Addison-Wesley, Reading (Mass.) London
Levin, B.V. (1965): Comparison of A. Selberg's and V. Brun's sieves. (Russian). Uspehi Mat. Nauk 20, 214–220
Li, H.-Z. (1994): Almost primes in short intervals. Science in China A 37, 1428–1441
van Lint, J.H., Richert, H.-E. (1965): On primes in arithmetic progressions. Acta Arith. 11, 209–216.
Liu, H.-Q. (1996): Almost primes in short intervals. J. Number Theory 57, 303–322
Merlin, J. (1911): Sur quelques théorèmes d'arithmetique et un énoncé qui les contient. C. R. Acad. Sci. Paris 153, 516–518.
Merlin, J (1915): Un travail de Jean Merlin sur les nombres premiers. Bull. Sci. Math. 2, 121–136.
Mertens, F. (1874): Ein Beitrag zur analytischen Zahlentheorie. J. reine angew. Math. 78, 46–92
Miech, R.J. (1964): Almost-primes generated by a polynomial. Acta Arith. 10, 9–30
Möbius, A.F. (1832): Über eine besondere Art von Umkehrung der Reihen. J. für Math. 9, 105–123
Montgomery, H.L. (1971): Topics in multiplicative number theory (Lecture Notes in Mathematics 227). Springer, Berlin Heidelberg New York
Motohashi, Y. (1975): On some improvements of the Brun–Titchmarsh theorem. J. Math. Soc. Japan 27, 444–453
Motohashi, Y. (1981): Lectures on sieve methods and prime number theory. Tata Institute of Fundamental Research, Bombay.
Motohashi, Y. (1999): On the remainder term in the Selberg sieve. Number Theory in Progress, 2 (Zakopane, 1997) (K. Győry, H. Iwaniec, J. Urbanowicz, eds.), 1053–1064. W. de Gruyter, Berlin New York
Nicomachus (1926): Introduction to arithmetic (transl. by M. L. d'Ooge). University of Michigan Press.
Pisano, L. (1852): Il liber abbaci di L. Pisano (1202, revised 1238). Rome
Rawsthorne, D. (1982): Selberg's sieve estimate with a one-sided hypothesis. Acta Arith. 41, 281–289
Richert, H.-E. (1969): Selberg's sieve with weights. Mathematika 16, 1–22.
Salerno, S. (1991): Iwaniec's bilinear form of the error term in the Selberg sieve. Acta Arith. 59, 11–20
Salerno, S., Vitolo, A. (1994): Bilinear form of the error term of Buchstab's iteration sieve. Stud. Scient. Math. Hungarica 29, 47–54
Selberg, A. (1947): On an elementary method in the theory of primes. Kong. Norske Vid. Selsk. Forh. B. 19(18), 64–67
Selberg, A. (1952a): On elementary methods in prime-number theory and their limitations. 11 Skand. Math. Kongr., Trondheim 1949, 13–22. Johan Grundt Tanums Forlag, Oslo
Selberg, A. (1952b): The general sieve-method and its place in prime-number theory. Proc. Internat. Cong. Math. 1950 I, 286–292. Amer. Math. Soc., Providence, R.I.
Selberg, A. (1972): Sieve methods. Proc. Sympos. Pure Math 20 (Stony Brook, 1969), 311–351. Amer. Math. Soc., Providence, R.I.
Selberg, A. (1989): Sifting problems, sifting density, and sieves. Number Theory, Trace Formulas and Discrete Groups (Oslo 1987) (K.E. Aubert, E. Bombieri, D. Goldfeld, eds.), 467–484. Academic Press, San Diego London
Selberg, A. (1991): Lectures on sieves, in Collected Papers, vol. 2. Springer, Berlin Heidelberg New York
Siebert, H. (1983): Sieve methods and Siegel's zeros. Studies in Pure Mathematics, 659–668. Birkhäuser, Boston Basel Berlin

Snirel'man, L. (1933): Über additive Eigenschaften von Zahlen. Math. Ann. **107**, 649–690

Tenenbaum, G. (1990): Introduction à la théorie analytique et probabaliste des nombres. Institut Elie Cartan, Nancy. [English transl.: Introduction to analytic and probabilistic number theory. Cambridge University Press (1995)]

Titchmarsh, E.C. (1930): A divisor problem. Rend. Circ. Mat. Palermo **54**, 414–429

Tsang, K.-M. (1989): Remarks on the sieving limit of the Buchstab-Rosser sieve. Number Theory, Trace Formulas and Discrete Groups (Oslo 1987) (K.E. Aubert, E. Bombieri, D. Goldfeld, eds.), 485–502. Academic Press, San Diego London

Vakhitova, E.V. (1995): On the application of Buchstab functions. Mat. Zametki **57**, 121–125 [English transl.: Math. Notes **57**, 121–125]

Vakhitova, E.V. (1999): Selberg's one-dimensional sieve with Buchstab weights of new type. Mat. Zametki **66**, 38–49 [English transl.: Math. Notes **66**, 30–39]

Vaughan, R.C. (1973): Some applications of Montgomery's sieve. J. Number Theory **5**, 64–79

Wirsing, E. (1961): Das asymptotische Verhalten von Summen über multiplikative Funktionen. Math. Ann. **143**, 75–102

Wu, J. (1992): P_2 dans les petits intervalles. Sém. de Théorie des Nombres Paris 1989–90 (Progr. Math. **102**) 233–267. Birkhäuser, Boston Basel Berlin

Index

Adjoint equation   115–120, 138, 262
- standard solution   120, 122
Almost primes   181, 186
- in polynomial sequences   185
- in short intervals   184
Asymptotic expansion   120, 277
Auxiliary functions   135, 145

Bombieri-Vinogradov Theorem   18
Brun, V.   51, 85, 94, 96
Brun-Titchmarsh theorem   63
Buchstab transform   272, 285, 288
Buchstab's function   113

Cauchy
-- condensation   58, 226
-- convergence criterion   57
- inequality   46, 224, 251
- mean value theorem   144
Chebyshev's Theorem   12
Contour   119, 130
- integral   119, 120

Density of sieve   27
Dickman's function   113
Dirichlet series   59
Discontinuity   128, 264, 268
Discretisation   226, 229, 234
- fineness   236

Euler
- constant   12, 123, 126
- function   16, 18
Exponent pair   250
Extremal example   104–105, 158, 164
- for weighted sieve   175

Fourier series   223, 246, 249
Fourier transform   249
Fundamental
- identity   74, 77, 199

-- discretisation   232
- inequality   73
- lemma   92, 104, 114, 146, 151, 152, 157

Gamma function   120
Goldbach problem   15, 28, 65

Hankel's integral   120

Identity   80
- Buchstab's   72, 77, 80, 109, 161, 166, 189, 203, 270
- for main term   205
- fundamental   74, 77, 199
- in Selberg's method   286
- Legendre's   14–18
- reduction   202
Inductive argument   111, 149, 151, 161
Inequality   80, 82, 90
- Cauchy's   46
- fundamental   73
Inner product   116, 124, 139, 140, 262, 265, 267, 269
Integral equation   113, 262
- approximate   55, 261, 269
- polynomial solutions   117, 122

Laplace transform   117–119, 123–127, 262, 281
Legendre   8, 13
- formula   104
Level
- of distribution   21, 174
- of support   20, 21, 23
Liouville function   159

Maclaurin series   54, 82
Mean Value Theorem   83
- Cauchy's   144

Mertens' product   12, 53, 55, 95, 104, 160, 165
Möbius function   9, 10, 14, 159, 205, 230

Partial summation   30, 108, 213, 272
Poisson sum formula   247–250, 252
Prime
– in an interval   49
– twin   27, 65, 71, 84
Prime Ideal Theorem   67, 185
Prime Number Theorem   12–13, 159
– analogue   165
– for arithmetic progressions   63
Probability   11

Quadratic form   42

Rankin, R. A.   51, 53
Remainder   19
– bilinear form   224
– trivial treatment   21, 223
Riemann zeta function   59
Rosser, J. B.   71, 85, 103, 144

Saddle point method   277, 281
Selberg, A.   105, 113
Sieve
– bilinear structure   224
– Brun's   177
– combinatorial   71
– composition   37, 146, 151, 234
– dimension   19
– factorisation   226
– Legendre's   13, 20, 23, 51, 71, 72, 79, 177
– linear   158
– of Eratosthenes   7
– pure   71–72, 79, 85, 92
– Rosser's   85, 103, 194
– Selberg's   72, 79, 104, 174, 260, 273, 285
– weighted   77, 156
– – Buchstab's   188
– – improved   186
– – Laborde's   190
– – Richert's   180
Sifting
– density   27, 29, 37, 50, 83, 86, 97, 108, 260
– – alternative hypotheses   32, 265
– – hypothesis   28, 144, 150
– – two-sided   29, 156, 196, 210, 213
– limit   37
– – as $\kappa \to \infty$ 130, 277, 285
– – as $\kappa \to \frac{1}{2}$ 145
– – as zero of function   113
– – in Brun's sieve   93, 94, 99
– – in Rosser's sieve   86, 105, 114
– – weighted   174
– – when $\kappa = 1$ 117, 125
Stationary point   132
Stieltjes integrals   30

Taylor
– coefficients   120
– expansion   120, 282
– theorem   120
Trigonometrical sums   225, 250

Printed in the United States
38565LVS00001BA/19